# Earthquake Research and Analysis - Seismology, Seismotectonic and Earthquake Geology

# Earthquake Research and Analysis - Seismology, Seismotectonic and Earthquake Geology

*Editor*

Hubert O'Kelly

scitus
academics

**Earthquake Research and Analysis - Seismology, Seismotectonic and Earthquake Geology**
Edited by **Hubert O'Kelly**

ISBN: 978-1-68117-302-3
Library of Congress Control Number: 2016937487


SCITUS Academics LLC,
www.scitusacademics.com
Box No. 4766, 616 Corporate Way,
Suite 2, Valley Cottage,
NY 10989



Printed in United States of America on Acid Free Paper

# Preface

Earthquake studies play an important role in minimizing the inevitable human and material loss. Earthquake or seismic analysis is a sub-set of structural analysis that calculates the structure's response to earthquakes. This is necessary for the structural design, structural assessment and retrofitting of the structures in the regions of earthquakes. Seismic structural analysis requires different seismic data. You can access these data in two ways—deterministic or probabilistic form. Data in deterministic form are used for structural design, etc. While probabilistic data are used for seismic risk analysis, structural studies are subject to random vibration and structural damage assessments during specific earthquake movement. The main seismic inputs include soil displacement data, magnitude of the earthquake, peak soil parameters, etc. Earthquake engineering has developed a lot since the early days, and some of the more complex designs now use special earthquake protective elements either at the base (base isolation) or throughout the structure. Specialized explicit finite element computer code is required to analyze these types of structures, which divides time into very small sections and models of actual physics, similar to common video games that often have physics engines. This can shape very large and complex buildings.

This book contains statistical seismology studies, state-of-the-art techniques and advances in early detection, acquisition and interpretation of earthquake precursors. This book focuses not on a unique problem in earthquake processes, but on historical earthquakes and seismology in a variety of tectonic environments for geological studies with more application of earthquakes.

# Table of Contents

# 1

# CHAPTER

# ASSESSMENT OF SEISMIC HAZARD OF A TERRITORY

**V.B. Zaalishvili[1]**

*[1] Geophysics and Engineering Seismology Department of Geophysical Institute of Vladikavkaz Scientific Centre of RAS,, Russian Federation*

## 1. INTRODUCTION

The aim of this work is to analyze the modern concepts of the seismic hazard of the territory, its evaluation and development of an algorithm for such assessments. One of the main problems is the adequacy of such assessments. In order to analyze the physical mechanisms of different approaches the evolution of methods and techniques on one side, and their comparison with the other side should be presented. It is also necessary to show the physical validity of choice of approach and its accessibility and simplicity, which increases the reliability of the final results. The next task is to analyze the integral estimates of seismic hazard of a territory. It is necessary to consider the calculation of the expected seismic effects on the final stage of the algorithm.

Seismic hazard assessment is the basis of modern seismic design and engineering. Seismic hazard maps of different contents and scales are result of such investigations. It makes possible to take into account all the features of the territory forming mentioned hazard. In particular, expected on the territory typical seismic impacts and calculated on the basis of world achievements not only of the West but also of the East are taking into account features of possible seismic sources (active seismic faults, their seismic potential, etc.).

The new integrated method for seismic hazard assessment is presented. Probabilistic maps of seismic hazard are created in the result of this assessment. The following databases were formed to analyze seismic hazard and seismicity of the territory: macroseismic, seismologic and database of possible seismic source zones (or potential seismic sources-PSS). With usage of modern methods (over-regional method of IPE RAS-Russia) and computer program (SEISRisk-3 – USA) in GIS probabilistic seismic hazard maps for the Republic North Ossetia-Alania in intensity units (MSK-64) at a scale of 1:200 000 with exceedance probability of 1%, 2%, 5%, 10% for a period of 50 years that corresponds to recurrence period of 5000, 2500, 1000, 500 years were created. Furthermore, for the first time the probabilistic seismic hazard maps were made in acceleration units for territory of Russia. For the large scale building is likely to be used 5% probability map, i.e. for the major type of constructions, whereas for high responsibility construction should be used 2% probability map. The approach based on the physical mechanisms of the source is supposed to use to produce the synthesized accelerograms generated using real seismic records interpretation.

For each of zoning objects the probabilistic map of the seismic microzonation with location of different calculated intensity zones (7,8,9,9*) is developed (the zones, composed by clay soils of fluid consistency at quite strong loadings can be characterized by liquefaction are marked by the index 9*). Similar results are observed for maps in acceleration units.

The integrated approach is based on the latest achievements in engineering seismology. It can reduce measure of inaccuracy in earthquake engineering and construction and also significantly increase the adequacy or foundation for assessments.

Carrying out of investigations on mapping of seismic hazards such as detailed seismic zoning (DSZ), which is based on the most advanced field research methods and analysis of every subject of the sufficiently large region (for example Caucasus) in a scale of 1: 100 000 or 1: 200 000 allows to create a physically reasonable general DSZ map of any wide territory. Such maps are generated organically through summation or the imposition of the calculated seismic fields that are a reflection of seismic potential of the corresponding seismic sources.

## 2. ASSESSMENT OF SEISMIC HAZARD: GENERAL AND DETAILED SEISMIC ZONING

Expected seismic hazard assessment is reduced to seismic potential estimation of a particular seismic source or a combination of sources. Wherein the mentioned potential is formed by a number of factors, such as geodynamic (tectonic movements), geological and geophysical features of the territory, the catalog of strong earthquakes, local soil conditions (geotechnical, hydrogeological and geomorphological conditions), resonance properties, attenuation rates, and others. At the same time physical validity problem of hazard level is one of the important problems of engineering seismology. The involved assessment of seismic hazard, presented as seismic zoning maps, in fact is a long-ranged forecast of the earthquake strength and location unlike short-range and middle-range earthquake forecast.

There can be marked out three types of analysis, three consecutive stages of seismic zoning:

1. general seismic zoning – GSZ or SZ, which is realized in 1:5 000 000 or 1:2 500 000 scale;
2. detailed seismic zoning DSZ, which was originally carried out in 1:50 000 - 500 000 scale;
3. seismic microzonation – SMZ, in 1: 5 000 - 10 000 scale.

In the result of seismic zoning the appropriate maps of GSZ, DSZ and SMZ were created. The difference between DSZ and GSZ lies in investigation scale. During DSZ can be and has to be studied all potential sources of possible earthquakes that may be not taken in account because of their relatively small seismic potential during GSZ analyzing. It is necessary to note that in actual conditions of consequences of seismic hazard generation with that types of sources can have, if not great, but evident negative effect. At once both types of zoning are quite similar, not to mention minuteness.

Despite similar name with GSZ and DSZ the third stage or stage of seismic hazard assessment in SMZ type has absolutely other physical meaning. The usage of SMZ allows taking into account the seismic properties of site soils, including physicomechanical and dynamical properties of soil.

Seismic hazard assessment of any given territory on the initial stage was performed using mainly deterministic methods. In the deterministic methods, all parameters of a close particular source (magnitude), medium (particular site), the distance from the source to the site, the wave attenuation when propagating through a particular medium (the propagation paths of seismic energy in the form of waves) are mostly known or thoroughly investigated and finally form the medium (soils) reaction of a site in the form of maximum seismic effect. Medium is considered in a simplified form - combined horizontal layers. A number of such deterministic computer programs is known.

At present in developed countries the hazard assessment is increasingly performed using probabilistic methods. Probabilistic seismic hazard assessment techniques include various alternative models of earthquake sources (focuses), return periods of events, accounting of attenuation models of seismic energy and accordingly the seismic effect. Herein the probabilistic method takes into account uncertainty bounds of certain parameters that are mainly random and the most important - the probabilistic nature of the actual implementation of the seismic event that is formed by a number of random factors. The last circumstance significantly reduced the intensity of emotions that was in the 60-70 years of the last century about development of a reliable method for prognosis of strong earthquakes. Multifactorial nature of the earthquake, on the other hand made it even more complicated to obtain reliable prognoses. Although increasing depth of investigation made such a prognosis more reliable.

Almost all major seismic zoning maps in former USSR were deterministic. They include the first map of 1937 and map of 1978.

In 1947 S.V. Medvedev proposed to differentiate the seismic hazard zones depending on the return period of strong earthquakes and durability of various buildings and structures [9]. Thus, in 1947, an offer to take into account seismicity or seismic activity of the territory when creating maps was made. Later, the famous scientist of former USSR Yu.V.Riznichenko for the first time introduced the concept of seismic "shakeability" and developed the first algorithm and program for its calculation [16]. Unfortunately, these and other progressive ideas and developments have not been used in the process of creating new maps for a variety of reasons. With that, this approach was developed in the West [5]. Just then K.A. Cornell suggested differentiating between seismic hazard and risk. The risk began to be understood not as risk of seismic event occurrence, but only as economic and social losses. The probabilistic seismic zoning maps with the probability of seismic hazard exceeding (or not exceeding) in certain time periods were first created in the West.

It is well known that nature is essentially non-linear. There are always uncertainties in nature laws that completely eliminate the use of deterministic seismic zoning maps. In other words, physically-based maps of seismic hazard differentiation or seismic zoning maps must be probabilistic. Herein the risk that always exists should be assessed and has to be the minimum.

This has led to a situation where the deterministic approach is expelled or banished from consideration. At the same time, it should be noted that the deterministic seismic hazard maps sometimes give estimates that are close to reality. In this context for calculation of seismic hazard in the beginning were used both map types.

The map that was developed later than the map of 1978 to some extent was taking into account the probabilistic nature of the formation of zones of seismic hazard, although essentially still not very different from deterministic.

The probabilistic approach was implemented for the first time in the new, more advanced seismic zoning maps of Russian Federation. These maps are called general seismic zoning maps or GSZ-97.

General map GSZ -97 is presented on Fig. 1. The map generalization is enough for state overall planning, but is not enough for reliable estimation of real objects seismic conditions.

The DSZ process is a complex of very complicated and expensive geology-tectonical, geophysical and seismical investigations for quantitative estimation of seismic effect in any site of perspective region [1]. This investigation type estimated quantitatively the source seismic effects only on concerned site GSZ (more precisely for mean soil conditions). [28]

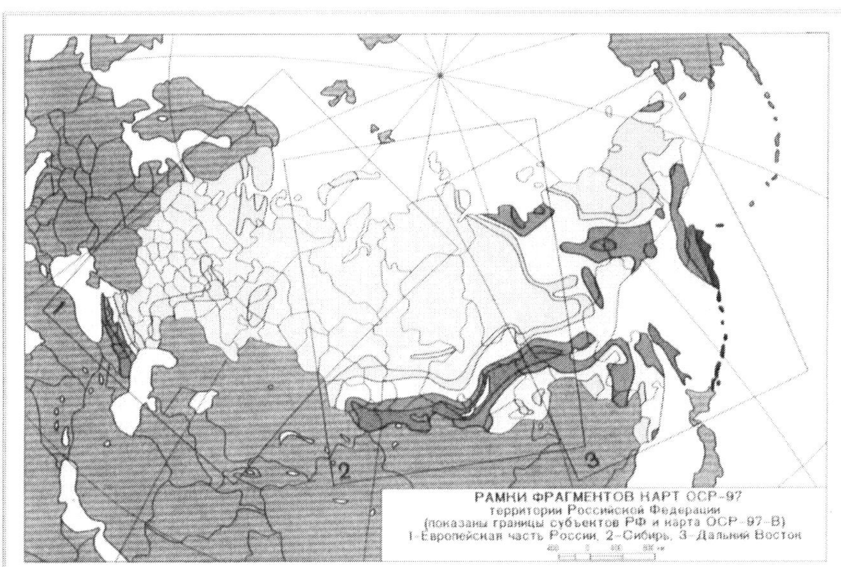


**Figure 1.** Map of General seismic zoning GSZ-97 of Russia

The modern DSZ has clear and argumented content. There are some methods that may be used in GSZ and DSZ for seismic generic structures (SGS) identifications. It is identification of hazardous earthquake occurrence zones. [12]

## 2.1. Seismogeological method

At the turn of XIX and XX centuries G.Abich, A.E.Lagorio and the other scientists found out as a result of earthquakes epicentral zones investigations that earthquakes are connected with tectonic structures. So, when describing the strongest Vern earthquake consequences (1887) Mushketov I.V. (1889) straightly connected earthquakes with fractures and faults. He believed that earthquakes have maximum effect on the lines of large and new faulting (Mushketov, 1891). He also discovered that some groups of earthquakes are connected with transverse structures. Later on the basis of the Kebi earthquake investigation results Bogdanovich K.I. introduced the concept of seismotectonic elements and showed migration of seismic shocks within seismically active zone.

Thus, seismogeological method determined the connection between strong earthquakes and tectonic structures. Such connections were named geologic criteria of seismicity. [12, 28].

## 2.2. Seismotectonical methods

By the results of Garm region between Pamir and Tien Shan in the late 40-es of the XX century Gubin I.E. for the first time introduced the term seismotectonical method. Herewith he connected strong seismic events with ruptured occurrence with the width up to several tens of kilometers. He supposed that within such a zone "seismogenity" remains and "seismogenity level" corresponds to other similar zones which are characterized by development processes of equal intensity. Gubin's law of seismotectonic says that in active structures of similar type and size maximum earthquakes originated as the result of rock displacement along active faults are characterized by equal values of focuses and magnitudes. The method at that connects geologic criterion of seismicity with a velocity of new fault displacements.

## 2.3. Seismostructural method

This method was offered in the middle of 50-es by Belousov V.V., Goryachev A.V., Kirillova I.V., Petrushevskiy B.A., Rezanov I.A., Sorskiy A.A. But the main development the method gained in the works of Petrushevskiy B.A. Such method connects seismic events with large structural complexes – blocks. Mentioned blocks were separated with the help of historical and structural analysis [12].

The analysis of blocks scale allowed connecting a considerable range of earthquake focuses depths with them and with deep faults. The deepest faults are located on the boundary of the Pacific Ocean and Eurasian and American continents. The concept about strong earthquake focuses connection with three-dimensional structures of Earth crust was later developed by Gorshkov G.P. (1984) [28]. Although the direction is considered as promising it needs to be fleshed out.

## 2.4. Tectonophysical method

The method was developed in the end of 50-es by Gzovsky M.V. [12]. The feature of the method is the connection of strong seismic event with the areas of maximum shearing stress. Such areas are characterized by faults and maximum gradients of average velocities of tectonic movements. Although Gzovsky M.V. made earthquake energy dependent on a number of factors it is practically impossible to make exact calculations. It was defined exceptionally by qualitative but not quantitative terms of mechanical properties of Earth crust and its viscosity in the region of maximum shearing stresses.

## 2.5. Method of quasi-homogeneous zones allocating

This method was first offered in the end of 50-es of the XX century as well as the previous method. The concept of the method consisted in detection of quasi-homogeneous areas where earthquake could be caused by the presence of geological and geophysical criteria. In this case it was supposed that some of them are connected with tectonic movements [12]. Since the number of characteristics is endless the number of situations with maximum magnitudes can also be endless. It stipulated an extension of characteristics and usage of all geological, seismological and geophysical data for allocation of zones with different seismic potential of tectonic nature. Data correlation allows giving prognosis of maximum magnitude $M_{max}$.

Today the techniques developed by Reisner and Ioganson are frequently used. At that the whole number of characteristics such as relief, crustal thickness, heat flow density, isostatic anomalies of gravity force, stratum depth of consolidated basement etc. is used for analysis. Realization of the techniques supposes usage of reference sites from all active regions of the world. In this connection such approach is named overregional [28].

## 2.6. Method of seismically active nodes

This method was also offered in the end of 50-es of the XX century. Data analysis of central Asia earthquakes allowed Reiman V.M. connecting disjunctive nodes to which strong earthquakes or seismogenetic nodes are confined. The method was significantly developed by Rantsman E.Ya. (1979). She used the approach to the number of world regions and connected focuses of strong earthquakes with these nodes. Herewith the author noticed that "focuses can exceed nodes in size and amount up to thousands kilometers". Structure differentiation on their seismicity is realized with the help of formalized criteria (relief type, maximum height, distance from node edges, volume of soft sediments etc.). It was determined that transverse uplands form structures which generate nodes [12].

## 2.7. Paleoseismological method

The method (Solonenko V.P., Khromovskikh V.S., Nikonov et al.) allows allocating PSS zones and assessing their seismic potential (maximum magnitude and seismic intensity) on the basis of study of paleosiesmodislocations spatial orientation. As a rule, seismotectonic dislocations are used for the assessment of these parameters. The number of formulas which connect statistic correlations between seismodislocation characteristics (length, shift amplitude) and seismologic parameters (magnitude, focus depth, seismic impacts intensity) of earthquake is known. Return period of earthquake is determined by paleoseismodislocations age. Herewith all existing approaches are compared: geologic-geomorphological, archeological, historical data and the results of radiocarbon dating, dendrochronology, connections of tree growth and earthquakes and lichenometry (lichen dating).

# 3. DETAILED SEISMIC ZONING

Analysis of the considered approaches shows that each method characterizes one or another aspect of the problem. Herewith some of them are developed straight on the assumption of practice demands. Other methods suppose a number of complimentary stages for achievement of final problem investigation. Overregional method is referred to the first group and we'll try to use it.

For avoidance of any doubt let us consider typical way of seismic hazard level determination and all other attributes necessary for goal achievement. So let's begin to realize the problem of assessment on the level of detailed seismic zoning (DSZ) by the example of the territory of the Republic North Ossetia-Alania (Zaalishvili, Rogozhin, 2011).

As was mentioned above we chose overregional seismotectonic method for allocation of PSS zones; such method permits to identify seismic or seismogenic source quite completely. Although there are some disadvantages on practice the method gives quantitative indices and it is characterized by the certain decision-making algorithm. The method was successfully used in works of prof. Rogozhin E.A. in many countries of the world (Russia, Israel, China, Kirghizia etc.). We were able to realize the method in North Ossetia exactly due to his active support. At the same time, this does not preclude obtaining reliable results and other known methods. The methodology used in most probabilistic seismic hazard analysis was first defined by Cornell and as usually accepted it consists of four steps [5, 15, 28]: 1. Definition of earthquake source zones (SSZ), 2. Definition of recurrence characteristics for each source, 3. Estimation of earthquake effect and 4. Determination of hazard at the site. The probabilistic hazard maps were created on the basis of investigation of corresponding territory main characteristic. 1-4 stages of such work realization are given further by the example of the territory of North Ossetia.

## 3.1. Definition of earthquake sources

As it was mentioned previously probabilistic assessment gain the largest extension at seismic hazard assessment of territory in the world in recent years. Various investigations had shown that such assessment gives physically more proved results. Indeed, at such assessment it is possible to forecast more sequentially a location of expected seismic events, their intensity etc. [14, 17]. The main difference from other methods consists in the fact that special processing of the existing data allows accounting uncertainties caused by our half knowledge of cause-and-effect relations and random factors at formation of our understanding about spatiotemporal realization of one or another seismic event. It is necessary to mention the computer program EQRISK of McGuire [8]. This program noticeably simplified assessments of seismic hazard what caused its wide extension. It came to a point where seismic hazard assessment with the help of probabilistic approach sometimes is called Cornell - McGuire's method.

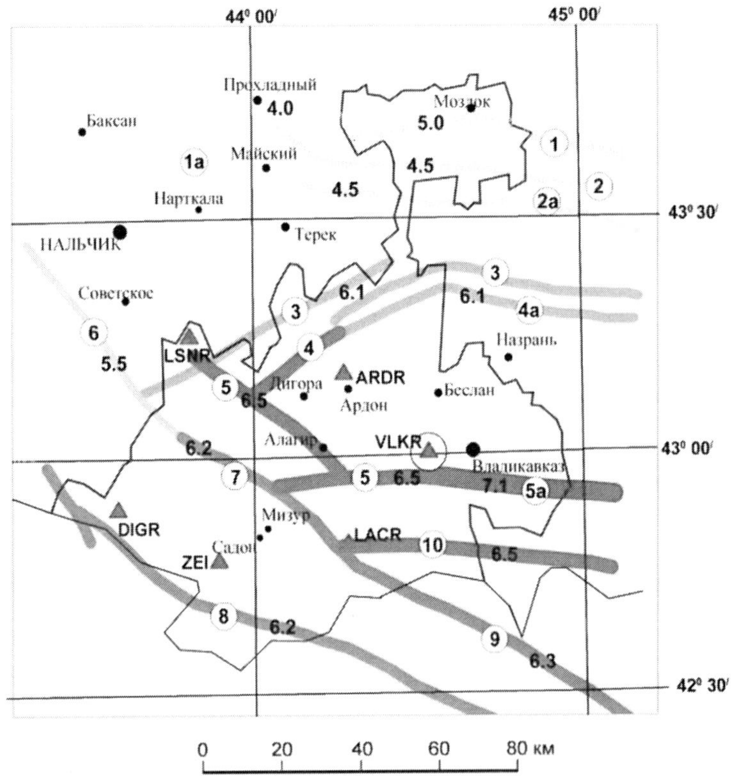


**Figure 2.** Map of PSS zones of the territory of the Republic North Ossetia-Alania. Red triangles – basic seismic stations in the region. Blue and black lines are the state borders of North Ossetia.

The investigations on determination and parameterization of the seismic source zones in recent decades has been realized by V.P.Solonenko, V.S.Khromovskikh, E.A.Rogozhin, V.I.Ulomov, V.G.Trifonov, I.P.Gamkrelidze and others [6, 12, 17, 20, 21]. On the basis of investigation results of the active faults located southward of the Great Caucasian ridge, parameters of seismic source zones were chosen according to data of I.P.Gamkrelidze work [6]. We used the investigation results of Ulomov V.I., Rogozhin E.A., Trifonov V.G. for accounting the indices of Northern Caucasus in our work. On the first stage of seismic hazard map making there was carried out an approximate or expert calculation of seismic potential (Mmax) of different determined active faults in the form of zones of possible seismic sources (PSS zones) for the territory of the Republic North Ossetia-Alania. Further the map of allocation (zoning) of such PSS-zones was made.

Maximum magnitude of expected earthquakes (seismic potential, $M_{max}$) was assessed on the results of usage of the over-regional seismotectonic method of seismic hazard assessment, offered by G.I.Reisner..

**Table 1.** PSS zones for North Ossetia characteristics (numbers in the rings on Fig.2)

| No | PSS zone | Magnitude | H, km | Kinematics |
|---|---|---|---|---|
| 1 | Mozdok eastern | 5.0 | 10 | reverse faulting |
| 1a | Mozdok western | 4.0 | 5 | strike-slip |
| 2 | Tersk | 4.5 | 5 | reverse faulting |
| 3 | Sunzha northern | 6.1 | 15 | reverse faulting |
| 4 | Sunzha southern (western branch) | 6.5 | 15 | strike-slip |
| 4a | Sunzha southern (eastern branch) | 6.1 | 15 | reverse faulting |
| 5 | Vladikavkaz (western branch) | 6.5 | 15 | reverse faulting |
| 5a | Vladikavkaz (eastern branch) | 7.1 | 20 | reverse faulting |
| 6 | Nalchik | 5.5 | 10 | strike-slip |
| 7 | Mizur | 6.2 | 15 | strike-slip |
| 8 | Main ridge | 6.2 | 15 | reverse faulting |
| 9 | Side ridge | 6.3 | 15 | reverse faulting |
| 10 | Karmadon | 6.5 | 15 | reverse faulting |

Usage of this method showed that the Northern Caucasus is the region of very high seismic hazard. In 2007 it was determined on data of field investigations that for the urbanized territories of North Ossetia (Fig.2), (table 1), [14, 17, 28]It is necessary to notice that on the next stage each possible seismic source was subjected to more exact assessment of its seismic potential. This stage is named a parameterization of seismic sources. The $M_{max}$ was determined by the data of the above mentioned authors.

On the next stage of investigations it was necessary to determine a range of strong earthquake focuses depths. There are no such deep hypocenters as in other regions. The range varies a little and the depth amounts 20-25 km. Herewith it is necessary to notice that hypocenters depths obviously increase monotonically to Grozny city and Caspian Sea. With a lack of data about earthquake distribution the average value of depth was assumed equal to 10 km (see Table 1).

## 3.2. Seismicity of North Ossetia-Alania and its connection with PSS zones

The catalogue of strong earthquakes was defined further. It supposes to check the existing data by correlating it with catalogues of certain authors and results of international investigations. Seismicity data for each zone were defined by the following catalogues: New Catalogue... 1982, Corrected Catalogue of Caucasus, Institute of Geophysics Ac. Sci. Georgia (in data base of IG), the Special Catalogue of Earthquakes for GSHAP test area Caucasus (SCETAC), compiled in the frame of the Global Seismic Hazard Assessment Program (GSHAP), for the period 2000 BC - 1993, N.V. Kondorskaya (editor), (Ms>3.5) Earthquake catalogues of Northern Eurasia (for 1992-2000), Catalogue of NSSP Armenia, Special Catalogue for the Racha earthquake 1991 epicentral area (Inst. Geophysics, Georgia) and also the Catalogue of North Ossetia 2004–2006.

Corrected Catalogue of Caucasus contains data for more than 61000 of earthquakes, including 300 historical events [26, 28]. The issue is seismic events which are referred to historical period meaning last 20 centuries or a new era. So the catalogue was defined. In certain cases when needed a depth of event focuses was recalculated.

Values *a* and *b*of law of frequency and the largest magnitude were determined only for large zones. This fact was caused by absence of such data. Well known formula of Gutenberg-Richter was applied in calculations:

$$\lg(N / T) = a - bM \tag{1}$$

where: N-number of events for a time period T, a and b are parameters of inclination and level of recurrence graph at M=0, respectively.

Analysis of recurrence graph dependence on distance allowed determining representativity and weighted contribution of each seismic zone. Herewith the opportunity of taking into account curve deviation of density distribution became available.

## 3.3. Estimation of earthquake effect

Thus seismic effect was estimated in intensity units and peak ground acceleration (PGA). Observation data and instrumental data on 43 significant earthquakes that occurred in Caucasus were revised to obtain the necessary information [19].

Data on 37 earthquakes was selected and in some cases the maps in the 1:500 000 scale were created. It was determined at the formation of the maps that within three isoseists for events with magnitude Ms>6 coefficient values are quite high (ν~4.5-5.0). For small and moderate earthquakes such values are less (ν~3.4). Data analysis allowed obtaining two different formulas for strong and weak earthquakes:

$$I = 1.5M_S - 3.4\lg(\Delta^2 + h^2)^{1/2} + 3.0 \text{ for small events} \quad (2)$$

$$I = 1.5M_S - 4.7\lg(\Delta^2 + h^2)^{1/2} + 4.0 \text{ for large events} \quad (3)$$

where: $M_S$ – the surface-wave magnitude, Δ – epicentral distance, h – focal depth

We used the second formula at the assessment of seismic hazard. Maximum intensity in the focus at that was 9 points (for magnitudes 6.5-7.0) and for 8 points the magnitudes were 5.5-6.0.

Due to the fact that on Caucasus in the period between June 1990 and September 1998 about 500 accelerograms were obtained for 300 relatively strong earthquakes [19] they became the basis for the formulas of macroseismic fields. The data of temporary and constant instrumental systems on Southern Caucasus and nearby areas were used during the data analysis. They included 84 corrected accelerograms from 26 earthquakes with magnitudes between 4.0 and 7.1. Correlations were obtained with the help of two step regression model (Joyner and Boore). The resulting equation for larger horizontal values of peak horizontal acceleration is:

$$LogPHA = 0.72 + 0.44M - \log R - 0.00231 + 0.28p,$$
$$R = (D^2 + 4.52)^{1/2}, \quad (4)$$

where:

PHA – peak horizontal acceleration in cm/sec$^2$,

M – surface-wave magnitude,

D – hypocentral-distance in km;

p is 0 for 50-percentile values and 1 for 84-percentile.

It is necessary to mention that coefficients of attenuation for horizontal accelerations are similar to the coefficients for Western part of North-America. Herewith analogous indices in Europe are less than on Caucasus and adjacent areas.

## 3.4. Determination of hazard

The probabilistic seismic hazard maps in intensity units (MSK-64) and in acceleration units for the territory of North Ossetia were worked out in a scale 1:200000 with exceedance probability for a period of 50 years (standard time of building or construction durability!) with 1%, 2%, 5%, 10%. All works were carried out in GIS technologies. At that return periods are equal to 5000, 2500, 1000 and 500 years correspond to the given probabilities (Fig. 3, 4).

The essence of the approach consists in the fact that the longer a considered time period the higher an intensity of design earthquake is. The recurrence is changing depending on the level of design intensity. So, expected intensity equal to 10 points corresponds to a period of 5000 years, 9 points – 2500 years, 8 – 1000 years, and finally, intensity level equal to 7 points corresponds to return period of 500 years [18, 26-28].

For calculation of design seismic effect we used computer program SEISRisk- 3 [5, 8]. When choosing an intensity map it is necessary to correlate obtained maps with real intensity effect of expected earthquake on considered territory. At that we must choose the map which will correspond most of all to a real situation particularly to territory features [10].

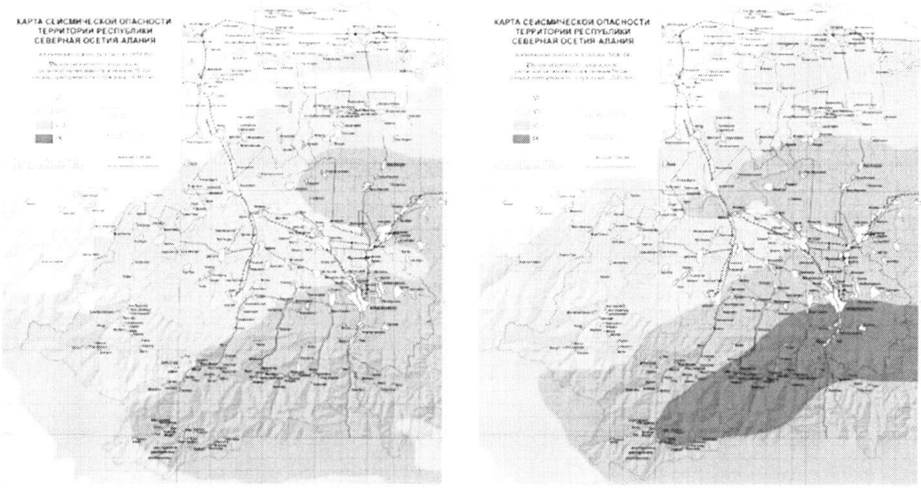


(a) (b)

**Figure 3.** Probabilistic maps of seismic hazard (DSZ) in the intensity units (MSK-64) with the exceedance probability 5% (a) and 2% (b) for North Ossetia territory and adjacent areas [27].

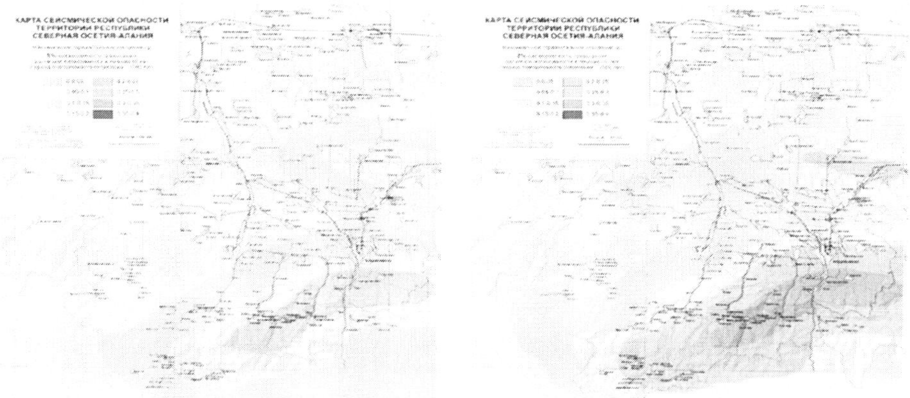


(a) (b)

**Figure 4.** Probabilistic map of seismic hazard (DSZ) in acceleration units (PGA) with exceedance probability 5% (a) and 2% (b) for North Ossetia territory [27, 28].

Thus, for the first time in Russia we realized seismic hazard assessment quite detailed. In this connection it is necessary to remind that seismic hazard map that is included in building code of Russia has the scale M 1:2 000 000 and our map has the scale M 1:200 000! Besides it must be noted that some part of the map of General seismic zoning of Russia has the scale M 1:8 000 000! It, undoubtedly, can characterize the maps in a scale M 1:200 000 as the maps of detailed seismic zoning. It should be noted that we simultaneously worked out seismic hazard maps both in traditional for Russia intensity units and in acceleration units.

Further analysis of the data which characterize the territory allowed concluding quite reasonably that 5 % map with exceedance probability of 50 years is the most appropriate for the Republic territory. It corresponds to earthquake recurrence of 1000 years. Exactly such recurrence was revealed for seismic events with 8 points intensity with the help of special field investigations of the previous earthquakes on the territory of North Ossetia. Thus, we recommended this map for mass construction. At calculation of responsible buildings and constructions to seismic impacts where risk of expected economic and social losses reaches higher values we recommend 2 % map for seismic hazard level with exceedance probability of 50 years (Fig.3). The zone of increased 9 intensity is clearly seen on intensity map. Such intensity is caused by presence of powerful Vladikavkaz fault (Fig.2). Correlation of seismic hazard maps in intensity and acceleration units shows that for acceleration map we have smoother change of hazard level (Fig.3, 4). Indeed, offset distance with any level of hazard can't show such abrupt intensity changes only in cases of appearance of additional hazards or changing of some other conditions, for example, soil conditions, which we don't take into account. That is why seismic hazard maps in acceleration units are more preferable for engineers. The history of forming of seismic processes observation networks stipulated quite full absence of acceleration records on the territory of Russia. On the other hand acceleration maps usage allows reaching any monotonous transition between different zones boundaries with change of consideration stage. It is very important under the conditions of increased seismic hazard. New approach where nonintegral values instead of whole-number values (for example, 8.1 or 7.5) are considered as intensity level of seismic hazard was worked out in recent years in Russia. It is hard characterizing and even understanding an assessment of seismic effect equal to 0.1 or 0.7 intensity. In spite of its conditional and uncertain characteristic this approach was used on practice widely at Olympic objects constructing in Sochi. It is obvious that authors used the so-called "expert" assessments here.

The way out is independent creation of seismic hazard maps in units of peak ground accelerations (PGA). We have worked out such maps for exposition of 50 years with exceedance probability. They have the scale M 1:200 000 and probability 1%, 2%, 5%, 10% (Fig. 4). Rapid change of acceleration levels on those maps can be excluded within reasonable rates. Exactly these maps (which were the first on the territory of Russia and CIS) while being probabilistic maps of detailed seismic zoning will be the basis for creation of probabilistic maps of seismic microzonation. Maps of seismic microzonation as a rule are made in scales 1:2000, 1: 5000 or 1: 10 000 [28]. It is possible to made maps in larger scale but it doesn't have practical sense. Indeed, soil conditions do not generally change faster and at construction of certain building or structure we always know soils on more detailed level. In this connection it is necessary to remember that we create maps of microzonation in the form of schemes for typical soils of investigated territory.

At the same time hydroelectric stations or atomic power stations may need higher level of protection. That's why considerably larger return periods can be considered for such objects. According to the accepted approaches for investigated territory the period will be 5000 years with the intensity 10 points. At the same time this period easily can be 2500 years with intensity 9 points for majority of other responsible objects (high-rise building etc.). Herewith very interesting consequence appears. It consists in the fact that level of territory security under otherwise equal conditions depends on a level of economic potential of this or that country. Developed country within common sense can presume any economically accessible level of design intensity. And we can see that earthquake consequences of the same level are always more drastic and catastrophic for poor hindward country. On the other hand in authoritarian state it all depends on a good will of bodies of government, which know that fact. Exactly due to that at strong consequences of natural and anthropogenic processes they can be concealed truckle to bodies of government. For example, the fact that death toll after strong Ashkhabad earthquake in 1948 amounted to 80 000 people was brought to light only many years on. In all fairness it must be noted that such and more egregious facts also took place later in other countries.

# 4. SEISMIC MICROZONATION OF TERRITORY

The sites with reference soil conditions corresponding to specified seismic hazard level are specified in the process of seismic microzonation. In Russia as reference soil conditions for a given territory are traditionally considered the soils with mean seismic properties (usually soils with shear wave velocity of 250–700 m/s). For example in Georgia depending on specific engineering-geological situation for a given territory the reference soils in their seismic properties can be worst or mean. In USA firm rocks are referred to reference soils. Seismic microzonation concludes the computation of intensity increments caused by soil condition differences. Seismic microzonation is carried out with the help of instrumental and calculational methods [28].

## 4.1. Instrumental method of seismic microzonation

The main method of seismic microzonation is an instrumental method. Exactly this method urges to solve a forecast problem of forming earthquake intensity. However the calculational method which allows modeling any definite conditions of area and impact features is often characterized as more reliable. It is very important for high power soil stratum. Usage of both methods together significantly increases the results validity.

### 4.1.1. Seismic microzonation on the basis of strong earthquakes instrumental records

A number of corresponding international projects were worked out for receiving a data about different soils behavior at strong earthquakes. It is necessary to mention that the data of permanent systems of instrumental observations on the island Taiwan (the groups SMART-1 and SMART-2.) [25, 26, 28] in our opinion were characterized by the most reliable background. It is also necessary to notice that obtaining of even strong earthquake record cannot guarantee its adequate proper usage. Therefore working out of approaches which can increase physical validity of the data is very topical problem. In particular one can create design impacts with taking into account seismic sources features (for example, accelerograms) for the level of DSZ [25, 26, 28]. Further accounting a distortion of wave field caused by soil condition change we can receive new implementation of each site. Herewith usage of even single record of strong earthquake will give very important results for testing of strong earthquake design records.

### 4.1.2. Seismic microzonation with the help of weak earthquakes records

It is obvious that the number of earthquakes is limited except the fact that you live in Japan. That's why registration of soil vibrations caused by weak earthquakes in absence of strong earthquakes became quite important factor. Formation of weak earthquake database became a reliable basis for data testing. Herewith that single virtual record of strong earthquake which at first glance was neglected will take an important place again. It is clearly seen that linear deformations during weak earthquakes must be transformed into nonlinear-elastic and even into nonelastic links "deformations – tensions". Amplitude-frequency characteristics of areas and recording form significantly vary when changing soils (Fig. 5) [25, 28].

Increase of the soil stratum depth (alluvium) considerably changes the character of earthquake records in the process of approaching the city.

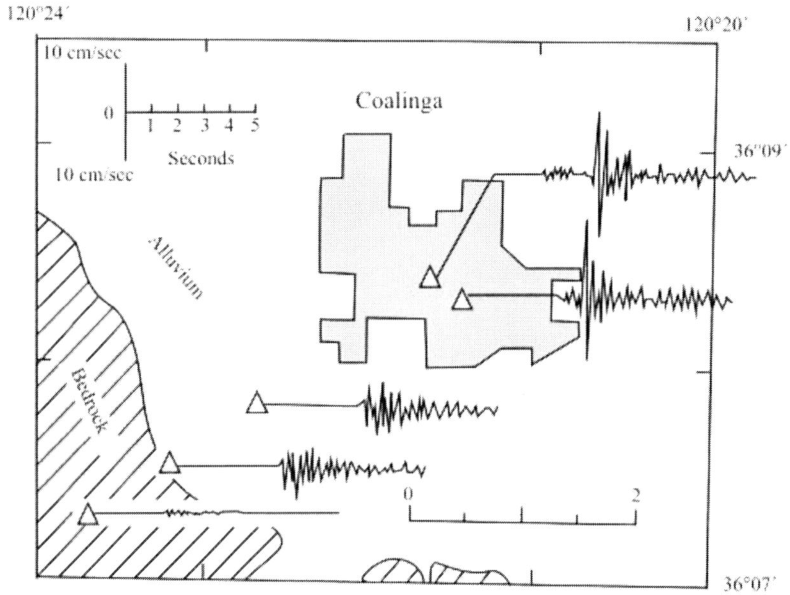


**Figure 5.** Scheme of California earthquake in Koaling city

Calculation of intensity increment with the help of weak earthquakes is realized by the formula [25]:

$$\Delta I = 3.3 \lg A_i / A_0, \tag{5}$$

where: *Ai,A0* mare the vibration amplitudes of investigated and reference soils, respectively.

The usage of tool in the form of registration of strong and weak earthquakes needs the organization of instrumental observations in a waiting mode.

### 4.1.3. Seismic microzonation with the help of weak earthquakes records

Strong earthquake extremely seldom occur on territories with small and moderate seismic activity. At that hazard level not only decreases but it can even increase due to impossibility of timely unloading of high tensions. In this connection when calculating intensity increment of weak earthquake records the dependence “deformation – tension” is linear. It causes inaccuracies in soil behavior assessments at expected strong earthquakes when dependence is nonlinear.

Calculation of intensity increment with the help of weak earthquakes is realized by the formula [25]:

$$\Delta I = 3.3 \lg A_i / A_0, \tag{6}$$

where: *Ai,A*0 are the vibration amplitudes of investigated and reference soils, respectively.

The usage of tool in the form of registration of strong and weak earthquakes needs the organization of instrumental observations in a waiting mode.

### 4.1.4. Seismic microzonation using microseisms

The results of microseisms observations [25] are used as subsidiary instrumental tool of SMZ. Strictly speaking, the reference of microseism on their origin to the purely natural phenomena is not quite correct. Numerous artificial sources, influence degree of which can’t be controlled, undoubtedly, take part in their forming along with the natural sources (Fig. 6)

Intensity increment for strong earthquakes on microseism is calculated by the formula [25]:

$$\Delta I = 2 \lg A_i / A_0, \tag{7}$$

where: *Ai,A*0 are the maximum amplitudes of microvibrations for investigated and reference soils, respectively.

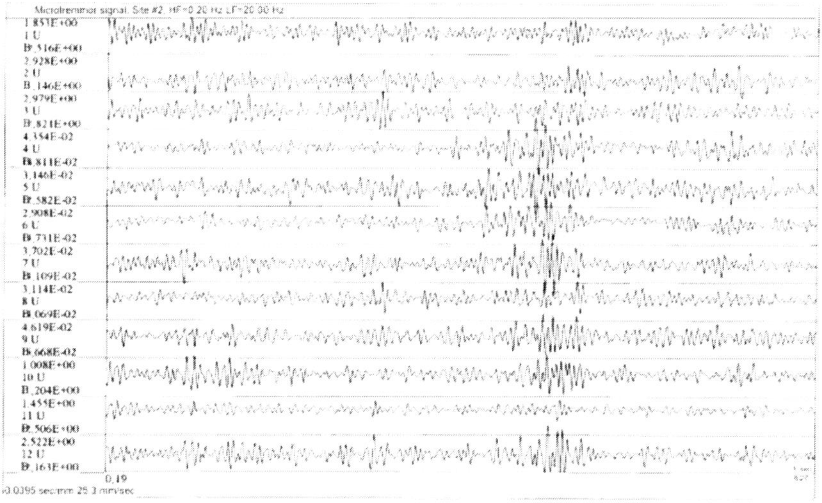

**Figure 6.** Microseisms records (10.07.1996, Voronezh Region, Russia)

Impossibility of the compliance of necessary standard conditions of microseism registration and large spread in values of maximum amplitudes limit the usage of microseism for calculation of soil intensity increment.

The above mentioned causes the application of microseism tool only in complex with other instrumental tools. Spectral features for different sites are estimated by means of H/V-rations [11, 25].

### 4.1.5. Seismic microzonation using explosive impact

The intensity increment ΔI of the soils of the zoned territory is calculated by the formula [25] at usage of weaker explosions:

$$\Delta I = 3.3 \lg A_i / A_0, \tag{8}$$

where: $A_i$, $A_0$ are vibrational amplitudes of the investigated and reference soils, respectively.

Execution of powerful explosions on the territory of cities, settlements or near the responsible buildings is connected with large and often insurmountable obstacles (technical and ecological problems, safety problems, labouriousness and economical expediency) and practically isn't used nowadays. This leads to the wide spreading of nonexplosive vibration sources [28].

#### 4.1.6. Seismic microzonation using nonexplosive impulse impact

The features of SMZ methods development led to the situation when the tool of elastic wave excitation with the help of low-powered sources (for example, hammer impact with m = 8–10 kilograms) has become the most wide spread in the CIS countries, in order to determine S- and P-wave propagation velocities in soils of the typical areas of territory. Velocity values are used in order to calculate the intensity increment using the tool of seismic rigidities by S.V.Medvedev [25]:

$$\Delta I = 1.67 \lg \rho_i V_i / \rho_0 V_0 \tag{9}$$

where: $\rho_0 V_0$ and $\rho_i V_i$ is the product of the soil consistency and P-wave (S-wave) velocity – seismic rigidities of the reference and the investigated soil, respectively.

The intensity increment, caused by soil watering, is calculated by the formula

$$\Delta I = K e^{-0.04 h_{GL}^2} \tag{10}$$

where K = 1 for clay and sandy soils; K = 0.5 for large-fragmental soils (with sandy-argillaceous filler not less than 30%) and strongly weathered rocks; K = 0 for large-fragmental firm soils consisting of magmatic rocks (with sandy-argillaceous filler up to 30%) and weakly weathered rocks; $h_{GL}$ is the groundwater level.

The given approach of S.V.Medvedev gained unexpectedly wide extension in 70-es of the XX century due to its simplicity and efficiency (CIS countries and countries of Eastern Europe, USA, Chile, Italy, India). This approach could be realized on practice in a very short time etc. thanks to its territory seismic regime independence. To a certain extent this blocked the development of other alternative approaches. At the same time negative sides of such approach soon were revealed. In the case of watered soils absence formation of intensity increment exceptionally due to type of correlated soils didn't hold up against criticism and gave inconformity with displayed differences of correlated soils at strong earthquakes. Thereafter it brought to the approach disregarding almost in all countries besides CIS countries [25, 28]. On the other hand it is necessary to notice that in case of such work performance all opportunities must be used.

By means of the special investigations it was determined that the reliability of calculated intensity increments considerably increases at usage of modern powerful impulsive energy sources (Fig. 7).

**Figure 7.** Surficial gas-dynamical pulse source (SI-32)

The lowering of final results quality is to a certain extent caused by the fact that in the tool of "intensities" the seismic effect dependence in soils on frequency or "frequency discrimination" of soils [22] and also the origin of typical "nonlinear effects" at strong movements isn't taken into account. A.B.Maksimov tried to remedy this deficiency by developing the tool, where frequency peculiarities of soils were taken into account [25]:

$$\Delta I = 0.8 \lg \rho_0 V_0 f_0^2 / \rho_i V_i f_i^2 \tag{11}$$

where: $f_0$, $f_i$ are predominant frequencies of reference and investigated soils, respectively.

A.B.Maksimovs' tool didn't find wide distribution, as frequency differences of soil vibrations with sharply different strength properties (at usage of traditional for the seismic exploration of small depths low-powered sources) were insignificant and the calculation results on the formulas (9) and (11) were practically similar [22].

Intensity increment was determined by the following formula [25]:

$$\Delta I = 0.8 \lg \rho_0 V_0 f_{wa0}^2 / \rho_i V_i f_{wai}^2 \quad (12)$$

where: $f_{wa0}$, $f_{wai}$ are weighted-average vibration frequencies of reference and investigated soils, respectively.

Weighted-average vibration frequency of soils was calculated at that on the formula [22, 28]:

$$f_{wa} = \sum A_i f_i / \sum A_i \quad (13)$$

where: $A_i$ and $f_i$ are the amplitude and the corresponding frequency of vibration spectrum, respectively.

#### 4.1.7. Seismic microzonation using vibration impact

At usage of a vibration source (Fig. 8) the calculation of intensity increment is realized with the help of the formula [25]:

$$\Delta I = 2 \lg S_i / S_0, \quad (14)$$

where: $S_i$ and $S_0$ are the squares of vibration spectra of investigated and reference soils, respectively.

**Figure 8.** Vibration source (SV-10/100)

The developed tool was used at SMZ of the territories of cities Tbilisi, Kutaisi, Tkibuli, single areas of the Large Sochi city. The tools' feature consists in the fact that it allows to assess soil seismic hazard without any preliminary investigations: at realization of direct measurements of soil thickness response on standard (vibration or impulse) impact. Later the formula was successfully used at SMZ of the sites of Novovoronezh atomic power-plant (APP) with the help of an impulsive source.

### 4.1.8. Seismic microzonation on the basis of taking into account soil nonlinear properties

The comparison of the absorption and nonlinearity indices with the corresponding spectra of soil vibrations shows that at higher absorption the spectrum square prevails in LF field and at high nonlinearity it prevails in HF field of the spectrum. In other words, the presence of absorption is displayed in additional spreading of LF spectrum region, and the presence of nonlinearity – in spreading of HF range.

All the mentioned allowed to obtain the formula for calculation of intensity increment on the basis of taking into account nonlinear – elastic soil behavior or elastic nonlinearity (at usage of vibration source) [25, 28]:

$$\Delta I = 3\lg A_i f_{wai} / A_0 f_{wa0}, \tag{15}$$

where: $A_i f_{wai}$, $A_0 f_{wa0}$ are the products of spectrum amplitude on weighted-average vibration frequency of investigated and reference soils, respectively.

The formula (14) characterizes soil nonlinear–elastic behavior at the absence of absorption.

If the impulsive source is used at SMZ then the formula will have the form [25]:

$$\Delta I = 2\lg A_i f_{wai} / A_0 f_{wa0}, \tag{16}$$

### 4.1.9. Seismic microzonation based on accounting of soil inelastic properties

The estimation of potential soil nonelasticity adequately and physically proved at intensive seismic loadings is the most important problem of SMZ as soil liquefaction and differential settlement of the constructions are observed at strong earthquakes (Niigata, 1966; Kobe, 1995).

For direct assessment of soil nonelasticity the specific scheme of the realization of experimental investigations (fig. 9, a) with gas-dynamic impulsive source GSK-6M (with two radiators) was used. Chosen longitudinal profile location allowed making impact sequentially by two emitters from near and somewhat far radiation zones. The HF component that quickly attenuates with distance (Fig. 9, b) prevails in the spectrum of soil vibrations, caused by near emitter. In a case of distant emitter impact the LF component predominates in the spectrum of vibrations (Fig. 9, c). In other words, at nonlinear-elastic deformations the main energy is concentrated in the HF range of spectrum and at nonelastic – in the LF range. The signal spectrum has the symmetrical form in the far and practically linear-elastic zone.

Elastic linear and nonlinear vibrations are characterized for the given source by the constancy of the real spectrum square, which is the index of definite source energy value, absorbed by soil (which is deformed by the source). The analysis of strong and destructive earthquake records and also the analysis of specially carried out experimental impacts showed that at nonelastic phenomena spectra square of corresponding soil vibrations is not the constant value. It can decrease and the more it decreases, the less the soil solidity and the greater the impact value is [28].

At usage of vibratory energy source, the whole number of new formulas [25] was obtained in order to assess soil seismic hazard with taking into account the amount of their nonelasticity:

$$\Delta I = 2.4\left[\lg\left(S_{ri}\right)_n\left(S_{r0}\right)_d / \left(S_{ri}\right)_d\left(S_{r0}\right)_n\right], \qquad (17)$$

where: $(S_{ri})_{n,d}$ $(S_{r0})_{n,d}$ are the squares of real spectra of soils under investigation and reference soils in near and distant zones of the source, respectively.

$$\Delta I = 3.3\lg[(A_i f_{awi})_n (A_0 f_{aw0})_d / (A_i f_{awi})_d (A_0 f_{aw0})_n], \qquad (18)$$

where: $(A_i f_{awi})_{n,d}$ and $(A_0 f_{aw0})_{n,d}$ are the amplitudes and weighted-average frequencies of soils under investigation and reference soils in near and distant zones of the source, respectively.

If a powerful impulsive source is used the offered formulas will be as following:

$$\Delta I = 1.2[\lg(S_{ri})_n (S_{r0})_d / (S_{ri})_d (S_{r0})_n], \qquad (19)$$

where: $(S_{awi})_{nd}$ and $(S_{aw0})_{nd}$ are the squares of real spectra of soils under investigation and reference soils in near and distant zones of the source, respectively;

$$\Delta I = 2\lg[(A_i f_{awi})_n (A_0 f_{aw0})_d / (A_i f_{awi})_d (A_0 f_{aw0})_n], \quad (20)$$

where: $(A_i\ f_{awi})_{n,d}$ and $(A_0\ f_{aw0})_{n,d}$ are the amplitudes and weighted-average frequencies of soils under investigation and reference soils in near and distant zones of the source, respectively.

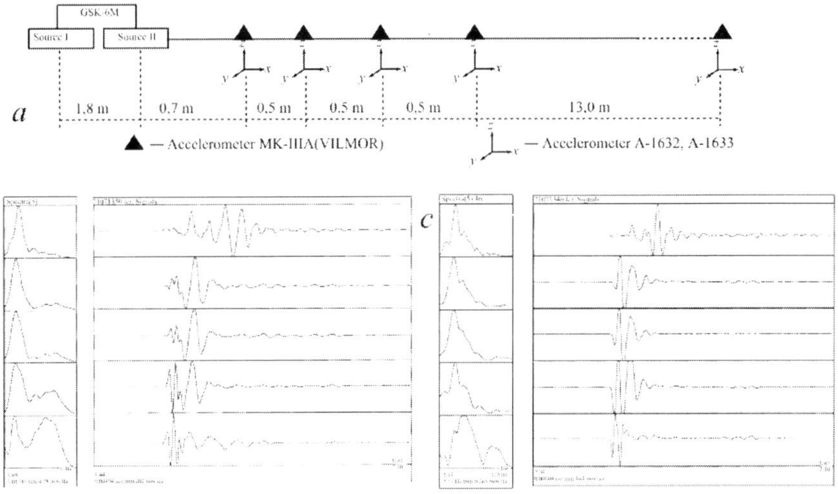


**Figure 9.** Investigation of site spectral features by means of GSK-6M seismic source: a) experiment scheme; b) records of first source impact; c) records of second source impact

The formulas (17) and (18) are adequate only for loose dispersal soils. The formulas (17) and (18) were used at SMZ of Kutaisi city territory. Besides, using the formulas (19) and (20) nonelastic deformation properties of soils in full-scale conditions on Novovoronezh APP-2 site were defined more accurately [25, 28]. The formulas were obtained based on physical principle that underlies the scheme used at the soil looseness assessment.

## 4.2. Calculational method of seismic microzonation

In order to analyze the features of soil behavior with introduction of definite engineering–geological structure characteristics of investigated site as initial data the calculational method of SMZ is used: values of shear wave velocities, modulus of elasticity, index of extinction, power of soil layers, their consistency etc. Calculational method includes the following techniques: thin-layer medium, multiple-reflected waves, finite-difference method, finite-elements analysis (FEA) and others.

Nonlinear soil properties can be taken into account in the problems of earthquake engineering with the usage of instrumental and calculation methods. The main method of seismic microzonation is the instrumental method. However, it is often necessary to use calculational method for solving such problems. Calculational method allows modeling virtually any conditions that are observed in the nature. The requirements of practice however reduced to the necessity of calculation of soil vibrations for nonlinear-elastic and nonelastic deformation conditions. Solving such a problem it is assumed that elastic half-space behaves as linear-elastic medium and at intensive seismic or dynamic impacts the covering soil stratum displays strong nonlinear properties.

Received instrumental stress-strain dependences can be applied, for example, for plastic clay soil shown in Fig. 10. Offered by A.V.Nikolaev [13] conception of the so-called soil bimodularity is taken into account in that dependence [25]. Considerable differences in “weak” soils behavior at compression and extension underlie in the phenomenon. Such soil is characterized at extension by very small modulus of shearing.

Solving of the given nonlinear problem for soils in the analytic form is usually based on considerable assumptions due to the complication of adequate accounting of behavior features of such complicated system as the soil. Thus, the numerical solving of nonlinear problems on the present-day stage of knowledge is the most proved under the condition that the data of field or laboratory investigations are considered in these or those connections [23, 24, 28].

So, the basis for solution of calculation nonlinear problems is the correlation determined using experimental investigations. Otherwise stated, programs for solving of calculation nonlinear problems are in essence analytical-empirical. Such programs like SHAKE, NERA etc. are the most adequate.

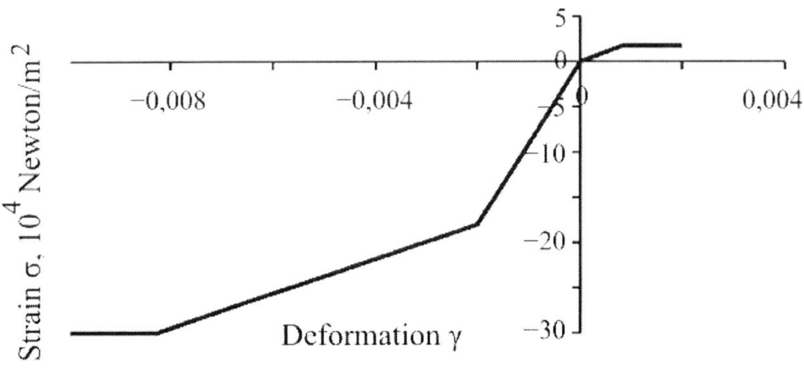


**Figure 10.** Instrumental stress-strain curve, showing property of soil bimodularity

### 4.2.1. Equivalent linear model – SHAKE and EERA programs

One of the first models which take into account nonlinear soil behavior is equivalent linear model. Equivalent linear approximation involves Kelvin–Voight model's modification (for taking some types of nonlinearity into account) and, for example, is realized in the programs SHAKE [2] and EERA [27]. Equivalent linear model is based on the hypothesis that shear modulus G and attenuation coefficient ξ are the functions of shearing strain γ. At calculations in both programs the parameters of soil multilayered structure were accounted in "natural occurrence" through the introduction of values of shear modulus G and attenuation coefficient ξ for each layer in thickness structure. With the help of layer combination it allowed receiving a necessary deformation level which fully corresponds to real thickness deformation at hard loads.

### 4.2.2. NERA program

It is necessary to notice that approach used in the EERA work out became later the basis of new computer program of NERA (Nonlinear Site Response Analysis) [2]. This program allows calculating nonlinear reaction of soil thickness on seismic impact. It is based on the medium model that was offered by Iwan (1967) and Mroz (1967). For short this model is often called the IM model. It is demonstrated that this model supposes strain-deformation simulating of nonlinear curves, using a number of n mechanical elements, which have different sliding resistance $R_j$ and stiffness kj, where $R_1 < R_2 <... < R_n$. Initially the residual stresses in all elements are equal to zero. At monotonically increasing load the element j deforms until the transverse strain τ reaches $R_j$. After that the element j keeps positive residual stress, which is equal to $R_j$. The equation, describes dynamics of soil medium, is solved by the method of central differences.

### 4.2.3. Calculation of nonlinear absorptive ground medium vibrations using multiple reflected waves' tool of seismic microzonation

Let's suppose that we have the seismic wave, which falls on the soil thickness surface. Let's assume that soil thickness is nonlinear absorptive unbounded medium with the density ρ and S-wave propagation velocity $v_S$. At small deformations the value of shear modulus G will be maximum for the given soils:

$$G = G_{max} = \rho v_S^2 \qquad (21)$$

At the deformation increase the value G remains constant at first but at reaching some value (which is definite for each material or soil) the value G considerably changes, i.e. the soil begins to display its nonlinear properties. At the continued deformation increase the growth of stresses decelerates and then can remain unchanged until material destruction or hardening, i.e. until structural condition change.

As the main soil index, which characterizes its type and behavior at intensive loads, the value of plasticity PI was chosen. The parameters, which are necessary for calculations, are determined on basis of empirical ratios [7, 25]:

$$k(\gamma, PI) = 0.5\left\{1 + \tanh\left[\ln \frac{0.000102 + n(PI)}{\gamma}\right]^{0.492}\right\} \qquad (22)$$

where

$$n(P_I) = \begin{cases} 0.0 & for \quad \mathrm{PI} = 0, \\ 3.37 \cdot 10^{-6} PI^{1.404} & for \quad 0 < PI \le 15, \\ 7.0 \cdot 10^{-7} PI^{1.976} & for \quad 15 < PI \le 70, \\ 2.7 \cdot 10^{-5} PI^{1.115} & for \quad PI > 70\ ; \end{cases}$$

$$d = 0.272\left\{1 - \tanh\left[\ln\left(\frac{0.000556}{\gamma}\right)^{0.4}\right]\right\} e^{-0.0145 PI^{1.3}}.$$

Then the change of shear modulus is determined on basis of the ratio

$$\frac{G}{G_{max}} = k(\gamma, PI)(\sigma)^{d}, \tag{23}$$

where G is the current shear modulus, σ is normal stress.

Seismic energy absorption is calculated by the formula

$$\xi = 0.333\frac{1+\exp\left(-0.0145PI^{1.3}\right)}{2}\left[0.586\left(\frac{G}{G_{max}}\right)^{2} - 1.547\frac{G}{G_{max}} + 1\right] \tag{24}$$

On the basis of the given ratios and introduced by us ratios for determination of necessary indices (normal stress, deformation etc), nonlinear version of the program ZOND was worked out [25]. From the database of strong motions AGESAS, which was formed by us [26, 28], the accelerogram, which was recorded on rocks in Japan, with the characteristics (magnitude, epicentral distance, spectral features etc.) similar to the territory of Tbilisi city, was chosen as the accelerogram, given into the bedrock.

The analysis of the results of linear and nonlinear calculations models of definite areas of Tbilisi city territory confirms the adequacy of calculations to the physical phenomena, which were obtained in soils at intensive loads (Fig. 11). With the increase of seismic impact intensity the nonlinearity display increases. Absorption grows simultaneously. Hence the resulting motion at quite high impacts levels can be lower than the initial level. It corresponds to the fact, which is known on the results of analysis of strong earthquake consequences, which happened in recent yares (for example, Northridge earthquake, 1994).

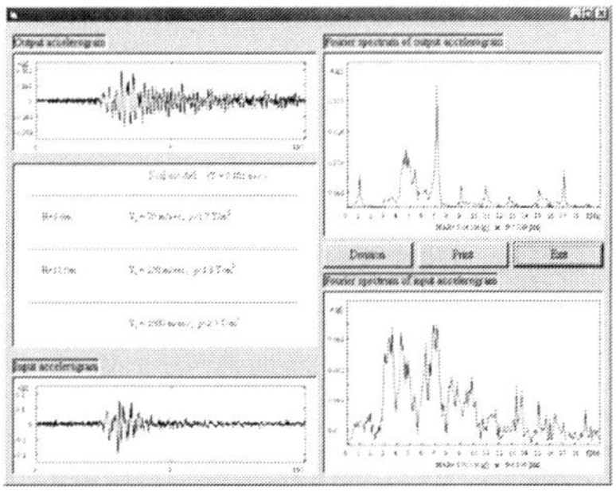
(a)

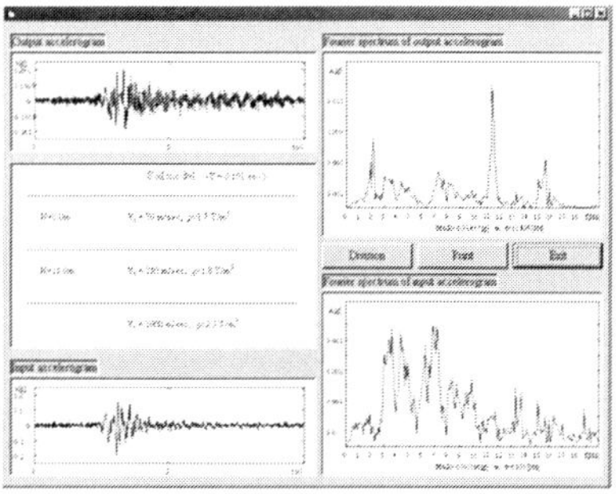
(b)

**Figure 11.** Results of calculations using multiple reflected waves' tool in linear (a) and nonlinear (b) cases.

### 4.2.4. Calculation of nonlinear soil response using FEM tool of seismic microzonation

The problem of the determination of soil massif response on dynamic impact with taking soil nonlinear properties into account can be solved by usage of finite element method (FEM) in the following way [25].

Soil medium is represented in the form of two-dimensional massif, which is approximate by triangular finite elements. The net, which consists of triangular elements, allows to describe quite accurately any relief form and form of the layer structure of soil massif with its physics-mechanical parameters. Within finite element the soil is homogeneous with inherent to its characteristics, which vary in time depending on impact intensity. Earthquake accelerogram of horizontal or vertical direction, which is applied, as a rule, to the foundation of soil massif, is used as the impact. Soil is in the conditions of plane deformation and it is considered as an orthotropic medium. Axes of the orthotropy coincide with the directions of main strains [28]. The problem of nonlinear dynamics of soil massif is solved by means of the consecutive determination of mode of deflection of the system on the previous step. The system is linear-elastic on each step.

## 4.3. Instrumental-calculational method of seismic microzonation

In recent years a new «instrumental-calculational» method of SMZ (per se simultaneously having the features of both instrumental and calculational method) which includes tool of «instrumental-calculation analogies» has been developed in Russia in recent years [25]. Its usage is based on direct usage of modern databases of strong movements.

As a basis at realization of tool instrumental database of strong movements, registered in definite soil conditions, is used. As a result of given database with the help of numerical calculations it is possible more or less safely to forecast behavior of these or those soils (or their combination) for strong (weak) earthquakes with typical characteristics for the investigated territory (magnitude, epicentral distance, focus depth etc.).

## 4.4. Relief influence on the earthquake intensity in SMZ problems

The correlation analysis of the dependence of seismic intensity increment on true altitude, slope steepness and relief roughness showed that the main factors, which change the value of seismic intensity, are the first two indices

[25]. It conforms well to the investigation results of V.B.Zaalishvili, who introduced the new parameter of the relief coefficient (Fig. 12):

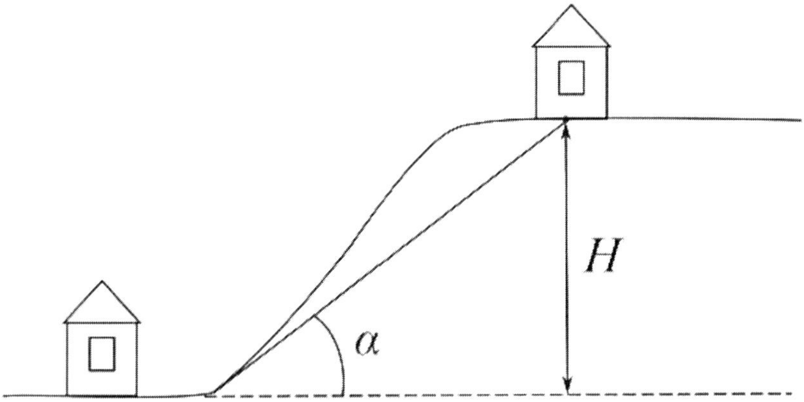


$$R = \alpha \times H$$

**Figure 12.** Relief coefficient R

Later the data analysis allowed to offer the empirical formula for the possible amplification calculation K and intensity increment ΔI, which are caused by the relief [25]:

$$K = -0.1 + 0.68 \lg R \tag{25}$$

where R= α × H is the relief coefficient; α is the relief slope angle, degree; H is height, m.

The analysis of the experimental data shows that intensity increment can vary at that independently of the type of rocks, from 0 to 1.5 degree.

Finally, let's try to assess the amplification of vibrational amplitude, which is caused by relief, with the help of the calculational method of FEM (Fig.13).

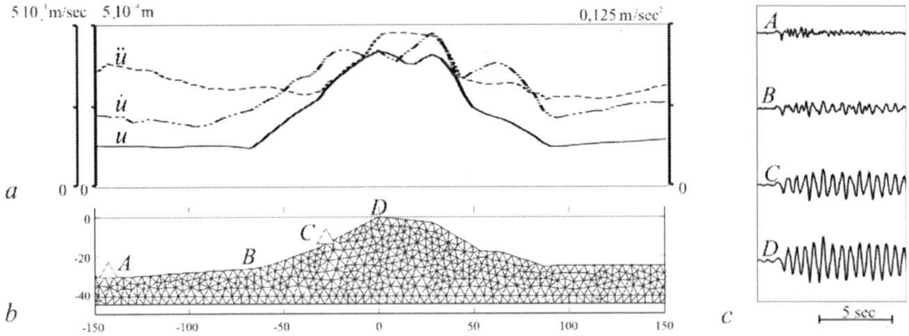

**Figure 13.** Final elements analysis (FEA) application example: a) Variation of amplitudes of displacement, velocity and acceleration along surface; b) calculational model; c) seismograms, calculated in points A, B, C, D.

It was determined that the vibrational amplitude considerably chances with the relief. The given dependence at that is various for the displacements, velocities and accelerations. The largest value of the amplification is observed for displacements and the maximum ratio of vibrational amplitudes, for example, in the point C to the point A, is 2.1 and for the point D – 3.2. It satisfies well the results of experimental observations where the ratio in the point C for the S-wave is equal to 2.3 and in the spectral region the maximum values are 1.8 (at T = 0.4 s) and 3.2 (at T = 0.7 s) for P- and S-waves accordingly. Spectral analysis also shows the resonance increase of vibrational amplitudes in the top part of the slope on the frequency 1.6 Hz (i.e. T=0.6 s).

Considerably fewer investigations are dedicated to the influence of the underground relief on the intensity. At the vee couch of the rocks, which are covered by sedimentary thickness, the ratio between wave length and the sizes of vee stripping have influence on seismic intensity change. Seismic intensity increment in the given case is formed by the wave interference and can be 1.5–2.0 degree.

Thus, at the execution of SMZ works in the mountain regions or under the conditions of billowy relief, it is necessary to pay special attention to the influence of surface or underground relief on the intensity forming. It is necessary to continue the investigations in order to obtain statistically proved ratio for the calculation of intensity increment, caused by relief.

## 4.5. Seismic microzonation of Vladikavkaz city

If we consider 5% DSZ map as basis for seismic microzonation so seismic intensity of 8 points corresponds to reference grounds for whole territory. Then, maps of seismic microzonation of cities must be created. According to the above mentioned maps of detailed zoning the maps of seismic microzonation with probability 1%, 2%, 5% or 10 %, correspondingly, were made up (Fig. 14).

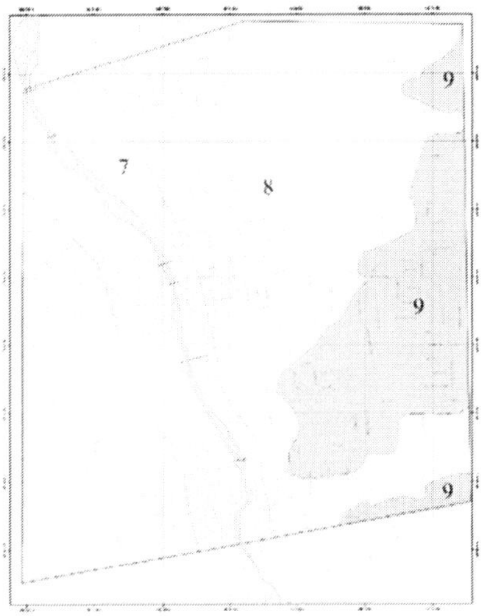


(a)

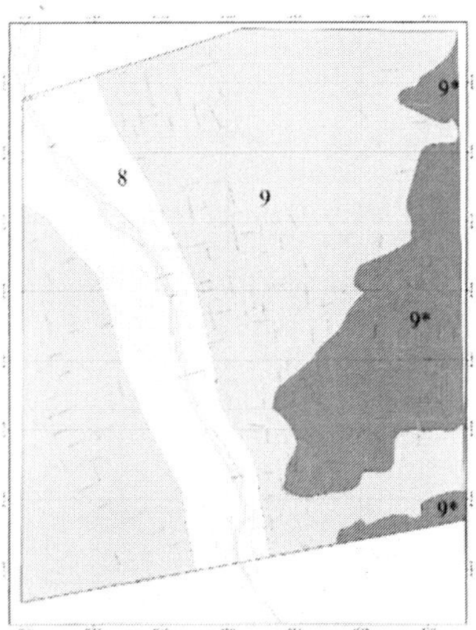


(b)

**Figure 14.** The maps of seismic intensity microzonation for probabilities of 5% (a) and 2% (b) for the central part of Vladikavkaz city territory [26, 28].

Though, that definitions of the word «zoning» are similar, actually they are quite different in essence. Unlike the maps of detailed seismic zoning, which give seismic potential (Mmax) and source features, the maps of seismic microzonation give assessments of soil condition influence (sands, rocks, pebbles, clays etc., their combination; watering; relief (as underground as surface); spectral distribution of incoming wave; predominant vibration frequencies on city square etc.) on forming of future earthquake intensity. It should be noted that as a basis the maps of different probability of exceedance will be used and as the initial intensity, the value of which corresponds directly to the intensity of the sites, composed by average soils or characterized by average soil conditions and, therefore, the maps will be referred to the 7, 8 or 9 points (and similarly for acceleration). The zones, composed by clay soils of fluid consistency, which can be characterized by liquefaction at quite strong impacts, are marked by the index 9*. Intensity calculation here supposes the usage of special approaches in the form of direct taking soil nonlinearity into account [25]. The maps in accelerations units show the similar results. As a rule, the scale of such maps is 1:10 000, in order to have the opportunity of taking them into account at building.

Engineering-geological zoning of territory is the basis of seismic microzonation. It assumes detailed investigations of features of the territory. Such works are quite expensive and labor-consuming. In the same time they are characterized by locality. During design of seismic microzonation maps it is important to adjust and refine data of engineering-geological conditions of investigated territory which are always exists in one or another form and can be used as approximate basis. And finally general view of these conditions must be given. Proven typification of main factors (thickness of quarternary deposits, velocity profile, groundwater level, surface and underlying soils slope angle, etc.). Actual examples of such investigations will be given– (Fig. 15):

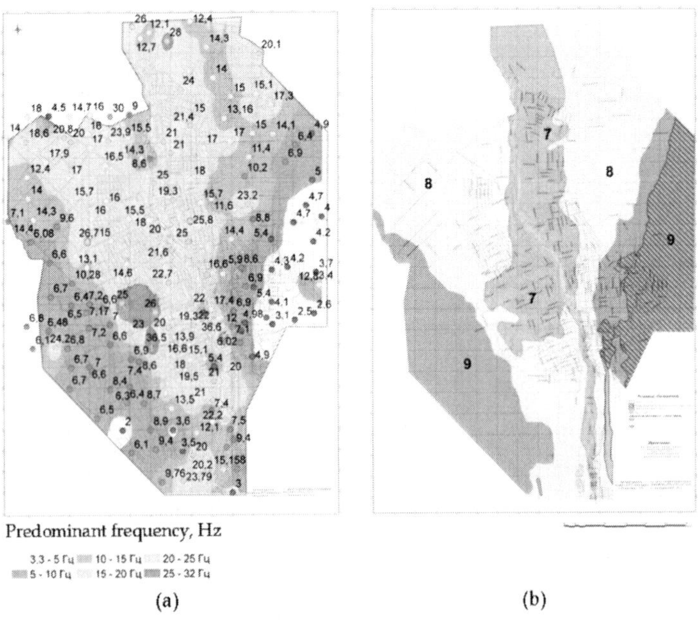


(a) (b)

**Figure 15.** Verification of engineering-geological conditions by means of H/V technique (a) and final map of seismic microzonaton of Vladikavkaz city (b).

Seismic microzonation maps, as the direct basis of earthquake design and practical construction, are maps of seismic hazard. Maps of seismic microzonation not just show where earthquake-resistant buildings are necessary. They show for what intensity buildings and constructions must be designed: 6, 7 or 8 or 9 points. This supposes the attachment of various financing for the implementation of the anti-seismic measures. It should be noted that the seismic zonation maps and microzonation can and should be constructed as well in units of acceleration. This will help to implement a more smooth transition from the borders of one seismic zone to another, thereby increasing the reliability of their allocation.

All types of maps and in particular maps of general and detailed seismic zonation as well as microzonation are nowadays formed in GIS technologies. The use of GIS technologies allows to lay a variety of information on a particular area or the whole territory, for example the city in the form of layers and to investigate their integral effect on the characteristics of seismic hazard occurrence.

# 5. SPECIFIED SEISMIC FAULT AND DESIGN SEISMIC MOTION

Analysis and consequent account of initial accelerograms transformation will become the basis for site effect analysis at strong seismic loadings (Fig. 16) [25, 28].

Methods of such modelling are based on accordance of spectral properties of modelled and real earthquake. In a whole modelling accuracy depending on the purposes of total motion usage and what characteristics defining structural system behaviour must be reproduced.

Earthquake source that is a region of rupture can be considered as point source only for much larger distances than fault size. At close distances effects of finite fault size become more significant. Those phenomena are mainly connected with finite rupture velocity, which causes energy radiation of different fault parts in different times and seismic waves are interference and causes directivity effects [3, 4].

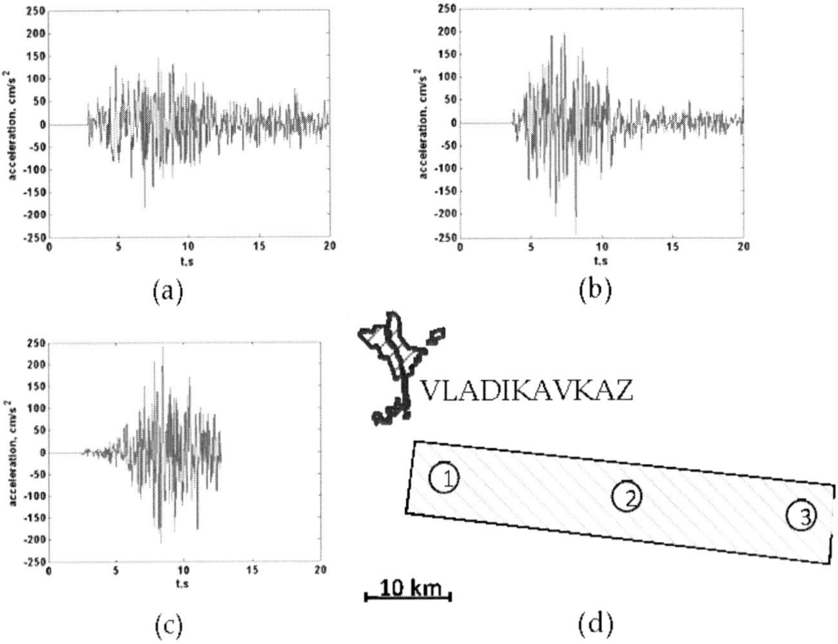


**Figure 16.** Synthetical accelerograms for different source locations: a – western part of fault; b – middle part of fault; c – eastern part of fault; d – scheme of sources of scenarios earthquakes

Let's compare amplitude spectra of obtained design accelerograms with spectrum of real earthquake from considered fault. Data analysis (Fig. 17 and Fig. 18) shows that spectra of calculated and real earthquakes in a whole are similar in their main parameters. It must be noted that spectrum of vertical component of real earthquake is closer to design spectra. The last fact is quite obvious and is explained by proximity to earthquake source. Indeed, close earthquakes in general are characterized by predomination of vertical component. Record of TEA station (located in theatre) was selected due to its location on dense gravel and has a minimal distortions caused by soil conditions.

Analysis of spectrum of weak earthquake shows that peaks are observed on 1.3 and 5.6 Hz (Fig. 17). In spectra of synthesize accelerograms mentioned amplitudes are also observed. At the same time medium response on strong earthquake, undoubtedly, differ from weak earthquake response (Fig. 18).

Usage of maps of detailed seismic zoning in units of accelerations at seismic microzonation level is possible only for calculation method giving results in units of accelerations. Today traditional instrumental method of seismic microzonation does not allow obtaining intensity increments in accelerations due to traditional orientation on macroseismic intensity indexes. The exclusion is the case of investigation of strong earthquakes accelerations when instrumental records are obtained. At the same time investigations are conducted and the problem is supposed to be solved.

On the other hand in recent years a new instrumental-calculation method was developed [25]. New method is based on selection from database (including about 5000 earthquake records) soil conditions which are the most appropriate to real soil conditions of the investigated site. Then the selection of seismic records with certain parameters or their intervals follows (magnitude, epicentral distance, and source depth). Then maximal amplitudes are recalculated for given epicentral distances. Absorption coefficient can be calculated by attenuation model for given region.

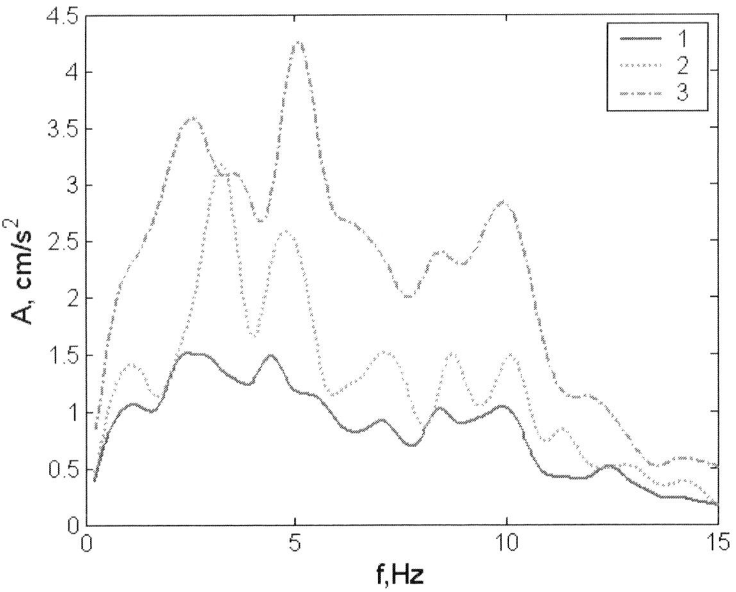


**Figure 17.** Spectra of design accelerograms at different source locations of earthquake M=7.1: 1 – western part of fault; 2 – middle part of fault; 3 – eastern part of fault

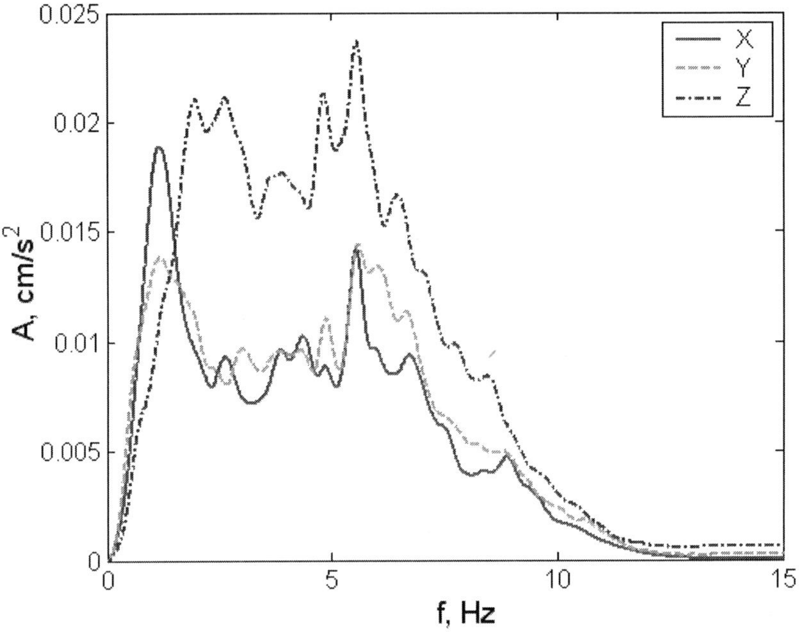


**Figure 18.** Spectra of accelerograms of weak earthquake with epicentre in the zone of Vladikavkaz fault. (25.08.2005 10:25 GMT, H = 8 km M= 2.5).

Thus, a new complex method of seismic hazard assessment providing probability maps of seismic microzonation, which are the basis of earthquake engineering, is introduced. Undoubtedly such approach significantly increases physical validity of final results.

Today, we have conditions for detailed seismic zoning maps development like the above mentioned but for all the territory of the Northern Caucasus on the basis of the modern achievements of engineering seismology. Thus algorithm of seismic hazard assessment of the territory that is taking into account multiple factors forming seismic intensity was considered. Forms of typical seismic loadings for firm soils are given, which will be changed from site to site in dependence of differences in ground conditions (engineering-geological, geomorphological and hydrogeological conditions).

# 6. CONCLUSIONS

1. The goal of the work was to analyze the modern concepts of the seismic hazard of the territory, its evaluation and development of an algorithm for such assessments. One of the main problems is to account a level of possible result errors, or adequacy of such assessments. The evolution of methods and techniques for seismic hazard assessments is presented. The algorithm of direct account of certain characteristics of the territory is given. The integrity of seismic hazard assessments is specified. Calculation of the expected seismic effects is considered.
2. The physical basis for creation of probability maps of general seismic zoning on the example of Russia (occupying a huge area) which needs small-scale mapping of seismic hazard of type 1:2 000 000-8000 000 for the management purposes are considered. On the other hand, seismic hazard assessment of a particular territory involves a need for a more detailed account of geological and geophysical features of the investigated territory.
3. Further the physical principles and methods for creating of probability (1%, 2%, 5%, 10%) detailed seismic zoning maps of a particular region in GIS technologies, characterized by greater large-scale: 1:25 000 – 200 0000, which corresponds to return period of maximum probable earthquake of 5000, 2500, 1000 and 500 years are considered. At the same time, geological and geophysical features are accounted here undoubtedly more accurately. The future creation of detailed seismic zoning maps of vast territories is substantiated.
4. The physical basis, methods and techniques for creation of seismic microzonation maps, including the use of modern high-power non-explosive sources (vibration and impulsive action) are considered. The physical formation mechanisms of algorithms of direct account of a number of soils indicators under heavy loads, which are the basis of relevant computer programs, are considered. It identifies changes or distortion of the amplitude-frequency characteristics of the original or the incoming wave field of seismic impact caused by the interaction of absorption and nonlinearity (or inelasticity) phenomena in different typical soils of the territory.
5. Despite the similarity of names, the physical mechanisms of seismic microzonation are fundamentally different from the general and detailed seismic zoning. Seismic microzonation takes into account

soil conditions and the territory is a direct basis of earthquake engineering and construction. At the same time to detailed zoning map of certain probability (for example, 10%) will correspond seismic microzonation map of the same probability obtained on its basis, i.e. likelihood of recurrence of 10% or 500 years.

6. The possibility of successful differentiation of soil conditions on the basis of the analysis of the relationship of the horizontal vibration spectrum of the initiated signal to the vertical spectrum and the predominant frequency of the ground motion is shown. The process of formation of seismic microzonation map of modern urban territory is considered.

7. At the end, based on the direct account of the features of the seismic source (for example, characteristics of the seismogenic fault) the formation of the expected seismic effect is shown. Comparison of calculated and real impacts showed their good similarity.

8. Complex of probabilistic maps of detailed seismic zoning, seismic microzonation maps and type of expected seismic impact is full and physically justified integral assessment of seismic hazard of territory.

## NOMENCLATURE

ACCELEROGRAM - record of ground acceleration changes in time, obtained by accelerographs.

BEDROCK - a relatively hard, solid rock that commonly underlies soil or other softer unconsolidated sedimentary materials.

DEFORMATION - a change in the original shape and/or volume of a material due to stress and strain.

DENSITY - either (1) the quantity of something per unit measure such as unit length, area, volume, or frequency (see, for example, power spectral density), or (2) the mass per unit volume of a substance under specified conditions of pressure and temperature.

DETAILED SEISMIC ZONING (DSZ) – type of seismic hazard assessment. The main tasks of DSZ – define seismic generating zones, assess focal parameters and their effects.

EARTHQUAKE - a shaking of the Earth that is either tectonic or volcanic in origin or caused by collapse of cavities in the Earth. A tectonic earthquake is caused by fault slip.

EARTHQUAKE ENGINEERING - the field of earthquake engineering is defined as encompassing man's efforts to cope with the harmful effects of earthquakes.

ENGINEERING SEISMOLOGY - that part of seismology which aims primarily at providing seismological data for earthquake engineering, earthquake hazard and earthquake risk applications.

EPICENTER - it is the point on the surface of the Earth, vertically above the place of origin (Hypocenter or Focus) of an earthquake. This point is expressed by its geographical coordinates in terms of latitude and longitude

EPICENTRAL DISTANCE - distance from a site (usually a recording seismograph station) to the epicenter of an earthquake. It is commonly given in kilometers for local earthquakes, and in degrees (1 degree is about 111 km) for teleseismic events.

FOCAL DEPTH - the conceptual "depth" of an earthquake focus.

GENERAL SEISMIC ZONING (GSZ) - type of seismic hazard assessment of a vast territories through the allocation of large seismic generating zones that determine area seismicity. In the result of GSZ are constructed maps in a scale 1:250 000 – 1:8 000 000, allowing rational planning of the development of different areas, to assess the total cost required for the anti-seismic measures on a national scale.

PEAK HORIZONTAL ACCELERATION - the maximum acceleration amplitude measured (or expected) in a strong-motion accelerogram of an earthquake, abbreviated PHA.

SEISMIC MICROZONATION (SMZ) – type of seismic hazard assessment. Initial seismicity or region intensity is set by maps of general and detailed seismic zoning (GSZ and DSZ). SMZ takes into account soil conditions that increase or decrease initial intensity of seismic vibrations.

VELOCIGRAM - record of ground velocity vibrations in time, obtained by velocigraphs.

## REFERENCES


1. Aptikaev, F.F. et al. (1986) Methodological recommendations on detailed seismic zoning. Questions of engineering seismology. Issue 27. Moscow, 1986. 184-212. (in Russian)
2. Bardet J.P., Tobita T., NERA, A computer program for Nonlinear Earthquake site Response Analyses of layered soil deposits. Univ. of Southern California, Los Angeles, 2001. 44 p.
3. Beresnev, I. A., Atkinson, G. M. (1997). Modeling finite fault radiation from ωn spectrum. Bull. Seism. Soc. Am., 87, 67–84.
4. Beresnev, I. A., Atkinson, G. M. (1998). FINSIM – a FORTRAN program for simulating stochastic acceleration time histories from finite faults. Seismological Research letters. Vol. 69. No. 1.
5. Cornell C. A. (1968) Engineering risk in seismic analysis. Bull. Seism. Soc. Am. 54 1968, pp. 583-1606
6. Gamkrelidze, I., T. Giorgobiani, S. Kuloshvili, G. Lobjanidze, G. Shengelaia (1998). Active Deep Faults Map and the Catalogue for the Territory of Georgia // Bulletin of the Georgian Academy of Sciences, 157, No.1, pp. 80-85.
7. Ishibashi, I. and Zhang, X. (1993). "Unified dynamic shear moduli and damping ratios of sand and clay," Soils and Foundations, Vol. 33, No. 1, pp. 182-191.
8. McGuire R. (1995) Probabilistic Seismic hazard analysis and design earthquakes: closing the loop. vol. 83, No. 5, pp.1275-1284
9. Medvedev, S.V. (1947) On the question of taking into account seismic activity of region at construction. Procs. of seismological institute of AS USSR. No 119, 1947(in Russian)
10. Musson R. (1999) Probapilistic seismic hazard maps for the North Balkan region. 1999. Annali di Geofisica. vol. 42, No. 6, pp. 1109-1124.
11. Nakamura Y, A Method for Dynamic Characteristics Estimation of Subsurface using Microtremor on the Ground Surface. QR of RTRI, Volume 30, No. 1, 1989
12. Nesmeyanov, S.A. (2004) Engineering geotectonics. Moscow: Nauka, 2004. 780 p. (in Russian)

13. Nikolaev, A.V. (1987) Problems of nonlinear seismics. Moscow: Nauka, 1987. p. 5-20. (in Russian)

14. Reisner, G. I., Ioganson, L. I. Complex typification of earth crust as basis for fundamental and applied tasks solution. Article 1 and 2. Bull. MOIP, 1997. Geology dept., vol. 72. issue 3. pp. 5-13 (in Russian).

15. Reiter L. Earthquake hazard analysis. New York: Columbia Univ. Press, 1991. 245 p.

16. Riznichenko, Yu.V. (1966) Calculation of points of Earth surface shaking from earthquake in surrounding area. Bull. of AS of USSR. Physics of the Earth. 1966. 5. pp. 16-32. (in Russian)

17. Rogozhin, E.A., Reisner, G.I., Ioganson, L.I. (2001) Assessment of seismic potencial of Big Caucasus and Apennines by independent methods // Modern mathematical and geological models in applied geophysics tasks: selected scientific works. Moscow: UIPE RAS, 2001, pp. 279-300 (in Russian).

18. Schnabel, P. B., Lysmer, J., and Seed, H. B. (1972) " SHAKE: A Computer Program for Earthquake Response Analysis of Horizontally Layered Sites", Report No. UCB/EERC-72/12, Earthquake Engineering Research Center, University of California, Berkeley, December, 102p.

19. Smit P., V. Arkanian, Z. Javakhishvili, S. Arefiev, D. Mayer-Rosa, S. Balassanian, T. Chelidze (2000). The Digital Accelerograph Network in the Caucasus. In: "Earthquake Hazard and Seismic Risk Reduction". Kluwer Academic Publishers, pp. 109-118.

20. Trifonov, V. G. (1999) Neotectonics of Eurasia. Moscow: Nauchniy Mir, 1999, 252 p. (in Russian)

21. Ulomov, V. I., Shumilina, L. S., Trifonov, V. G. et al. (1999) Seismic Hazard of Northern Eurasia // Annali di Geofisica, vol. 42, No. 6, pp. 1023-1038.

22. Zaalishvili, V.B. (2000) Physical basics of seismic microzonation. Moscow: UIPE RAS, 2000. 367 p. (in Russian)

23. Zaalishvili, V.B., Otinashvili, M.G. (2000) Analysis of ground strong motions by means of numerical methods. Theory of constructions and earthquake engineering. No 1, 2000. Tbilisi: ISMIS AS of Georgia. Pp. 67-71. (in Georgian)

24. Zaalishvili, V., Otinashvili, M., Dzhavrishvili, Z. (2000) Seismic hazard assessment for big cities in Georgia using the modern concept of seismic microzonation with consideration soil nonlinearity. INTAS/Georgia/97-0870. Periodic report. 2000. 170p.

25. Zaalishvili, V. B. (2009) Seismic microzonation of urban territories, settlements and large building sites. Moscow: Nauka, 2009, 350 p. (in Russian).

26. Zaalishvili, V.B., Rogozhin E.A. (2011) Assessment of seismic hazard of territory on basis of modern methods of detailed zoning and seismic microzonation. The Open Construction and Building Technology Journal, 2011, Volume 5, pp. 30-40.

27. Zaalishvili, V., Dzeranov, B., Gabaraev, A. Assessment of seismic hazard of territory and creation of probability maps /Geology and geophysics of South of Russia. 2011. # 1. pp. 48-58.

28. Zaalishvili, V.B. Assessment of seismic hazard territory. Earthquake Engineering. Published by InTech. Edited Halil Sezen, 2012, P. 25-64

# 2 CHAPTER

# SEISMIC PERFORMANCE OF COMPOSITE SHEAR WALLS CONSTRUCTED USING RECYCLED AGGREGATE CONCRETE AND DIFFERENT EXPANDABLE POLYSTYRENE CONFIGURATIONS

**Wenchao Liu, Wanlin Cao *, Jianwei Zhang, Qiyun Qiao and Heng Ma**

*College of Architecture and Civil Engineering, Beijing University of Technology, Beijing 100124, Ch*

## ABSTRACT

The seismic performance of recycled aggregate concrete (RAC) composite shear walls with different expandable polystyrene (EPS) configurations was investigated. Six concrete shear walls were designed and tested under cyclic loading to evaluate the effect of fine RAC in designing earthquake-resistant structures. Three of the six specimens were used to construct mid-rise walls with a shear-span ratio of 1.5, and the other three specimens were used to construct low-rise walls with a shear-span ratio of 0.8. The mid-rise and low-rise shear walls consisted of an ordinary recycled concrete shear wall, a composite wall with fine aggregate concrete (FAC) protective layer (EPS modules as the external insulation layer), and a composite wall with sandwiched EPS modules as the insulation layer. Several parameters obtained from the experimental results were compared and analyzed, including the load-bearing capacity, stiffness, ductility, energy dissipation, and failure characteristics of the specimens. The calculation formula of load-bearing capacity was obtained by considering the effect of FAC on composite shear walls as the protective layer. The damage process of the specimen was simulated using the ABAQUS Software, and the results agreed quite well with those obtained from the experiments. The results show that the seismic resistance behavior of the EPS module composite for shear walls performed better than ordinary recycled concrete for shear walls. Shear walls with sandwiched EPS modules had a better seismic performance than those with EPS modules lying outside. Although the FAC protective layer slightly improved the seismic performance of the structure, it undoubtedly slowed down the speed of crack formation and the stiffness degradation of the walls.

**Keywords:** recycled aggregate concrete; composite shear wall; seismic performance; expandable polystyrene modules; theoretical analysis; experimental study; numerical simulation

# 1. INTRODUCTION

Along with rapid urbanization, the construction industry has boomed significantly. As a result, construction waste has also increased tremendously. Removed concrete (construction waste) is often considered useless and disposed as demolition waste [1]. The reuse of discarded concrete has become a hot topic in the world leading to studies on waste concrete in many countries. Generally, the aggregate obtained from waste concrete by cleaning, crushing, and grading is called recycled concrete aggregate (RCA), and the concrete made from RCA is called recycled aggregate concrete (RAC). Since the 1950s, the possibility of RAC production from waste concrete has been studied [2]. In the 1970s, the United States reintroduced the use of RCA in nonstructural applications such as in filling materials, foundations, and base course materials [3]. In China, studies on RCA are still in their infancy. The country started to implement regulations [4] promulgated by the Ministry of Construction in 2005; the principle of construction waste reduction, harmless and resource-oriented environmental philosophy was proposed in the regulation. With the development of science and technology, more in-depth studies have been carried out. Kang et al. [5] conducted the flexural testing of reinforced concrete beams with RCAs; the results showed that the flexural behavior was not affected significantly by the use of up to 30% RCA. The fire performance of sustainable RCAs was studied by Gales et al. [6], and ten buildings were constructed using RCA concrete, thus playing an important role in the promotion of RCAs. RAC is no longer considered as a new technology in the world; so far, many studies have mainly focused on its applications [7,8,9,10,11,12].

Shear wall is one of the important structures that can bear a horizontal shear force; many studies have been carried out on shear walls [13,14,15,16,17]. With the increasing number of environmental problems in the world due to the construction industry, it has become necessary to use energy-efficient techniques for construction. Conventional shear walls cause several adverse effects on the environment; therefore, it is not appropriate to use them for constructions. Expanded polystyrene (EPS) composed of polystyrene, pentane, and a fire retardant has been used extensively because of its high chemical stability and excellent heat insulation property [18]. Therefore, new composite shear walls with energy conservation and heat preservation properties are highly desired. Wang et al. [19] conducted several tests on insulating masonry walls with eight different shear-span ratios and a solid wall to evaluate the seismic performance of the walls under a low-cycle repeated loading. Dou et al. [20] conducted four shear walls with different heights and width ratios of polystyrene insulation board, forming a heat insulation bearing integrated wall. Li studied the high shear-span ratio of glazed hollow bead insulation composite shear under a low cyclic loading [21]. Most of the previous studies were based on the combination of ordinary concrete and EPS modules, and mainly focused on the mechanical properties of the structure. However, studies on composite shear walls using a combination of recycled concrete and EPS module are rare. Cao et al. developed a new seismic-energy-saving shear wall structure using recycled concrete and EPS insulating cavity with a single row reinforcement net and surface mortar layer [22]. The combined single row of reinforcement and EPS heat preservation technology was applied to the recycled concrete shear wall. The surface mortar layer significantly affected the seismic performance of the shear wall. Because of the poor fire resistance of EPS modules, the fine aggregate concrete (FAC) outer protective layer was very necessary for the fire prevention in high-rise buildings. Therefore, it is very important to understand the behavior of a composite shear wall formed using EPS modules, recycled concrete, and FAC as the protective layer. In particular, the effect of a FAC protective layer and EPS module on the seismic performance of a composite wall should be evaluated.

Therefore, in this study, six recycled concrete wall specimens with EPS modules were designed for low-cycle reverse-loading tests. The effects of different shear-span ratios, EPS module layout, and FAC protective layer on the load-bearing capacity, stiffness, hysteresis characteristics, and seismic energy consumption of composite walls were analyzed, and the role of EPS thermal insulation modules and outer protection layer during the seismic process was verified. A calculation formula considering the role of FAC protective layer for the load-bearing capacity of the EPS module composite wall is proposed. Thus, this study provides a reference for the seismic design of this type of composite walls.

# 2. EXPERIMENTAL

## 2.1. Material Properties

The RAC, used to prepare the specimens, was collected from an old building in Beijing, China. The discarded concrete was cleaned, crushed, and screened by grade before preparing the finely ground RCAs needed for the specimens as shown in Figure 1a. The RCA was tested according to GB/T 25177-2010 [23]; sieving, water absorption, and hazardous materials tests were carried out. Because the RCA was produced by demolition, components such as bricks were inevitable (~1.62 wt %). The RCA used in the study had an aggregate content of 40.1%, mortar aggregate content of 34.9%, residual mortar content of 23.42%, and toxic material content of 1.62%. The density of the RCA was 2290 kg/m$^3$. Its water absorption capacity was 3.23%. Figure 1b shows the sieve-curve of the coarse aggregate in RCA.

According to JGJ 55-2011 [24], the material consumption amount needed by the recycled concrete was determined based on the mix proportion design method of ordinary concrete. The strength grade of the six recycled concrete specimens was RC40, with the designed mix proportion (cement:water:sand:recycled-pebble) = 1:0.83:4.01:4.01. To be specific, the replacement rate of the coarse aggregates in RAC was 100%, and the fine aggregate of RAC was ordinary sand. Moreover, according to JGJ 55-2011, FAC is a type of ordinary concrete, and the maximum size of coarse aggregate is less than 15 mm. The designed mix proportion (cement:water:sand:pebble) of the FAC was 1:0.58:2.46:3.27.

Table 1 shows the measured mean compressive strength (28 day) of cubes and mean elastic modulus of recycled concrete blocks [25]. Notably, all the tested blocks were naturally cured under the same conditions with the specimens produced in this study.

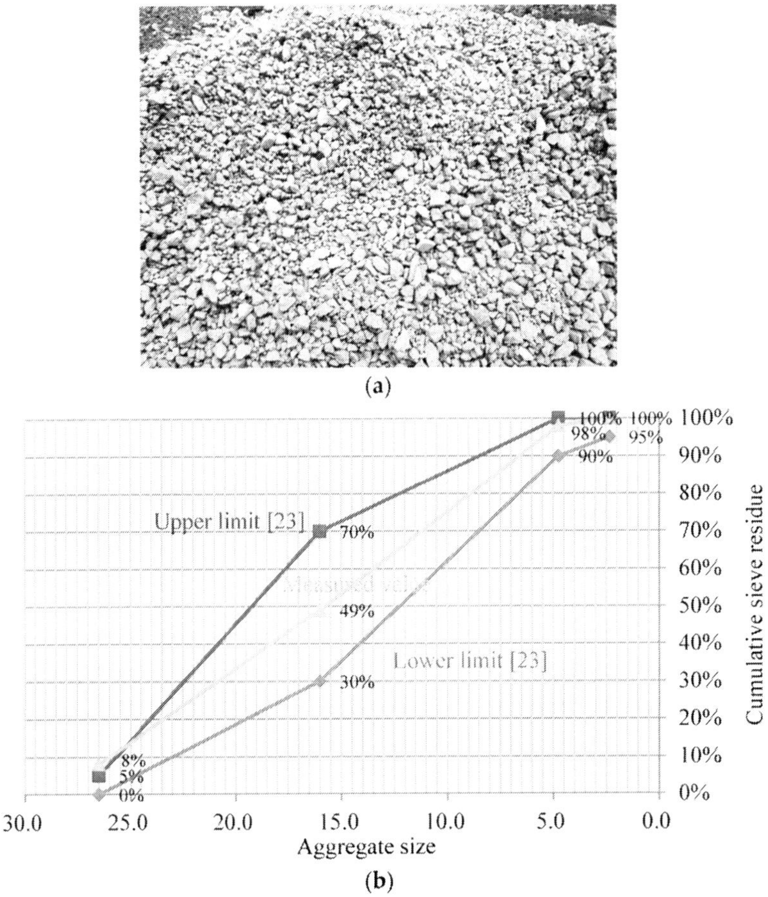


**Figure 1.** The aggregate gradation of the recycled aggregate concrete (RAC). (**a**) The coarse aggregate of RAC (aggregate size is 5–25 mm); (**b**) The sieve-curve of the coarse aggregate.

**Table 1.** Experimental results of concrete mechanical properties. RAC = Recycled aggregate concrete, FAC = Fine aggregate concrete.

| Type of Concrete | Concrete Strength | Cube Compressive Strength (MPa) | Elastic Modulus (MPa) |
|---|---|---|---|
| RAC | RC40 | 42.6 | 32,958 |
| FAC | C40 | 41.8 | 31,553 |

EPS is an excellent energy-saving and thermal insulation material that can be manufactured into EPS module components by industrialization processing [18]. EPS modules were assembled and lap-jointed together by tongue-and-groove modules as shown in Figure 2. The properties of the EPS module are listed in Table 2.

(a) (b)

**Figure 2.** Expanded polystyrene (EPS) module. (**a**) Side elevation of EPS module; (**b**) Plan view of EPS module.

**Table 2.** Physical properties of expanded polystyrene (EPS) module.

| Physical Properties | Value |
|---|---|
| Apparent density (kg/m$^3$) | 18–22 |
| Thermal conductivity (W/(m·K)) | ≤0.039 |
| Compressive strength (MPa) | ≥0.10 |
| Tensile strength of the plate direction (MPa) | ≥0.10 |
| Dimensional stability (%) | ≤0.50 |
| Water absorption (%) | ≤4 |

The rebar diameters used in walls were 4, 6, 8 and 12 mm. The measured mechanical properties of each type of steel are listed in Table 3, where $f_y$ is the yield strength, $f_u$ is the ultimate strength, δ is the ductility, and $E_s$ is the elastic modulus.

**Table 3.** Mechanical properties of steel bars.

| Grade Level | Diameter (mm) | $f_y$ (MPa) | $f_u$ (MPa) | $\delta$ (%) | $E_s$ (MPa) |
|---|---|---|---|---|---|
| HPB235 | 4 | 253.35 | 359.36 | 34.1 | $1.95 \times 10^5$ |
| HPB300 | 6 | 386.33 | 578.33 | 21.6 | $2.05 \times 10^5$ |
| HRB400 | 8 | 462.33 | 699.67 | 17.6 | $2.01 \times 10^5$ |
| HRB400 | 12 | 423.00 | 598.33 | 23.1 | $2.02 \times 10^5$ |

## 2.2. Details of Test Specimens

Six recycled concrete composite shear wall specimens with EPS modules were designed. Three of the six specimens were used to construct mid-rise walls with a shear-span ratio of 1.5, and other three were used to construct low-rise walls with a shear-span ratio of 0.8. The mid-rise and low-rise shear walls consisted of an ordinary recycled concrete shear wall, composite wall with an FAC protective layer (EPS modules external insulation layer), and composite wall with sandwiched EPS module insulation layer. The parameters of the specimens are listed in Table 4.

**Table 4.** Dimension and reinforcement details of specimens.

| Specimen | SW1.5-1 | SW1.5-2 | SW1.5-3 | SW0.8-1 | SW0.8-2 | SW0.8-3 |
|---|---|---|---|---|---|---|
| Height (mm) | 2100 | 2100 | 2100 | 1050 | 1050 | 1050 |
| Thickness (mm) | 200 | 320 | 320 | 200 | 320 | 320 |
| Shear-span ratio | 1.5 | 1.5 | 1.5 | 0.8 | 0.8 | 0.8 |
| Steel ratio | 0.25% | 0.25% | 0.25% | 0.25% | 0.25% | 0.25% |
| Vertical web rebar | D8-200 | D8-200 | D8-160 | D8-200 | D8-200 | D8-160 |
| Horizontal web rebar | D8-200 | D8-200 | D8-160 | D8-200 | D8-200 | D8-160 |
| Concealed column rebar | 4D12 | 4D12 | 4D12 | 4D12 | 4D12 | 4D12 |
| Steel net | - | D4-10 | - | - | D4-10 | - |
| Stirrup | D6-150 | D6-150 | D6-150 | D6-150 | D6-150 | D6-150 |
| Connecting bar | D6 | D6 | D6 | D6 | D6 | D6 |

Specimens SW1.5-1 and SW0.8-1 were recycled concrete shear walls of 200 mm in thickness, and the web rebar of wall had double rows, double directions, and with a steel ratio of 0.25%. The reinforcement of the concealed columns at both the ends of the wall corner was 4D12, and the stirrup was D6-150 mm.

Specimens SW1.5-2 and SW0.8-2 were recycled concrete composite shear walls of 320 mm in thickness, including a FAC fire wall of 50 mm in thickness, EPS modules of 70 mm in thickness, a recycled concrete wall of

200 mm in thickness, and a single layer D4-10 mm bidirectional steel net arranged in the middle of the FAC. D6 connecting bars were used to connect the thermal insulation modules. The steel ratio in the recycled concrete wall was the same as that in specimens SW1.5-1 and SW0.8-1.

Specimens SW1.5-3 and SW0.8-3 were recycled concrete composite walls of 320 mm in thickness with EPS module and sandwiched thermal insulation, including EPS modules of 70 mm in thickness in the middle and a recycled concrete composite wall of 125 mm in thickness on both sides. The web rebar of the recycled concrete composite wall on both the sides of the modules was single-row bidirectional D8-160, see the detailed information about the reinforcement of the component in Figure 3.

## 2.3. Construction of Test Specimens

The sample was prepared outside the laboratory of Beijing University of Technology (Beijing, China); the detailed preparation steps were as follows: (1) Binding the foundation reinforcing bars, followed by positioning and binding column bars and vertical reinforcement in the wall; (2) pouring foundation concrete; (3) joining the EPS thermal insulation modules, followed by strapping horizontal reinforcement and steel hoops; (4) pouring the recycled concrete wall and FAC as the outer fire protection layer, and making the test module; (5) strapping loading beam steel, and pouring loading beam concrete. Figure 4 shows the preparation of the specimens.

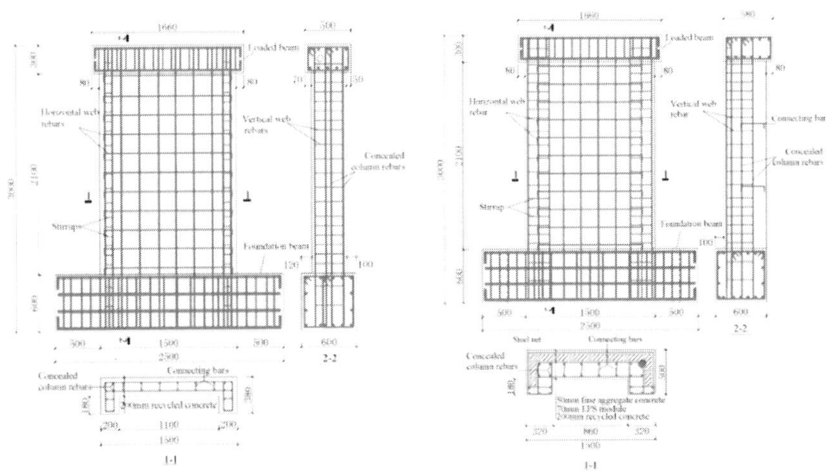

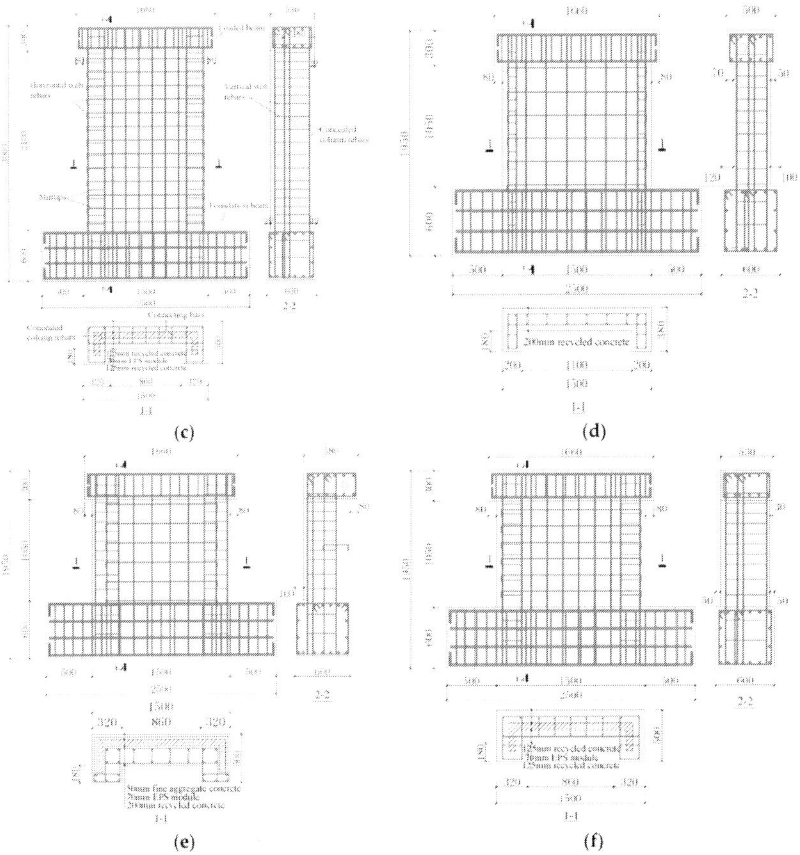


**Figure 3.** Detail information of the reinforcement of test specimens. (**a**) SW1.5-1; (**b**) SW1.5-2; (**c**) SW1.5-3; (**d**) SW0.8-1; (**e**) SW0.8-2; (**f**) SW0.8-3.

## 2.4. Test Set-Up and Loading Programme

The repeated low-cycle loading method was used in the experiments. First, a 1600 kN vertical load with an axial compression ratio of 0.15 was applied on the loading beam of the specimens, controlled by two vertical jacks with a measuring range of 1000 kN. Then, the horizontal load was applied through the horizontal tension and compression of the jacks; the distances (h) from the horizontal loading points to the foundation top surface of the specimen were 2250 mm (mid-rise wall with a shear-span ratio of 1.5) and 1200 mm (low-rise wall with a shear-span ration of 0.8), respectively. The loading device is shown in Figure 5.

Figure 4. Preparation of specimens. (a) Lashing foundation reinforcement; (b) Assembling EPS module; (c) Pouring recycled concrete; (d) Pouring fine aggregate concrete (FAC).

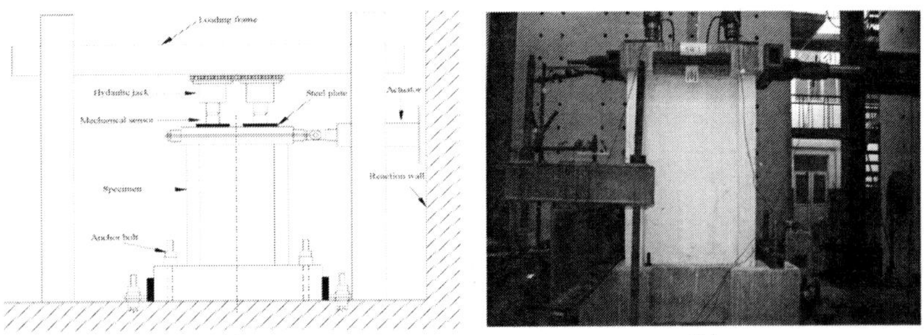


**Figure 5.** Test set-up. (**a**) Loading device; (**b**) Loading field.

The horizontal load was applied using a combined control of load and displacement. For the wall with a shear-span ratio of 1.5, in the first stage, the load was controlled by the force, and when the horizontal displacement attained $(\Delta = \frac{1}{375}h = \frac{1}{375} \times 2250 \leq 6\ mm)$, the force was applied with an increase of 30 kN force per cycle. In the second stage, a displacement controlled loading protocol with 1/750 displacement angle (θ) increment was used. On the other hand, for the wall with a shear-span ratio of 0.8, in the first stage, the load was controlled by the force, and, when the horizontal displacement attained $(\Delta = \frac{1}{200}h = \frac{1}{200} \times 1200 \leq 6\,mm)$, the force was applied with an increase of 30 kN force per cycle. In the second stage, a displacement controlled loading protocol with 1/600 displacement angle (θ) increment was used. The load course of the specimen is shown in Figure 6.

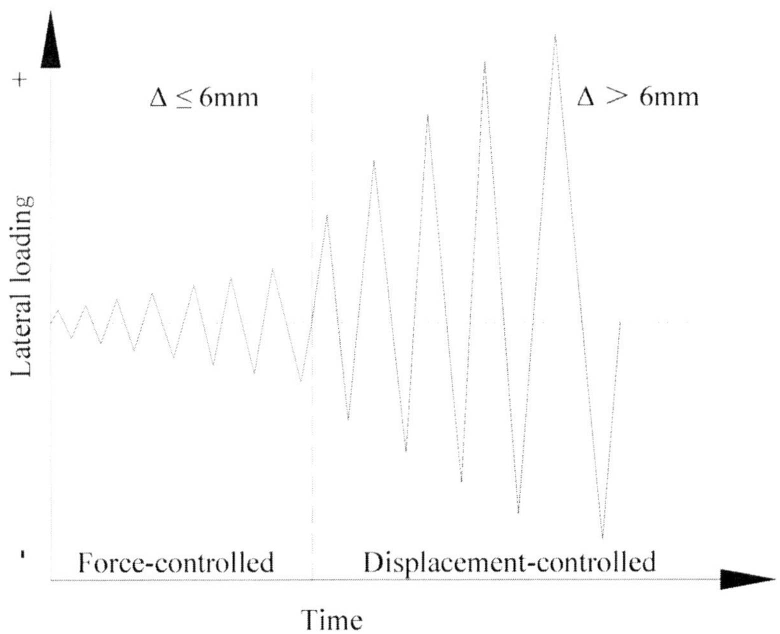


**Figure 6.** Loading history of specimens.

Differential displacement sensors were installed at the horizontal loading point and at the central height of the wall. Dial gauges were installed on the side of the foundation and on the top surface of the foundation for monitoring the foundation horizontal slip and the vertical turn-up of the foundation, respectively. The displacement sensor, dial gauges, and IMP (Isolated Measurement Pods) data collection system were butt-jointed as shown in Figure 7. During the tests, the cracks and failure at each position were recorded through the visual observation of experimental phenomena.

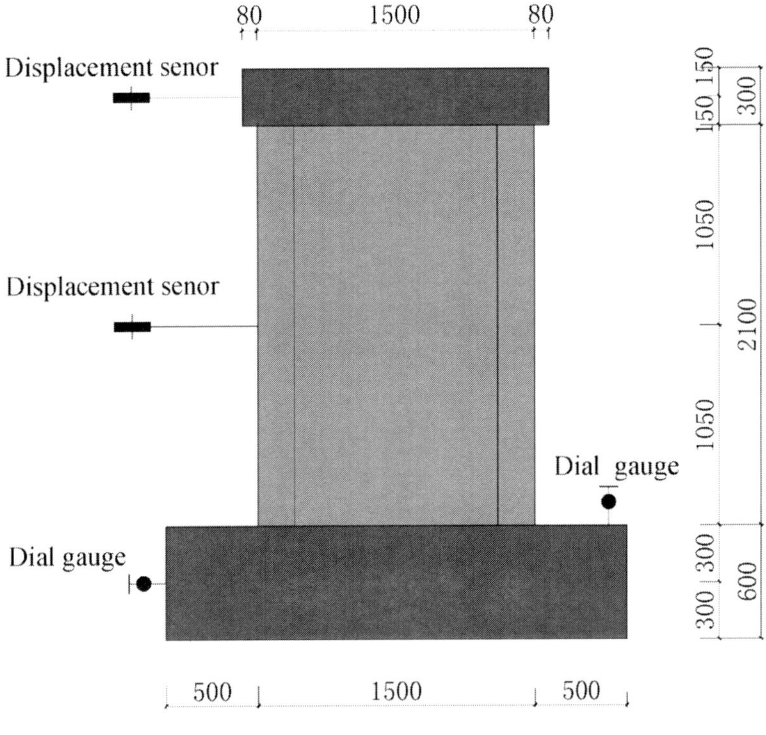


**Figure 7.** Test point arrangement of specimens.

# 3. RESULTS AND DISCUSSION

## 3.1. Failure Characteristics

When the displacement angle (θ) was ~1/50, the wall body with a shear-span ratio of 1.5 and that with a shear-span ratio of 0.8 failed. However, both the specimens showed different failure mechanisms, i.e., the former specimen showed signs of bending failure, whereas the latter specimen failed in shear. This indicates that the shear-span ratio plays a significant role in affecting the failure mechanism of shear walls. Figure 8 shows the failure mechanism of each specimen.

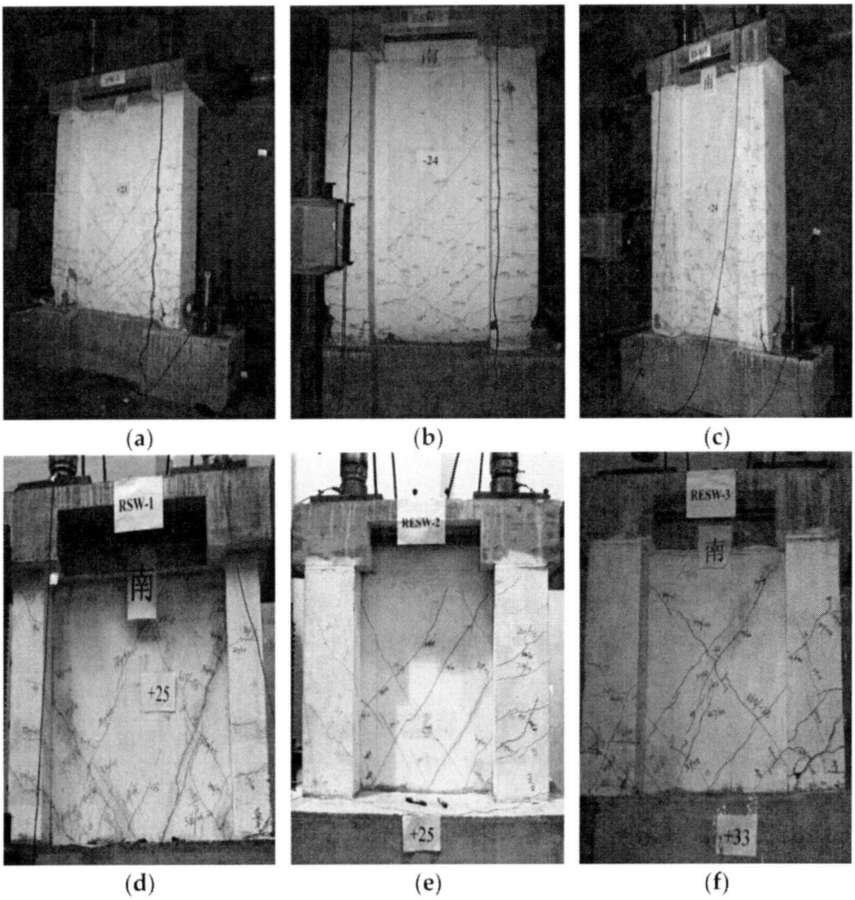


**Figure 8.** Failure characteristics of six walls. (**a**) SW1.5-1; (**b**) SW1.5-2; (**c**) SW1.5-3; (**d**) SW0.8-1; (**e**) SW0.8-2; (**f**) SW0.8-3.

The cracks on the specimens mainly appeared on both the flanges and on the web plates of the shear walls. Horizontal cracks appeared on the bottom as well as on the middle of the flanges, whereas 45° diagonal cracks and random diagonal cracks were observed on both sides of the web plates.

For the wall with a shear-span ratio of 1.5, the first crack appeared in the height range 200–300 mm of the south foundation top surface of the west flange for SW1.5-1, in the height range 600–700 mm of the east recycled concrete area of the east flange for SW1.5-2 and at the height of 830 mm of the south of the east flange for SW1.5-3 as shown in Figure 9. With the increase in horizontal load, the horizontal cracks on both the flanges of the shear walls increased.

(a) (b) (c)

**Figure 9.** The first crack of specimens (shear-span ratio is 1.5). (**a**) SW1.5-1; (**b**) SW1.5-2; (**c**) SW1.5-3.

Figure 10a,c show that when the displacement angle (θ) of specimens SW1.5-1 and SW1.5-3 reached 1/1406 and 1/1573, respectively, 45° cracks started to appear at the web angle of the shear wall. When the loading was carried forward to the mid-term, many "X"-shaped cracks formed with the reciprocal chiasma of diagonal cracks at the web plates in SW1.5-1; however, the diagonal cracks at the web plates in SW1.5-3 developed relatively slowly. When the displacement angle (θ) of the SW1.5-2 specimen reached 1/300, first, slant 45° cracks appeared in the FAC layer north of the web plates of the wall. With the increase in the displacement angle, the diagonal cracks at the north of the web plates did not increase noticeably; when the displacement angle (θ) reached 1/809, slant 45° cracks started to appear in the FAC layer south of the web plates of the wall body. Slant cracks at the south of the web plates were clearly less than those in SW1.5-1, and only a few "X"-shaped cracks were observed as shown in Figure 10b. This indicates that the crack development can be slowed down by the protection layer outside the FAC.

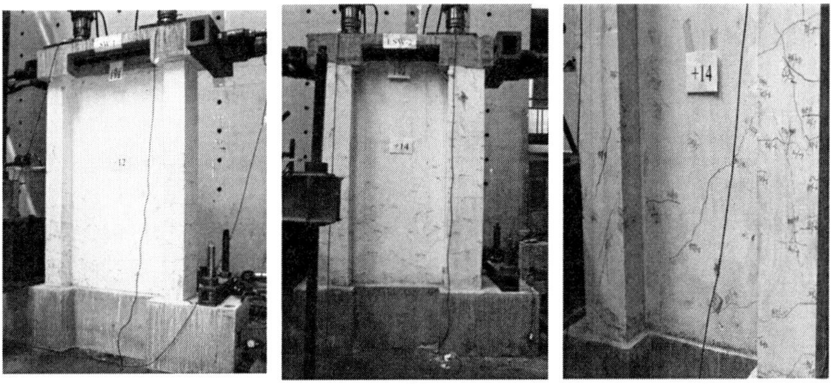

(a) (b) (c)

**Figure 10.** Web cracks at the middle stage of loading for specimens (shear-span ratio is 1.5). (**a**) SW1.5-1; (**b**) SW1.5-2; (**c**) SW1.5-3.

For the specimen with a shear-span ratio of 0.8, Figure 11a,b show that the first cracks in the SW0.8-1 and SW0.8-2 wall appeared at the north of east flange, 300–550 mm away from the foundation top surface. On the other hand, the cracks in the SW0.8-2 specimen appeared later compared to SW0.8-1 specimen. The first cracks in the SW0.8-3 specimen appeared at the south of the west flange, 250–300 mm away from the foundation, slightly earlier than that in SW0.8-1 as shown in Figure 11c. In the middle of the loading process, cut-through diagonal cracks appeared on the web plates of SW0.8-1, and diagonal inclined cracks appeared on the web plates of SW0.8-2 and SW0.8-3. Beyond this point, multiple "X"-shaped cracks appeared on the web plates of the SW0.8-1 wall, tightly close to both the flanges. For SW0.8-2, multiple diagonal cracks cut through the flanges on the web plates. Similarly, multiple "X"-shaped cracks appeared on the flanges and on the middle of the web plates in SW0.8-3.

In terms of failure mode, the damages in the web plates and flanges of ordinary recycled concrete specimens (SW1.5-1 and SW0.8-1) were most severe, followed by the sandwich thermal insulation composite walls (SW1.5-3 and SW0.8-3) and composite walls with the EPS modules and outer protection layer (SW1.5-2 and SW0.8-2). The failure process of the specimens was as follows: The vertically distributed steel bars yielded first or even snipped, a part of the vertically steel bars yielded, a large amount of concrete at the bottom of the flanges crushed, and a continuous crushing failure band formed and stretched to the web plates of the wall body as shown in Figure 12.

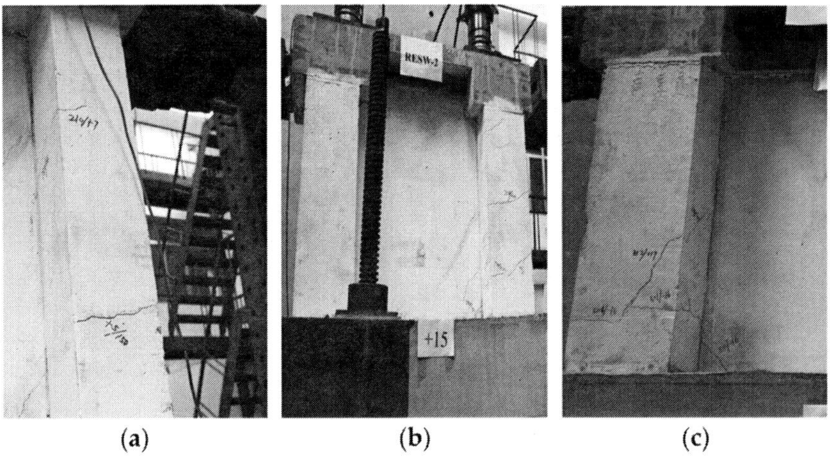


**Figure 11.** The first crack of specimens (shear-span ratio is 0.8). (**a**) SW0.8-1; (**b**) SW0.8-2; (**c**) SW0.8-3.

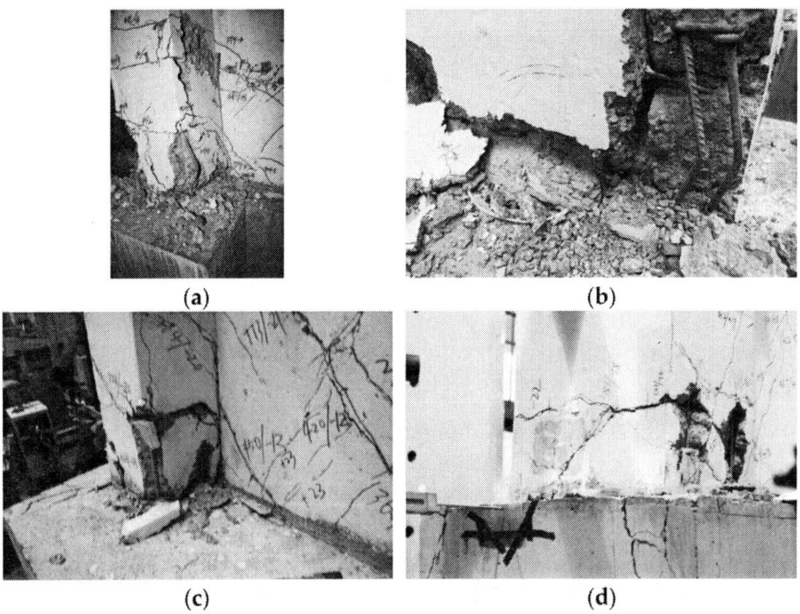


**Figure 12.** Failure details of specimens. (**a**) Reinforcement yielding (SW1.5-1); (**b**) Failure zone of concrete (SW1.5-3); (**c**) Concrete-flake-off (SW0.8-1); (**d**) Exposed reinforcement (SW0.8-3).

### 3.2. Load-Bearing Capacity

Table 5 shows the cracking load $F_c$, yield load $F_y$, peak load $F_u$, and their ratios ($\mu_{cu} = F_c/F_u$, $\mu_{yu} = F_y/F_u$). Each load has a measured value (MV) and relative value (RV). In this study, the cracking load was the load when the first visible cracks appeared, and the yield load was determined by the energy equivalence method [26] as shown in Figure 13. The loads were the mean absolute values of the positive and negative values.

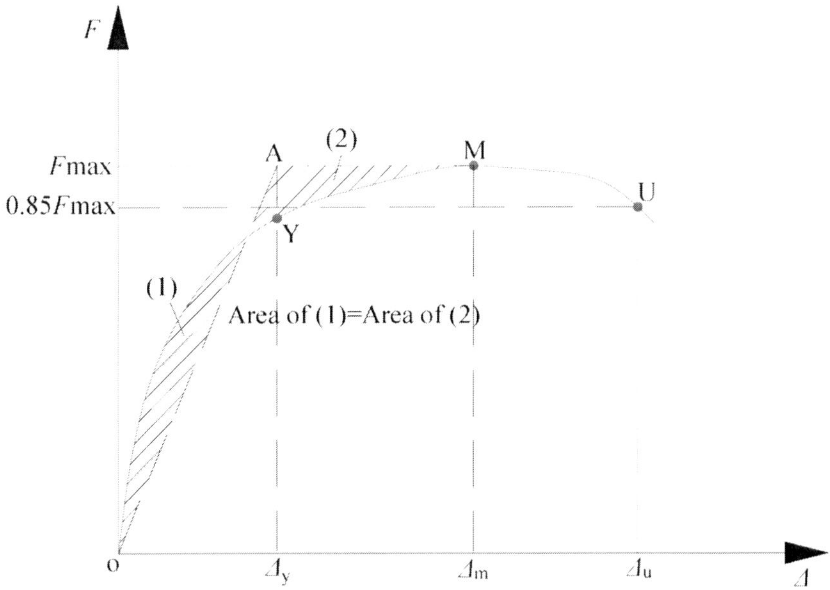


**Figure 13.** Determination of yield point.

**Table 5.** Experimental results of cracking load, yield load, and ultimate load.

| Specimen | $F_c$ (kN) | | $F_y$ (kN) | | $F_u$ (kN) | | $\mu_{cu}$ | $\mu_{yu}$ |
|---|---|---|---|---|---|---|---|---|
| | MV | RV | MV | RV | MV | RV | $F_c/F_u$ | $F_y/F_u$ |
| SW1.5-1 | 60.80 | 1.000 | 422.43 | 1.000 | 481.5 | 1.000 | 0.126 | 0.877 |
| SW1.5-2 | 59.05 | 0.971 | 388.28 | 0.919 | 450.5 | 0.936 | 0.131 | 0.862 |
| SW1.5-3 | 61.68 | 1.014 | 428.29 | 1.014 | 516.3 | 1.072 | 0.119 | 0.830 |
| SW0.8-1 | 180.47 | 1.000 | 688.26 | 1.000 | 830.43 | 1.000 | 0.217 | 0.829 |
| SW0.8-2 | 213.12 | 1.181 | 782.41 | 1.137 | 901.36 | 1.085 | 0.236 | 0.868 |
| SW0.8-3 | 204.96 | 1.136 | 739.84 | 1.075 | 892.62 | 1.075 | 0.230 | 0.829 |

Under the same configuration, the capacity of the specimens with a shear-span ratio of 0.8 was larger than that of those with a shear-span ratio of 1.5, indicating that when the shear-span ratio is smaller, the capacity of the component is higher. Compared to SW1.5-2, the yield strength of SW1.5-3 increased by 10.3%, and the peak load increased by 14.5%, indicating that the arrangement of the EPS modules in the middle of the shear wall is more beneficial for improving the capacity of recycled concrete composite shear walls than arranging them outside the wall. On the other hand, SW0.8-3 and SW0.8-2 had an almost equal peak load, indicating that the arrangement form of the EPS modules slightly affected the capacity of the shear walls. Compared to SW0.8-1, the yield strength of SW0.8-2 increased by 13.7%, and the peak load increased by 8.5%, indicating that the capacity of the walls was improved by the EPS modules and the outer protection layer of FAC.

## 3.3. Stiffness and Degradation

The secant stiffness of each specimen was computed from the skeleton curve of the hysteresis plot using Equation (1). F is the horizontal force, and the U is the horizontal displacement. The measured stiffness values and their degeneration factors are listed in Table 6. $K_0$ is the initial elastic stiffness, and $K_y$ is the yield stiffness. The relationship curve "stiffness K-displacement angle θ" is shown in Figure 14.

$$K_i = \frac{F_i}{U_i} \tag{1}$$

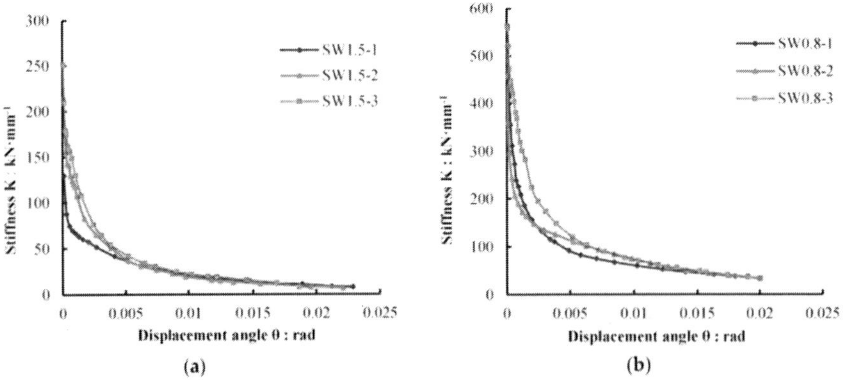


**Figure 14.** Stiffness degradation curves of specimens. (**a**) Shear-span ratio is 1.5; (**b**) Shear-span ratio is 0.8.

**Table 6.** Stiffness and degeneration coefficient of specimens.

| Specimen | $K_0$ (kN·mm$^{-1}$) | | $K_y$ (kN·mm$^{-1}$) | | $\beta_{y0} = K_y/K_0$ | |
|---|---|---|---|---|---|---|
| | *MV* | *RV* | *MV* | *RV* | *MV* | *RV* |
| SW1.5-1 | 129.8 | 1.000 | 35.8 | 1.000 | 0.276 | 1.000 |
| SW1.5-2 | 215.0 | 1.656 | 76.5 | 2.138 | 0.356 | 1.289 |
| SW1.5-3 | 251.2 | 1.935 | 72.0 | 2.010 | 0.287 | 1.038 |
| SW0.8-1 | 463.65 | 3.572 | 67.5 | 1.885 | 0.146 | 0.529 |
| SW0.8-2 | 524.33 | 4.040 | 103.1 | 2.880 | 0.197 | 0.714 |
| SW0.8-3 | 563.21 | 4.339 | 123.5 | 3.450 | 0.219 | 0.793 |

Under the same configuration, the stiffness of the specimens with a shear-span ratio of 0.8 was higher than those with a shear-span ratio of 1.5. Compared to SW1.5-1, the initial stiffness of specimen SW1.5-2 increased by 65.6%, and the yield stiffness increased by 113.8%. On the other hand, compared to SW0.8-1, the initial stiffness of SW0.8-2 increased by 13.1%, and the yield stiffness increased by 52.7%. This indicates that the EPS modules and protection layer outside the FAC improved the stiffness of composite shear walls and significantly reduced the rate of stiffness degeneration. By comparing the stiffness degeneration processes of SW1.5-2 and SW1.5-3 as well as SW0.8-2 and SW0.8-3, it was found that the arrangement of the EPS thermal insulation modules in the middle of the shear wall is more beneficial for improving the stiffness than arranging the EPS thermal insulation modules on one side of the wall.

## 3.4. Load-Displacement Response

The horizontal load–displacement relationships of the specimens are shown in Figure 15. The ultimate displacement angle of ordinary shear walls was in the order of 1/120, and in this experiments, the maximum displacement angle was 1/50.

At the initial stage of the loading, the displacement increased linearly with the force, and the residual deformations after the unloading were negligible. With the increase in repeated cyclic loading, the peak displacement as well as the residual displacement increased gradually, whereas the stiffness degradation became evident until failure.

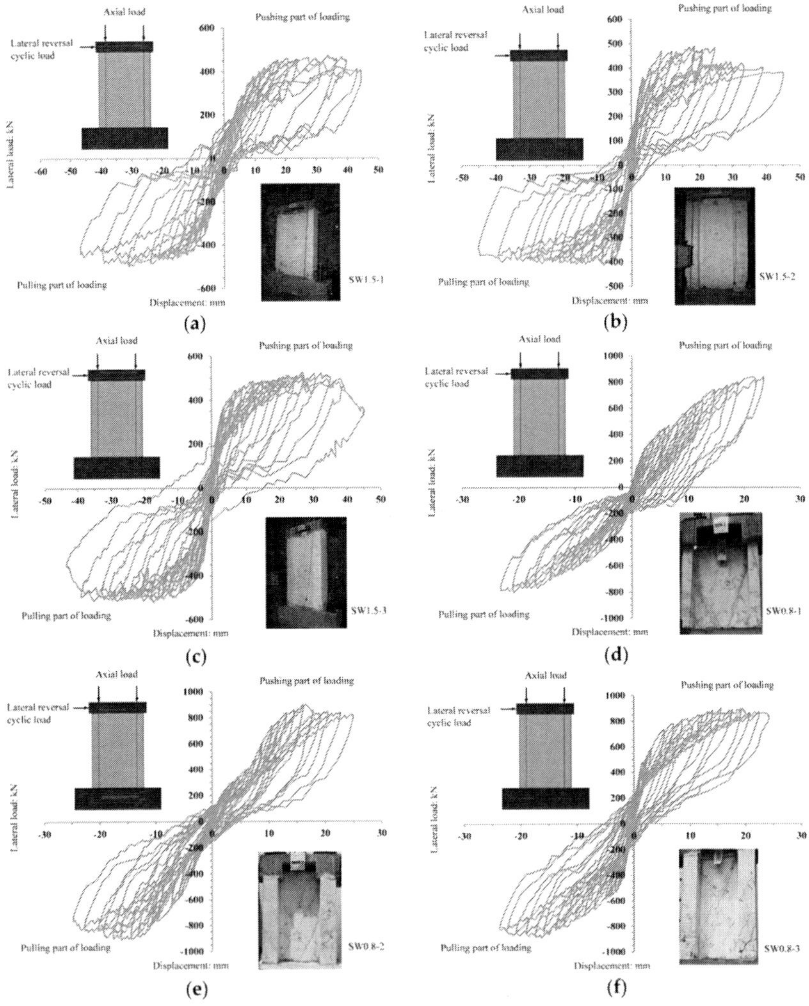


**Figure 15.** Hysteresis curves of specimens. (**a**) SW1.5-1; (**b**) SW1.5-2; (**c**) SW1.5-3; (**d**) SW0.8-1; (**e**) SW0.8-2; (**f**) SW0.8-3.

Compared to SW1.5-1, the residual strain of SW1.5-2 after the unloading became larger with a slightly smaller capacity. However, during the failure, the elastic-plastic deformation capacity and seismic-resistance and energy-dissipation capacity of the shear walls were improved by the combined effect of the EPS thermal insulation modules and FAC to some extent. Compared to SW1.5-2, the hysteresis loop of SW1.5-3 was full, and the capacity of the specimen improved. The adhesion between the EPS thermal insulation modules and the recycled concrete on both the sides of the EPS modules was better than that between the EPS thermal insulation modules and FAC. This not only improved the overall response of the specimen, but also enhanced the seismic-resistance and energy-dissipation capacity to a large extent.

On the other hand, compared to SW0.8-1, the hysteresis loop of SW0.8-2 was full, and the capacity improved. This indicates that the EPS modules and FAC participated satisfactorily in the seismic-resistance and energy-dissipation processes and improved the capacity of the specimens. Compared to SW0.8-2, the hysteresis loop of SW0.8-3 was fuller with a slow degradation of capacity, and its structural ductility was better. This indicates that the adhesion between sandwich-arranged EPS modules with recycled concrete on both sides of the modules was good, thus enhancing the overall response of the specimen and further improving the seismic response of the composite walls.

Figure 16 shows a comparison of the force–displacement skeleton curves of the specimens with different shear-span ratios. Under different shear-span ratios, the capacity of the specimens with a shear-span ratio of 1.5 was lower than that of the specimens with a shear-span ratio of 0.8. For the specimens with the same shear-span ratio, the recycled concrete wall with the EPS modules arranged in the middle had the highest capacity, and the strength degradation was the slowest. Compared to SW1.5-1, the yield stiffness of SW1.5-2 was relatively large with a relatively low capacity. After the wall cracking, the stiffness degeneration rate of SW1.5-2 was higher than that of SW1.5-1. SW1.5-3 had a force–displacement skeleton curve similar to that of SW1.5-2, but both its initial stiffness and yield stiffness improved. With the increase in displacement angle, the strength degradation of SW1.5-3 was slower than that of SW1.5-2. Compared to SW0.8-1, the initial stiffness of SW0.8-2 was relatively small, with a relatively higher yield stiffness and larger capacity. Compared to SW0.8-2, the initial stiffness and yield stiffness of SW0.8-3 improved. With the increase in displacement angle, the descent stage of the load-bearing capacity of the specimens became smooth and steady with a good structural ductility.

### 3.5. Energy Dissipation Capacity

The area under the hysteresis curve of the specimen under the action of low-reversed cyclic loading shows the magnitude of seismic energy dissipated by the structure. In this study, the area surrounded by the

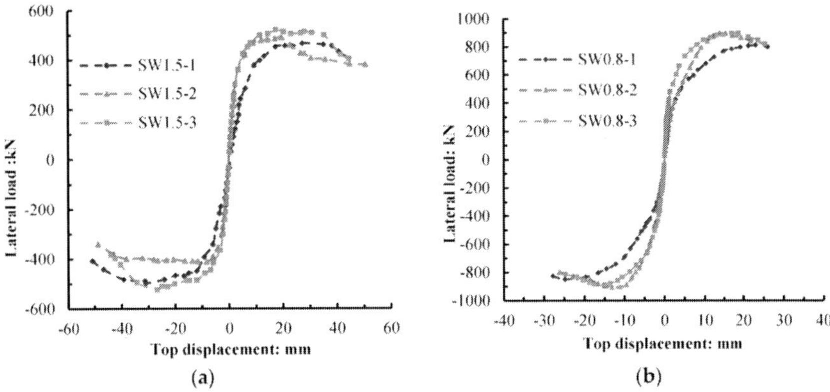


**Figure 16.** Force–displacement skeleton curves of the specimens. (**a**) Shear-span ratio is 1.5; (**b**) Shear-span ratio is 0.8.

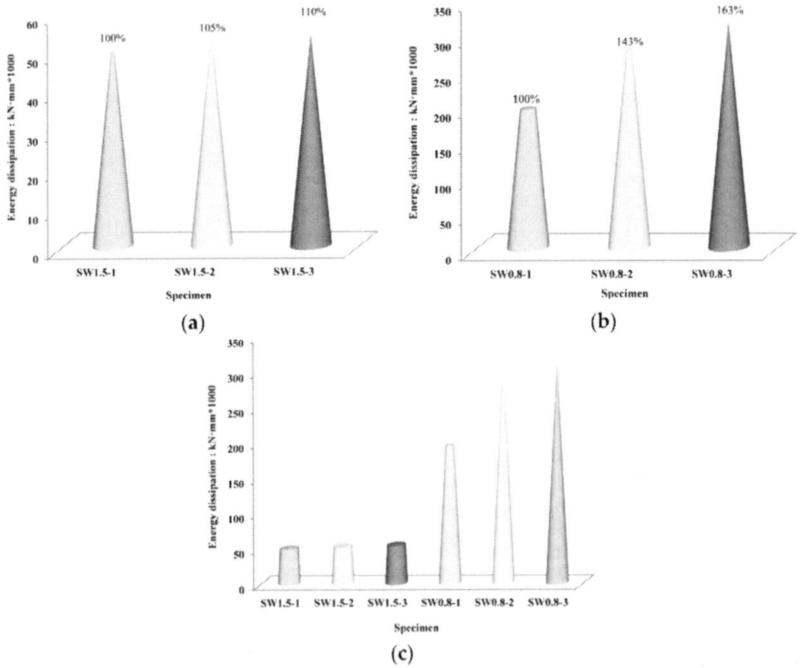


**Figure 17.** Energy dissipation histogram of specimens. (**a**) Energy dissipation for specimens (shear-span ratio is 1.5); (**b**) Energy dissipation for specimens (shear-span ratio is 0.8); (**c**) Energy dissipation of all specimens.

**Table 7.** Experimental results of energy dissipation.

| Specimen | $E_{0.02}$ (kN·mm) | *RV* | Specimen | $E_{0.02}$ (kN·mm) | *RV* |
|---|---|---|---|---|---|
| SW1.5-1 | 49,529 | 1.000 | SW0.8-1 | 195,206 | 1.000 |
| SW1.5-2 | 52,154 | 1.053 | SW0.8-2 | 278,420 | 1.426 |
| SW1.5-3 | 54,285 | 1.096 | SW0.8-3 | 319,057 | 1.634 |

skeleton curve (which was connected by the peak points of the hysteresis curve) of the specimen when the displacement angle reached 1/50 was used as the representative value of the energy dissipation of the shear wall. Table 7 lists the energy dissipation of each specimen, where $E_{0.02}$ is the energy dissipation value when the displacement angle of the specimen was 1/50. The energy dissipation of each specimen is shown in Figure 17.

For the same configuration, the energy dissipation of the specimens with a shear-span ratio of 1.5 was lower than that of the specimens with a shear-span ratio of 0.8. This is because the capacity of the medium-high wall was relatively smaller. For the specimens with a shear-span ratio of 0.8, compared to SW0.8-1, the total energy dissipation of SW0.8-2 increased by 42.6%. This indicates that the EPS modules and their outer FAC participated satisfactorily in the seismic-resistance and energy-dissipation processes and accordingly enhanced the energy dissipation capacity of the walls significantly. However, when the shear-span ratio was 1.5, the contribution of the EPS modules and FAC on the total energy dissipation was not very significant. When comparing SW1.5-3 with SW1.5-2, and SW0.8-3 with SW0.8-2, their overall energy dissipations increased by 4.1% and 14.6%, respectively. This indicates that the arrangement form of the EPS modules significantly affected the seismic-resistance and energy-dissipation of recycled concrete shear walls, and that, when the shear-span ratio was small, the sandwich arrangement significantly enhanced the seismic performance.

# 4. THEORETICAL ANALYSIS

## 4.1. Calculation of Load-Bearing Capacity

### 4.1.1. Analysis of the Load-Bearing Capacity of the Normal Section

The experimental study showed that the shear wall with a shear-span ratio of 1.5 suffered bending failure, and the failure mode was controlled by the bending moment in the bottom of the shear wall. Thus, the load-bearing capacity of mid-rise walls could be calculated by the normal section. The load-bearing capacity calculation for the normal section should be consistent with the following basic assumptions:

(1) The section meets the assumption of the plane section.
(2) Neglect the tensile effect of recycled concrete and FAC in the tension zone.
(3) The maximum compressive stress of recycled concrete and FAC is taken from the standard values of axial compressive strength $f_{ck,r}$ and $f_{ck,f}$.
(4) The double-fold line mode is used for the constitutive model of steel bar. The mode shows the linear elastic relationship before the steel bar yields, and the yield strength is used after the yielding.
(5) Ignore the vertical distribution force of reinforcement near the pressure area and neutral axis.
(6) Consider the tensile strength of the vertical bar in the range of $hw0-1.5x$ [27], where x is the height of the concrete compression zone.
(7) The calculation model for the load-bearing capacity of the normal section of the shear wall is shown in Figure 18. The calculation formula for the load-bearing capacity of the normal section is expressed by Equations (2) and (3).

$$\begin{aligned} N &= \alpha_1 f_c b_w x + \alpha_1 f_c (b'_f - b_w) h'_f + \alpha_1 f_a b_a (x + h_a + h_e) + \alpha_1 f_a h'_a (b'_f + b_e) + f'_y A'_s \\ &+ f'_{ya} A'_{sa} - f_y A_s - f_{ya} A_{sa} - f_{yw} b_w \rho (h_{w0} - 1.5x) - f_{ya} b_a \rho (h_{w0} - 1.5x + h_f + h_e) \end{aligned} \tag{2}$$

$$\begin{aligned} Ne &= \alpha_1 f_c b_w x (h_{w0} - \frac{x}{2}) + \alpha_1 f_c h'_f (b'_f - b_w)(h_{w0} - a_s) + \alpha_1 f_a b_a (x + h_a + h_e)(h_{w0} - \frac{x}{2} + \frac{h_a + h_e}{2}) \\ &+ \alpha_1 f_a h'_a (b'_f + b_e)(h_{w0} - \frac{x}{2} + \frac{h_a}{2} + h_e) + f'_y A'_s (h_{w0} - a_s) + f'_{ya} A'_{sa} (h_{w0} - a_s + a'_s + \frac{h'_a}{2} + h'_e) \\ &+ f_{ya} A_{sa} (a_s + \frac{h_a}{2} + h_e) - \frac{1}{2} f_{yw} b_w \rho (h_{w0} - 1.5x)^2 \\ &- \frac{1}{2} f_{ya} b_a \rho (h_{w0} - 1.5x + a_s + h_e)(h_{w0} - 1.5x - a_s - h_e) \end{aligned} \tag{3}$$

Among them, $e = e_0 + h_{w0} - \frac{h_w}{2}, e_0 = M/N, x \geqslant 2a'_s$ and the horizontal load-bearing capacity of the wall can be calculated as follows:

$$F = \frac{M}{H} = \frac{Ne_0}{H} \tag{4}$$

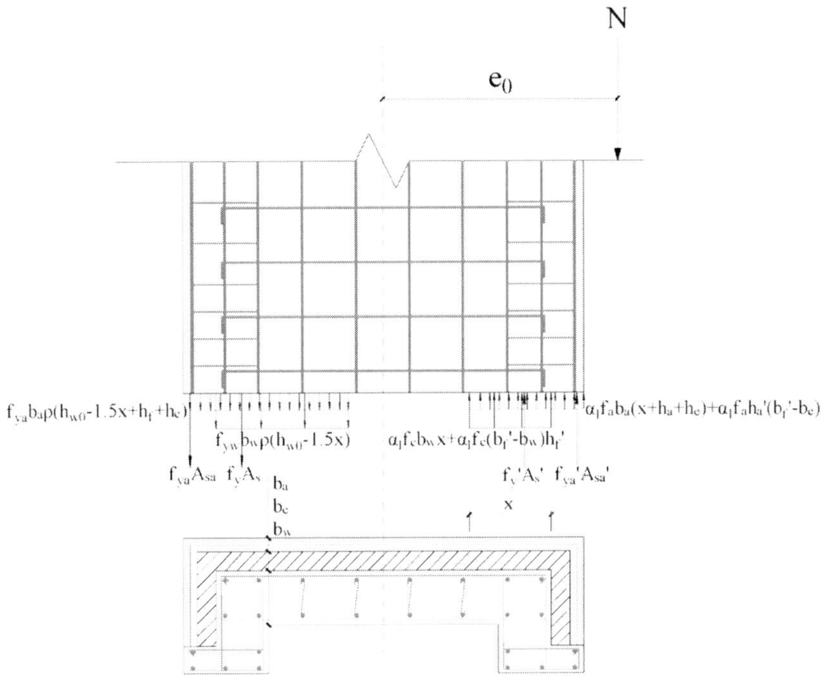


**Figure 18.** Mechanical model of load-carrying.

The wall of SW1.5-1 and SW1.5-3 did not have the FAC outer protective layer, and symmetrical reinforcement was used for specimens $(f_y' = f_y, A_s' = A_s, a_s' = a_s)$. Thus, the formula can be simplified to:

$$N = \alpha_1 f_c b_w x + \alpha_1 f_c (b_f' - b_w) h_f' - f_{yw} b_w \rho (h_{w0} - 1.5x) \quad (5)$$

$$Ne = \alpha_1 f_c b_w x (h_{w0} - \frac{x}{2}) + \alpha_1 f_c h_f' (b_f' - b_w)(h_{w0} - a_s) + f_y' A_s' (h_{w0} - a_s) - \frac{1}{2} f_{yw} b_w \rho (h_{w0} - 1.5x)^2 \quad (6)$$

The horizontal load-bearing capacity F of SW1.5-2 shear wall can be calculated using Equations (2)–(4). The horizontal load-bearing capacity F of SW1.5-1 and SW1.5-3 is calculated by the Equations (4)–(6). Table 8 shows the MV and calculated value of horizontal load-bearing capacity F of the shear wall. The results show that the error between MV and calculated value of SW1.5-1 and SW1.5-3 is <5%. However, the error in SW1.5-2 is relatively large, mainly because of the relatively low load-bearing capacity for the base cracking during the test. Overall, the calculation results are in good agreement with the experimental results.

**Table 8.** Mechanical and experimental results of ultimate load.

| Specimen | *MV* (kN) | Calculated Value (kN) | Relative Error |
|---|---|---|---|
| SW1.5-1 | 481.5 | 466.3 | −3.16% |
| SW1.5-2 | 450.5 | 514.5 | 14.21% |
| SW1.5-3 | 516.3 | 503.6 | −2.46% |

### 4.1.2. Analysis of Load-Bearing Capacity of the Cross Section

The shear features of inclined section were mainly reflected from the load-bearing capacity of the wall with a shear-span ratio of 0.8. Therefore, it was necessary to establish the calculation formula for the load-bearing capacity of low-rise walls. The basic assumption of the oblique section is as follows:

The shear force was undertaken by recycled concrete and the web distribution reinforcement of the wall [28],

$$V = V_C + V_S \tag{7}$$

For the shear force *VC*, which was undertaken by recycled concrete, the action of shear zone was considered,

$$V_c = 0.5 f_t b_w h_{w0} + \frac{0.28}{1 + 0.6(H/h_w)} N \frac{A_w}{A} \tag{8}$$

The shear force in the web of the low-rise wall was undertaken by the horizontal and vertical distribution reinforcement.

$$V_s = \frac{(H/h_w)^{4/3}}{1+(H/h_w)^{2/3}}\eta\rho_{sh}\omega_1 f_{yh} b_w h_w + \frac{1}{1+(H/h_w)^{2/3}}\eta\rho_{sv}\omega_2 f_{yv} b_w h_w \quad (9)$$

$$\omega_1 = \frac{1}{\sqrt{1+3(H/h_w)^{2/3}}} \quad (10)$$

$$\omega_2 = \frac{1}{\sqrt{3+(H/h_w)^{2/3}}} \quad (11)$$

In the formula, *ft* is the axial compressive strength of the recycled concrete; *bw* is the thickness of the web of the recycled concrete wall; *hw* is the total width of the recycled concrete wall along the shear direction; *hw*0 is the clear width of the recycled concrete wall along the shear direction; *N* is the vertical pressure of the wall section; *A* is the total area of the recycled concrete wall; *Aw* is the net area of the recycled concrete wall along the shear direction; *H* is the calculation value of the height of the recycled concrete wall; ρ*sh* and ρ*sv* are the reinforcement ratio of the horizontal and vertical distribution bar of the wall, respectively; *fyh* and *fyv* are the yield strength of the horizontal and vertical distribution reinforcement of wall, respectively; ω1 and ω2 are the stress coefficient of the horizontal and vertical distribution reinforcement, respectively; η is the shear friction coefficient among the aggregates on the diagonal section of the recycled concrete and is taken as 0.85 in this study.

The horizontal load-bearing capacity F of three shear walls was calculated using Equation (7). Table 9 shows the comparison between the MV and calculated value of the horizontal load-bearing capacity of the shear wall. Table 9 shows that the MV is in good agreement with the calculated value. The error between the MV and calculated value of all the three walls was <5%, and the accuracy of the calculation formula for the normal section was verified at the same time.

**Table 9.** Calculated and experimental results of ultimate load.

| Specimen | *MV* (kN) | Calculated Value (kN) | Relative Error |
|---|---|---|---|
| SW0.8-1 | 830.43 | 805.28 | −3.03% |
| SW0.8-2 | 901.36 | 860.36 | −4.55% |
| SW0.8-3 | 892.62 | 934.61 | 4.70% |

## 4.2. Finite Element Analysis

Further damage and evolution of recycled concrete was analyzed using a microscope. Under a low cyclic loading, the skeleton curve of the shear wall was similar to the load displacement curve under monotonic loading. Therefore, the performance of low cycle loading was simulated by using the monotonic loading. In this study, the damage and plastic strain of SW1.5-1 wall concrete was analyzed using the finite element software ABAQUS (ABAQUS Inc., Pawtucket, RI, USA), and the crack distribution, damage process, and mechanical characteristics of the specimen were discussed. The calculation results were compared with the experimental results.

### 4.2.1. Finite Element Parameter

The concrete plastic damage model was used to simulate the recycled concrete and FAC. The constitutive relationship between stress and strain under the uniaxial pressure of recycled concrete proposed by Xiao [29] was used as shown in Formula (12):

$$y = \begin{cases} ax + (3-2a)x^2 + (a-2)x^3, 0 \leqslant x < 1 \\ \dfrac{x}{b(x-1)^2 + x}, x \geqslant 1 \end{cases} \tag{12}$$

where $x=\varepsilon/\varepsilon 0'$,$y=\sigma/fc'$; $a$ is the coefficient at the ascending stage of the curve and $a$=2.2(0.748$r$2−1.231$r$+0.957), taken as a = 1.0428; $b$ is the coefficient at the descending stage of the curve and $b$=0.8(7.6483$r$+1.142), taken as b = 7.03224. The peak compressive strain of the recycled concrete prism is ε$r$0, 1.2 times the peak compressive strain of ordinary concrete [29], and $frc$ is the compressive strength of recycled concrete prism and taken as, $f_c^r = 0.67 f_{cu}$.

The model hypothesis of the tensile stress-strain law is concrete without damage before reaching the peak stress. The descending curve was calculated using Equation (13).

$$y = \frac{x}{\alpha_t(x-1)^2 + x}, x \geqslant 1 \tag{13}$$

where $x = \varepsilon/\varepsilon_{t0}^{r}$, $y = \sigma/f_t^{r}$; $\varepsilon_{t0}^{r}$ is the peak tensile strain of the recycled concrete prism. $\varepsilon_{t0}^{r} = f_t^{r}/E_c$; $f_t^{r}$ is the representative value of the uniaxial tension strength of concrete. $f_t^{r} = (-0.06r + 0.24)(f_{cu}^{r})^{2/3}$; $\alpha_t^{r}$ is the coefficient at the descending stage of the uniaxial tension stress-strain curve of concrete.

The "plasticity" model from the ABAQUS software was used to simulate the reinforcement law. The Mises yield and the kinematic hardening criteria were used in the numerical simulation. The stress-strain relationship of the steel bar was simplified as the oblique line. In this study, *E's*=0.01*Es* [27], and the Poisson's ratio was 0.3.

### 4.2.2. Finite Element Model

The 3D solid element C3D8R was used to simulate the recycled concrete, and the 3D truss element T3D2 was used for the rebar. The separating method was used to create the components from different steel, and the material and section properties were defined later. The reinforced skeleton was embedded into the recycled concrete wall, and then the interaction of the model was defined as shown in Figure 19.

The base of the wall was fixed on the ground, and the load surface of the loading beam was restrained. First, a vertical load was applied to the reference points on the load beam and remained unchanged during the process. Second, the horizontal displacement was controlled to simulate the horizontal loading process by defining different boundary conditions. The wall model was divided by the structure mesh method. The boundary conditions and the grid division of the model are shown in Figure 20.

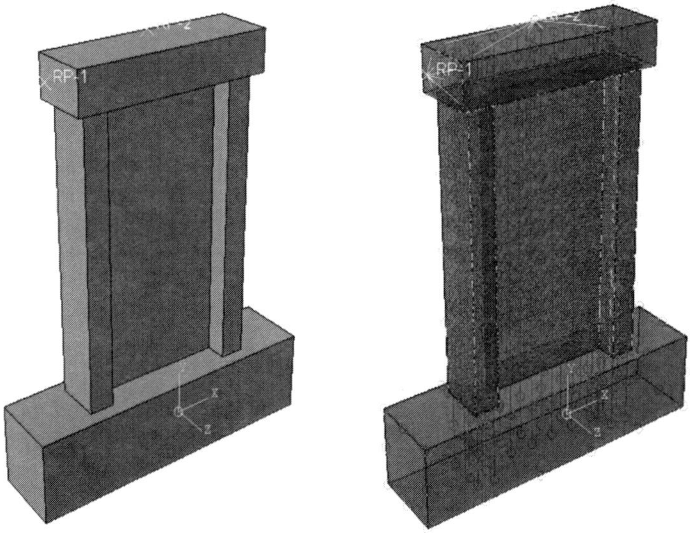


(**a**) (**b**)

**Figure 19.** Establishment of model. (**a**) Assembled model; (**b**) Interaction of model.

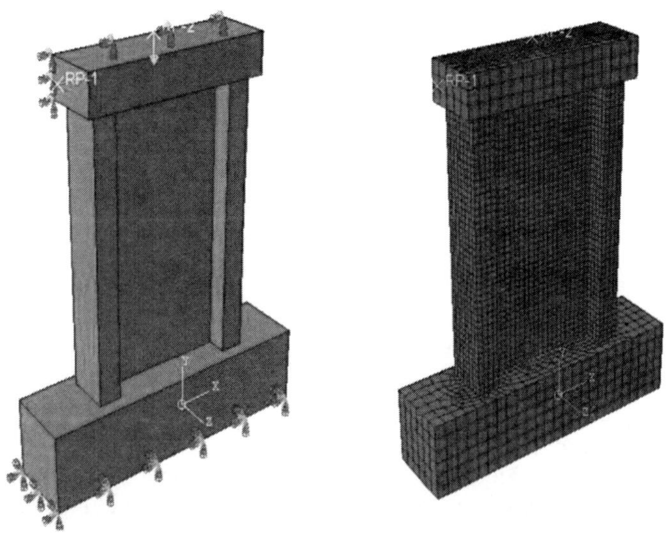


(**a**) (**b**)

**Figure 20.** Boundary condition and mesh of the model. (**a**) Boundary condition of model; (**b**) Element Mesh of model.

### 4.2.3. Calculation Results Analysis-Damage Evolution

When a tensile plastic strain appeared on the recycled concrete, it signified that the concrete element was cracked, and the width of the crack was reflected by the length of the vector arrow of the maximum plastic strain. The crack distribution was reflected by the vector map of the maximum plastic strain and the tensile damage of the concrete. Figure 21 shows the reinforced stress cloud and concrete stress cloud when the wall of SW1.5-1 reached the ultimate displacement. Figure 22a–c shows the pulling damage, compression damage, and actual damage of SW1.5-1. The vector map of the maximum plastic strain of SW1.5-1 is shown in Figure 22d.

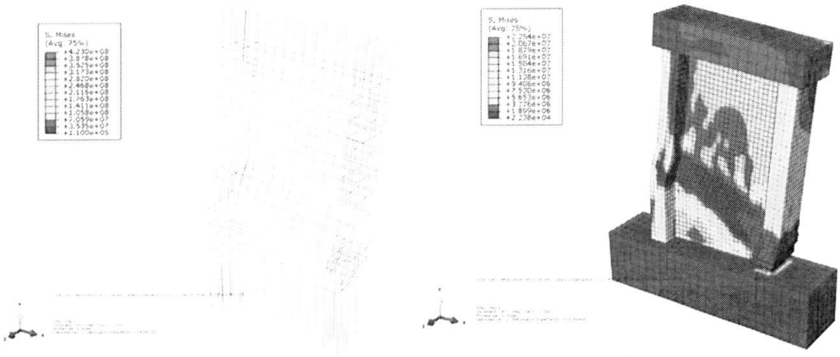

(a) (b)

**Figure 21.** Steel stress of the model. (**a**) Stress cloud of the steel; (**b**) Stress cloud of the concrete.

The following conclusions can be drawn from Figure 21 and Figure 22:

(1) The cracks on the wall of SW1.5-1 mainly appeared on the middle and lower part of the flange and web, and the horizontal and diagonal cracks appeared on the flange and web of the wall, respectively.
(2) When the wall of SW1.5-1 reached the ultimate displacement, compressional damage occurred at the bottom of the web and flange, whereas tensional damage occurred at the lower part of the flange, exhibiting the damage characteristics of bending failure. The simulation phenomena are consistent with the experiment.
(3) The maximum stress of the steel bar appeared at the bottom of the longitudinal reinforcement on the compression side when the wall of SW1.5-1 reached the ultimate displacement of the test. At the same time, the longitudinal steel of the pull side yielded.

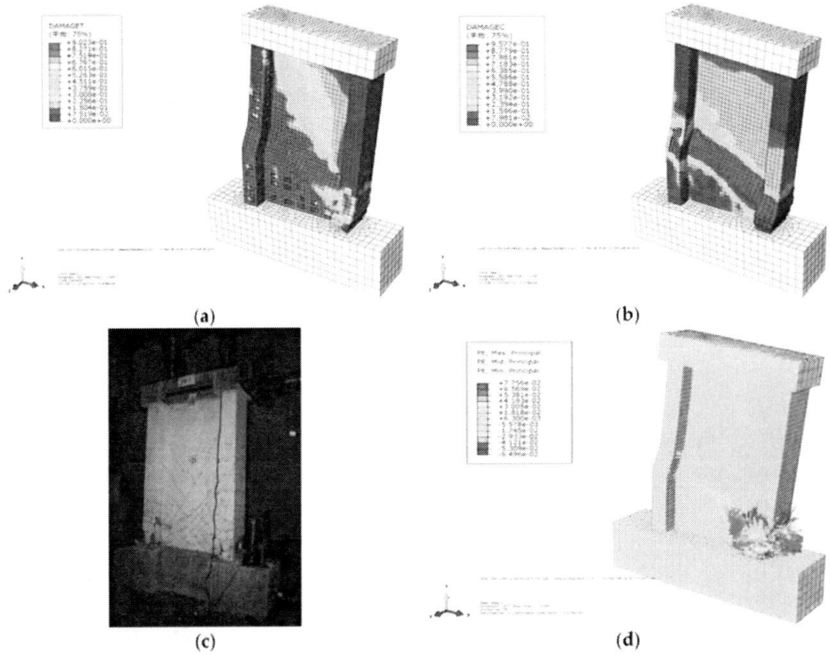

**Figure 22.** Concrete crack distribution of the model. (**a**) Pulling damage; (**b**) Compression damage; (**c**) Actual damage; (**d**) Vector map of maximum plastic strain.

#### 4.2.4. Calculation Result Analysis-Skeleton Curve

The model was calculated using the Newton–Raphson method. The calculated curves "F horizontal load–U horizontal displacement" was compared to the measured skeleton curve, as shown in Figure 23. The skeleton curve of the numerical simulation is in good agreement with the measured skeleton curve, and the calculated ultimate load-bearing capacity is slightly lower than that of the finite element model.

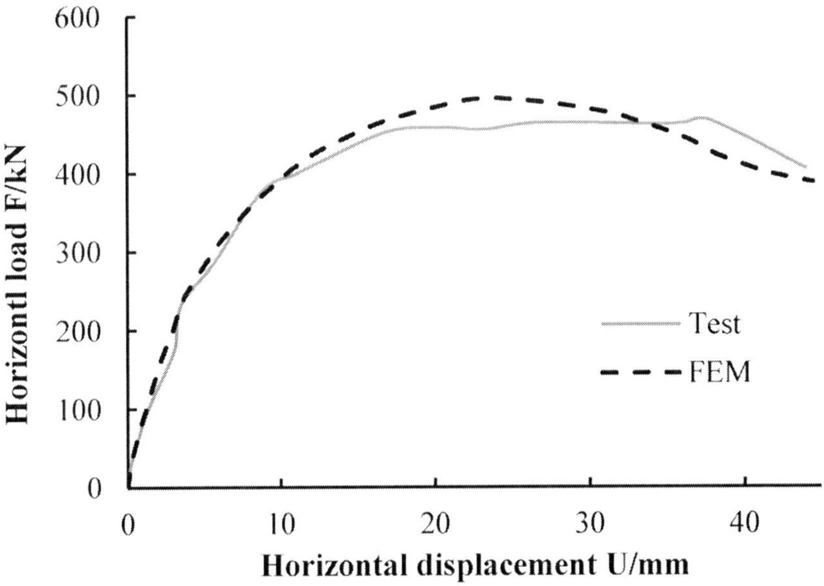


**Figure 23.** Comparison between measured and calculated curves of "F–U".

# 5. DISCUSSION AND SUGGESTIONS ON EXPERIMENTAL DESIGN

Because of the limitation of experimental conditions, the vertical and horizontal loadings in the entire test process were controlled artificially; thus, the data of the hysteresis curves fluctuated slightly. Therefore, it is recommended to select a controlling system to control the loading process if the condition allows, making the curve smoother. During the experiment, the early cracking of the foundation of the individual components significantly affected the test results. This is mainly because once the foundation cracked, the foundation became relatively slippery. Therefore, the horizontal load cannot be reflected in the real component, and the final result of the test data is less. Therefore, the foundation and loaded beam should be designed to be stronger than the component; thus, the test target can be damaged first. On the whole, the results of this experiment satisfy the requirements of accuracy and error.

The composite shear wall with the EPS module is suitable for cold regions with high requirements for heat preservation and energy saving, and the structure also has a good seismic performance. In the case of a low shear-span ratio, the seismic performance of the composite shear wall with the EPS module is better. Moreover, a high load-bearing capacity makes it suitable for the seismic design of low-rise buildings.

Considering fire prevention, the composite wall with a FAC protective layer is more reliable. The EPS module of the composite shear wall should be uniform and symmetrically arranged. For the design of a composite shear wall, a gap of 10–15 mm should be used if the EPS module cannot be split at the junction (such as the junction between the basis and EPS module), and then the gap should be sealed using a polyurethane foam.

For the composite shear wall with a FAC protective layer, a deeper dovetail groove for the EPS module would be better. The FAC protective layer and shear wall body should be well connected using a connecting bar, and the layer should be arranged as a wire mesh. The reinforcement ratio should be consistent with the ratio of the shear wall.

## 6. CONCLUSIONS

In this paper, the seismic performance of a recycled concrete composite shear wall with EPS modules was studied. The main conclusions are as follows:

(1) Compared to an ordinary recycled concrete shear wall, the shear strength of the composite shear wall with EPS modules significantly improved, the speed of stiffness degradation slowed down, the hysteresis loops were relatively full, and the energy dissipation capacity improved.

(2) The seismic performance of the shear wall with an EPS sandwich layer was better than that with an EPS module-FAC outer protective layer. The horizontal load-bearing capacity and stiffness of the shear wall with the EPS sandwich layer clearly improved, and the ductility and energy dissipation of the wall were also enhanced. Although the FAC outer protective layer contributed less to the seismic performance of the wall, it could slow down the speed of cracking and stiffness degradation of the wall.

(3) Based on the experimental study, considering the structural characteristics of an EPS module, calculation formulas are proposed for the load-bearing capacity of the normal/oblique section of the composite shear wall with an EPS module. The calculation results are in good agreement with the experimental results.

(4) Using the ABAQUS finite element software, a finite element model was developed for the shear wall without an EPS module. The damage evolution of the wall was analyzed. The skeleton curve of the finite element model is in good agreement with the experimental results.
(5) In the case of a low shear-span ratio, the seismic performance of the recycled concrete composite shear wall was better, and the load-bearing capacity was reliable. The symmetrical layout of EPS modules in the composite shear wall should be given priority, and a composite wall with a FAC protective layer would be better for fire resistance. The FAC protective layer should be arranged using a steel wire mesh, and it should be reliably connected to the wall.

## ACKNOWLEDGMENTS

The authors would like to acknowledge financial support from the topics of National Science and Technology Program (Grant No. 2015BAL03B01), the National Nature Science Foundation of China (Grant No. 51308011) and the Natural Science Foundation of Beijing, China (Grant No. 8144040).

## AUTHOR CONTRIBUTIONS

Wenchao Liu performed the experiments and analyzed the data, and wrote the paper. Wanlin Cao proposed the topic of this study and designed the experiments. Heng Ma performed the experiments, numerical simulations, and contributed to manuscript preparation. Jianwei Zhang and Qiyun Qiao participated in the manuscript preparation.

## CONFLICTS OF INTEREST

The authors declare no conflict of interest.

## REFERENCES

1. McNeil, K.; Kang, T.H.-K. Recycled Concrete Aggregates: A Review. Int. J. Concr. Struct. Mater. **2013**, 7, 61–69.
2. Lv, Z.Y. Study dynamics and development trend of recycled aggregate and recycled concrete. Concrete **2010**, 6, 77–82. (In Chinese).

3. Buck, A.D. Recycled concrete as a source of aggregate. ACI J. **1977**, 74, 212–219.
4. Ministry of Construction. Provisions on Urban Construction Waste Management; Law No. 139 of the Ministry of Construction of People's Republic of China; Ministry of Construction: Beijing, China, 2005.
5. Kang, T.H.-K.; Kim, W.; Kwak, Y.-K.; Hong, S.-G. Flexural Testing of Reinforced Concrete Beams with Recycled Concrete Aggregates. ACI Struct. J. **2014**, 111, 607–616.
6. Gales, J.; Parker, T.; Cree, D.; Green, M. Fire Performance of Sustainable Recycled Concrete Aggregates: Mechanical Properties at Elevated Temperatures and Current Research Needs. Fire Technol. **2015**.
7. Bhuiyan, M.Z.I.; Ali, F.H.; Salman, F.A. Application of recycled concrete aggregates as alternative granular infills in hollow segmental block systems. Soils Found. **2015**, 55, 296–303.
8. Pecur, I.B.; Stirmer, N.; Milovanovic, B. Recycled aggregate concrete for nearly zero-energy buildings. Mag. Concr. Res. **2015**, 67, 575–584.
9. Limbachiya, M.C.; Leelawat, T.; Dhir, R.K. Use of recycled concrete aggregate in high-strength concrete. Mater. Struct. **2000**, 33, 574–580.
10. Mohamad, N.; Samad, A.A.A.; Goh, W.I. The Use of Recycled Aggregate in a Development of Reinforced Concrete Container as a Retaining Wall: Preliminary Study. Advances in civil engineering and building materials III. Adv. Mater. Res. **2014**, 831, 153–157.
11. Jimenez, J.R.; Ayuso, J.; Lopez, M. Use of fine recycled aggregates from ceramic waste in masonry mortar manufacturing. Constr. Build. Mater. **2013**, 40, 679–690.
12. Aqil, U.; Tatsuoka, F.; Uchimura, T. Strength and deformation characteristics of recycled concrete aggregate as a backfill material. Soils Found. **2005**, 45, 53–72.
13. Ma, G.; Zhang, Y.; Liu, Y.Z. Seismic behaviour of recycled aggregate thermal insulation concrete (Ratic) shear walls. Mag. Concr. Res. **2014**, 67, 145–162.
14. Ahmadi, F.; Hernandez, J.; Sherman, J. Seismic Performance of Cantilever-Reinforced Concrete Masonry Shear Walls. J. Struct. Eng. **2014**, 140, 758–782.
15. Zhang, Y.Q.; Cao, W.L.; Zhang, J.W. Experimental study on seismic performance of coupled shear walls with upper part using recycled concrete. J. Earthq. Eng. Eng. Vib. **2010**, 30, 98–103.

16. Cao, W.L.; Liu, Q.; Zhang, J.W. Experimental study on seismic performance of low rise recycled aggregate concrete shear wall. World Earthq. Eng. **2009**, 25, 1–5. (In Chinese).
17. Zhang, J.W.; Dan, S.; Cao, W.L. Earthquake simulation test on mid-rise recycled concrete shear walls with concealed bracings. Eng. Mech. **2012**, 29, 101–106.
18. Gu, T.S.; Xie, L.Y.; Cheng, G. Building energy conservation and wall thermal insulation. Eng. Mech. **2006**, 23, 167–184.
19. Wang, Q.; Qian, J.R.; Ma, B.M. Experimental study on seismic behavior of concrete grillage walls. J. Build. Struct. **2005**, 25, 15–25. [
20. Dou, L.J.; Zhang, M.; Li, Y.C. Research on seismic behavior of EPS module shear walls. Build. Struct. **2013**, 43, 61–64. (In Chinese).
21. Li, Z.; Zhang, T.; Liu, Y.Z. Research on low-cyclic load test of large shear span ratio composite shear wall with GHB heat preservation formwork. Build. Struct. **2013**, 43, 5–8. (In Chinese).
22. Cao, W.L.; Cheng, J.; Zhang, Y.B. Experiment of seismic behavior of low-rise recycled aggregate concrete shear wall with insulation modules and single layer of reinforcement. J. Build. Struct. **2015**, 36, 51–58.
23. China Standards Publication. Recycled Coarse Aggregate for Concrete; GB/T25177-2010; Standard Press of China: Beijing, China, 2011.
24. China Industry Standard. Specification for Mix Proportion Design of Ordinary Concrete; JGJ 55-2011; Standard Press of China: Beijing, China, 2011.
25. China Standards Publication. Standard for Test Method of Mechanical Properties on Ordinary Concrete; GB/T50081-2002; Standard Press of China: Beijing, China, 2002.
26. Guo, Z.H.; Shi, X.D. Reinforced Concrete Theory and Analyze; Press of Tsinghua University: Beijing, China, 2003.
27. Bao, S.H. Design and Analysis of Tall-Building Structures; Press of Tsinghua University: Beijing, China, 2005.
28. Tang, X.R.; Jiang, Y.S.; Ding, D.J. Application of the theory of softened truss to low-rise steel fiber high strength concrete shear walls. J. Build. Struct. **1993**, 14, 2–11.
29. Xiao, J.Z. Experimental investigation on complete stress-strain curve of recycled concrete under uniaxial loading. J. TongJi Univ. Nat. Sci. **2007**, 35, 1445–1449.

# 3

# CHAPTER

# A SIMULATION AND EVALUATION SYSTEM ORIENTED TO THE EMERGENCY RESPONSE EFFECTIVENESS OF THE ABRUPT EARTHQUAKE DISASTER

**Yan-Yan Huang[1]**

*[1] Automation School, Nanjing University of Science and Technology, Nanjing, Jiangsu, China*

## 1. INTRODUCTION

The untraditional emergency disasters, like the 2008 Sichuan earthquake of China, break out with nothing indication, no time to prepare for. These emergency events always bring us the catastrophe with much casualty, homeless and immeasurable economic losses.

The emergency disasters response problem is an important thing to be researched all over the world. As we all know that we can not change the breaking out of the untraditional emergency event in present technology level, but it never means that we could do nothing, on the contrary, if we have enough response plans or preparedness, the disasters should be reduced to the minimum level. It requires us to prepare many valid emergency response plans by hands. However, how to get such valid emergency response plans?

Considering the emergency catastrophe like earthquake takes place so abruptly, it is difficult to make decisions to get the most suitable preparedness plan from many emergency response plans. In order to improve the emergency response capability for the abrupt earthquake disasters, it is necessary to build a decision-making support system to evaluate the response plans and to select a most suitable preparedness plans.

In fact, it is difficult to test and evaluate the response plans in the real conditions of a strong earthquake disaster. As far as we know, modeling and simulation methods are fit for evaluating the preparedness plans. Adopting the modeling and simulation methods, there are many advantages such as low cost, safety research, time-space easily convert, and so on.

Many emergency researches, authors in reference [1-3], show that the more preparedness for the emergency event, the less risk we suffer from the uncertainty emergency. As we all know that, the emergency plans are oriented to the future event, we have none of the real data and sense. So the modeling and simulation methods are utilized to the decision-make support system.

More researches have discussion about the modeling and simulation method in the emergency response management [4]. At the mean time, Many researcher and experts have noticed the importance of the simulation and evaluation method in the decision-making support system [5,6]. The emergency response needs the decision making support system [7,8].

In China, there are many learners to carry out their researches on the emergency response management, Professor Wang [9] puts forth the parallel simulation method to research the emergency management. Fan Weicheng [10] points to the public incidents, provides some useful suggestion and response methods for the emergency platform system.

In this chapter, we set the research object orient to the emergency earthquake disasters. Design a decision-making support system framework for the emergency response management, and then mainly research the simulation and evaluation methods and their application in the earthquake disasters.

# 2. SYSTEMATIC FRAMEWORKS FOR SIMULATING AND EVALUATING THE EMERGENCY PREPAREDNESS FOR EARTHQUAKE

Judging and analyzing the advantage and disadvantages of an emergency response plan needs a good decision making support system. To build up such a decision making system, a serial of technical methods and theory models are required to support the system. In this chapter, the author sums up the research methods for the emergency response plans, and put forth a whole research route map to design and develop the system. Such a research route map is a set of systematic framework which can simulate and evaluate the response plans for the untraditional emergency disaster like earthquake, seeing the Figure 1.

In Figure 1, there are 5 main parts (subsystems) in the set of systematic framework. These subsystems are built to support the whole decision framework. They are including: (A) emergency response effectiveness concept and the evaluation indices, (B) indices weights acquisition,(C)emergency response simulation theory based on OODA-DEVS, (D) simulation system and simulation data acquirement,(E) the integrated evaluation process and decision making support system.

The next research is to describe the related subsystems in the Figure 1. The following research will take emphasis on the main techniques of the simulation and evaluation in emergency response plans.

# 3. EVALUATION INDICES ORIENT TO THE EMERGENCY RESPONSE OF THE EARTHQUAKE DISASTER

In order to select a right plan from many emergency response plans, it is necessary to make a good decision to evaluate the emergency response plans. And therefore, build up a set of evaluation indices system is very important.

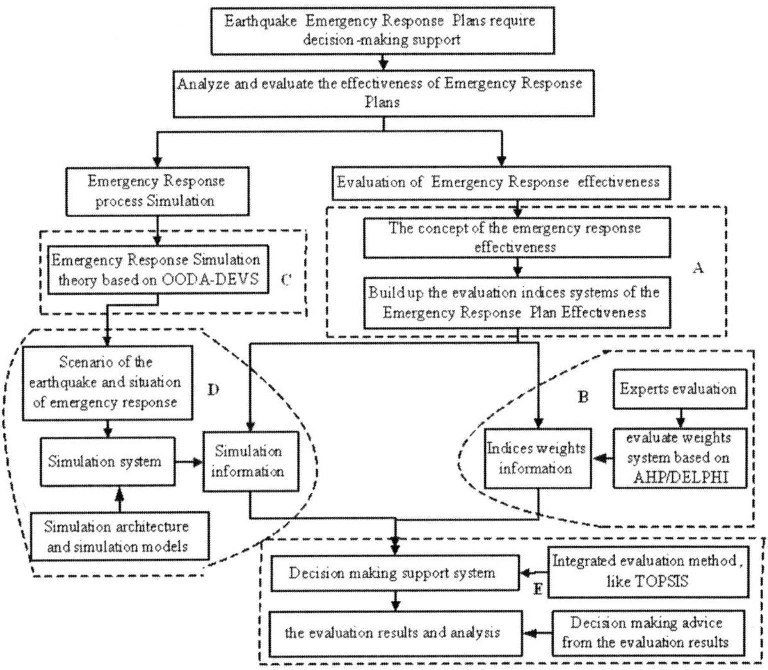


**Figure 1.** Systematic framework to simulate and evaluate the abrupt earthquake disaster

## 3.1. Measurements on the effectiveness for emergency response

To evaluate the emergency response plans, we need design a comprehensive evaluation index. Such a complex index is needed to measure what is the degree of an emergency response plan can perform.

The operational effectiveness is widely used in military evaluation. As we know that, a war is very similar to the fight for the earthquake disaster. And thus, we need to give a definition for the effectiveness. In fact, effectiveness means the degree to which objectives are achieved and the extent to which targeted problems are solved.

And therefore, the effectiveness of emergency response plan can be defined as the degree to which the emergency plans executed right by the decision makers are achieved, or the extent to which the emergency disaster risk to reduce and rapid response rescue.

By means of the definition of the emergency response effectiveness, we can put forth a comprehensive index to measure the degree of the emergency response.

Emergency response effectiveness concept includes: 1) it is a comprehensive index; 2) given a specific emergency conditions and specific mission, which degree does the emergency response plan perform in the mission.

According to the concept of the emergency response effectiveness, we can measure the emergency response plans of the earthquake disaster in some aspects including: rescue speed, capability, safety and cost. As we know that, judge whether a well earthquake emergency plan is valid need comprehensive evaluate the above factors.

## 3.2. Evaluation indices systems for the emergency response preparedness

As a comprehensive effectiveness index, it needs to decompose into detailed sub indices. Using the top-down analysis methodology of system engineering, we can break up the top effectiveness indices into the different sub-indices, seeing in the Figure 2.

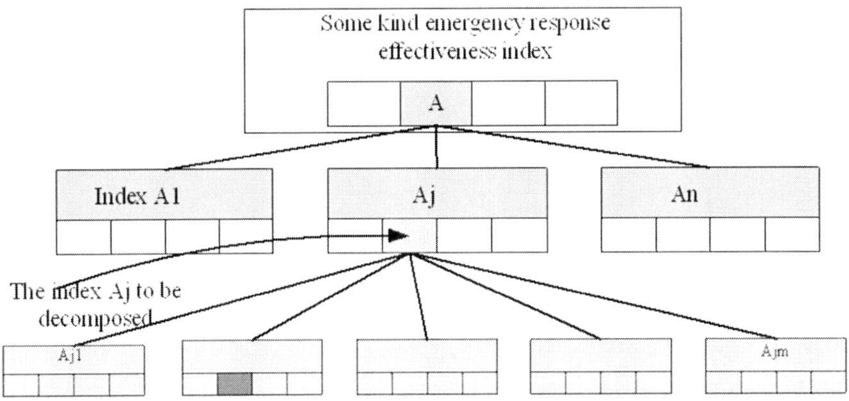


**Figure 2.** Effectiveness indices decomposed in top-down methodology

With the top-down methodology, we can design the response effectiveness indices system for the untraditional earthquake disaster, for example, seeing Figure 3.

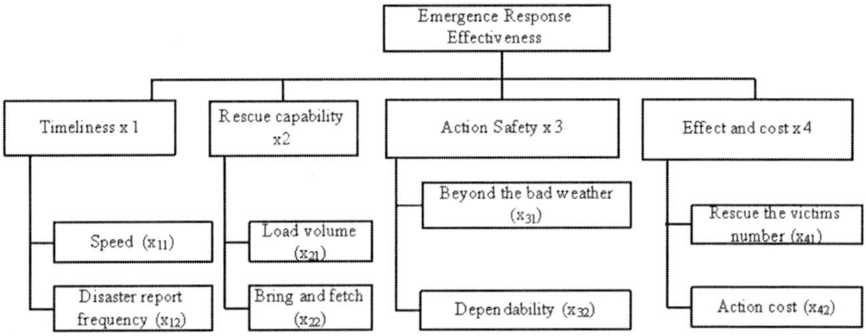


**Figure 3.**Response effectiveness indices of earthquake disaster

In order to describe the indices clearly and be easy to design for programming, we adopt the XML (Extensible Markup Language) file format to describe the effectiveness indices of the earthquake disaster. The emergency response effectiveness indices system can be written in XML file format as follows:

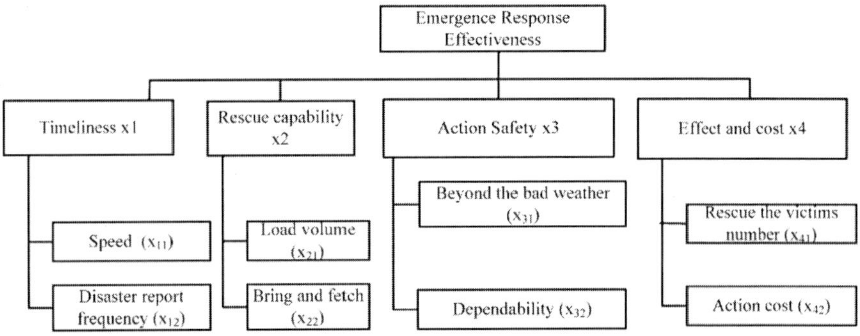


**Figure 4.** The emergency response effectiveness indices system

### 3.3. The weights acquirement for the emergency response effectiveness indices

For the evaluation indices system, we have to know how much the sub-indices make role in evaluating the emergency response preparedness. And therefore, it is essential to acquire and calculate the weights of the sub-indices, which support the whole response plans effectiveness. However, it is complicated to acquire the indices weights because the calculation full of experts' interaction with the evaluation systems.

In order to calculate the indices weights smoothly and make such an weights acquirement system reusable, we design a weights evaluation system based on MVC (Model View and Controller) mode. The MVC mode is a reusability mode in software engineering, seeing the Figure 5. The architecture of the weight evaluation system based on MVC is designed as the Figure 6.

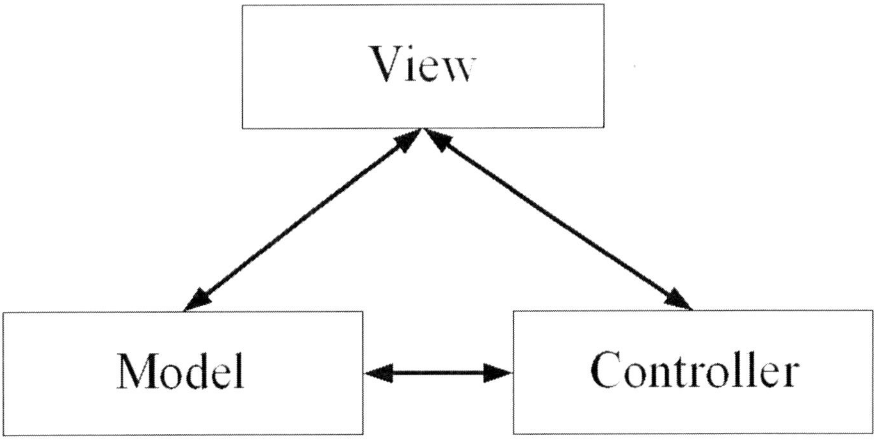


**Figure 5.** The Model-View-Controller mode in software engineering

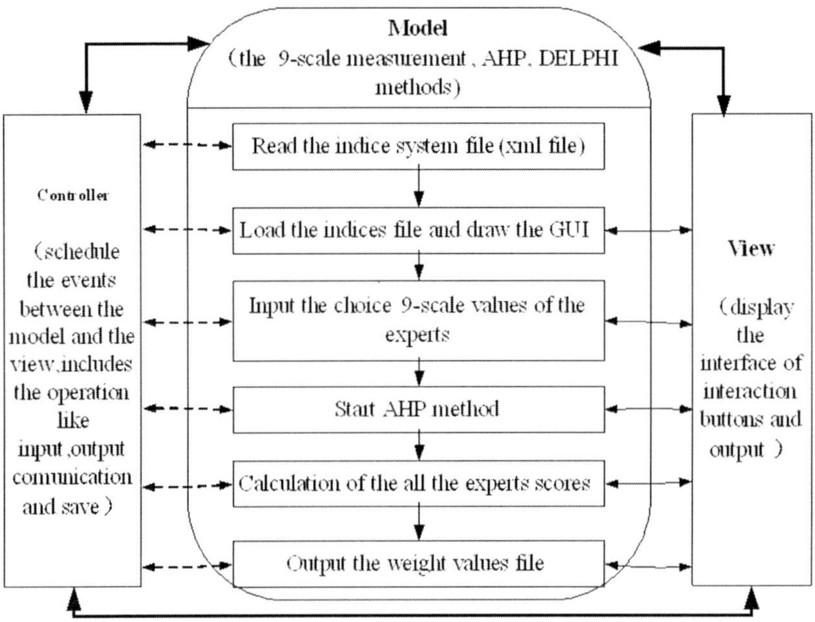


**Figure 6.** The architecture of the weight acquisition system based on MVC

In the Figure 6,the main parts are described as follows:

**Model**: it is a process algorithm of expert scoring information based on the idea of AHP (Analytic Hierarchy Process). Before the model starts, the response effectiveness indices system file (xml file) will be loaded in the weights acquisition system.

**View**: it is the graphic user interface (GUI) and it displays the visual information including the input windows of the experts, and the print of the weight values output.

**Controller**: it is the communication and control bridge between the Model and View, and it mainly includes the operations of the input, output, and calculation and saving.

As an application example, we load the xml file of the response effectiveness indices (seeing the response effectiveness indices system, Figure 3.), then the weight evaluation system with the graphic interface view is generated, (seeing the Figure 7). This system is developed in java programming language. Notice, Here the indicators are the same meaning of indices.

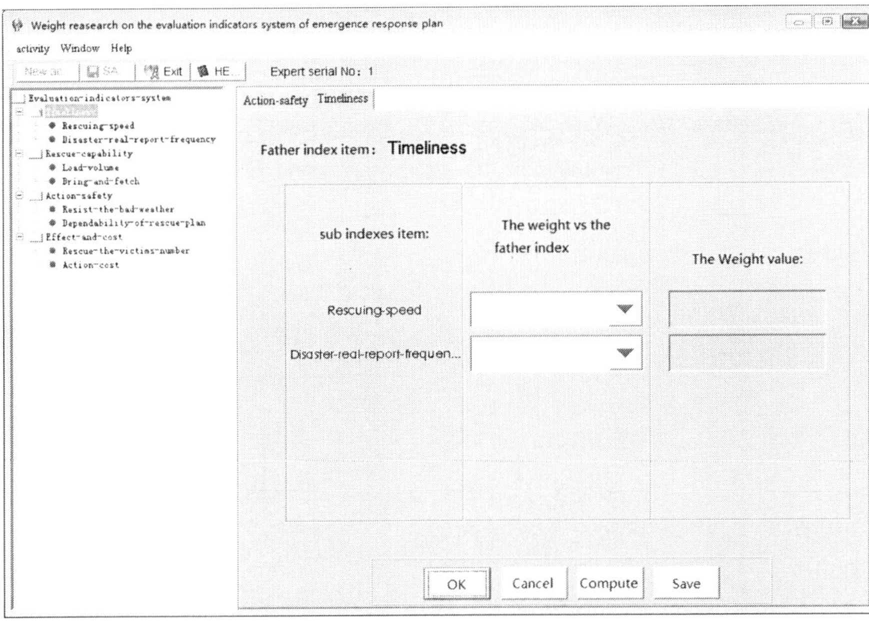


**Figure 7.** The weights calculation and the visual interaction system

By means of above system, we can get the indices weights W from index $x_1$ to $x_8$ : W=(0.142, 0.115, 0.091, 0.10, 0.092, 0.123, 0.159, 0.178). The weights W will be used in the integrated evaluation in later of this chapter.

## 3.4. The acquirement of the simulation information

The response plans of the emergency event are designed for the future emergency, so that we cannot collect the real data of emergency event, and therefore the simulation method are necessary to evaluate the future emergency plans. However, it is difficult to get the proper information from the simulation system because there is much difference between the simulation and evaluation. As we know that, the simulation system cannot get the proper information without the evaluation indices requirements. So we try to build a bridge between the two subsystems. This bridge is the so-called framework of menu of simulation data collecting, seeing the Figure 8.

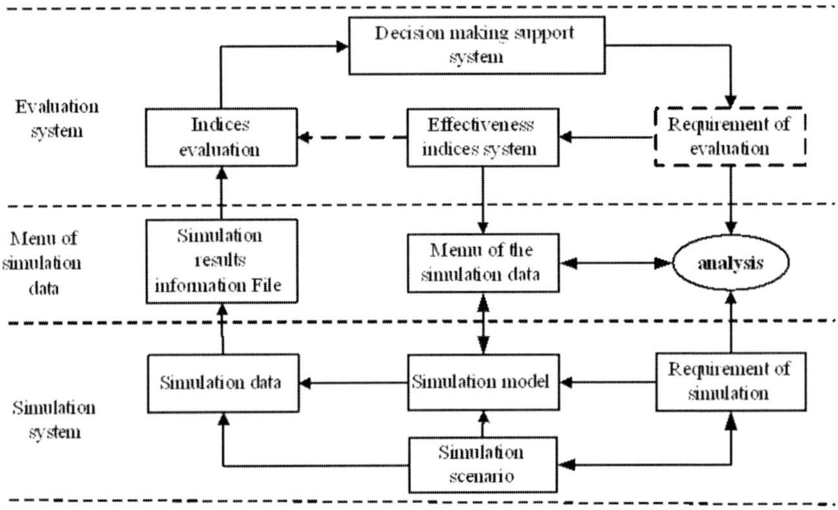


**Figure 8.** The framework of the menu of the simulation data collecting

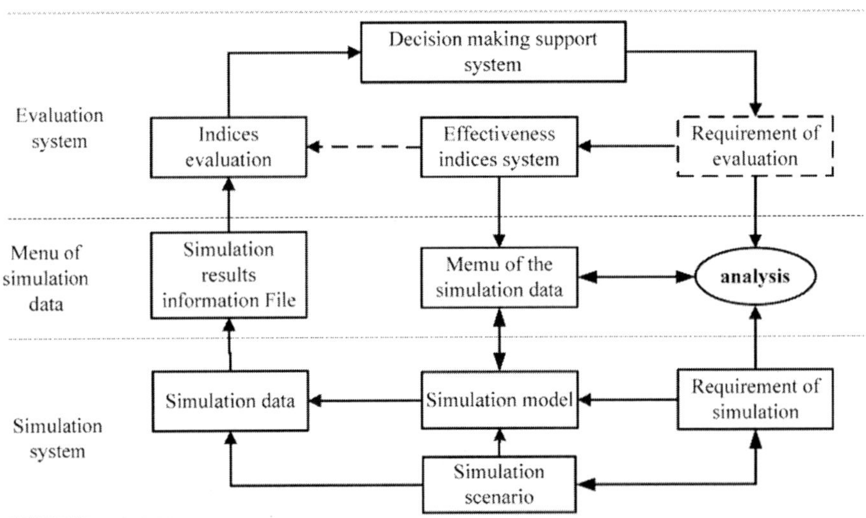


**Figure 9.** The menu of the simulation data

Because the simulation data is isomerous, it is difficult to save and transfer in form of the relational database format. In accordance with the format of emergency response effectiveness indices, the proper file format for the menu of the simulation data is the XML file. For the earthquake emergency response scenario, we design the menu of the simulation data to collect the simulation information as follows, seeing in Figure 9.

# 4. SIMULATION MECHANISM OF THE EMERGENCY RESPONSE PROCESS

As we know that, it is difficult to evaluate the emergency response effectiveness of the earthquake disaster in real environments. It is necessary to use simulation method to support the decision making. And therefore, a reasonable simulation method should be based on a scientific mechanism of emergency simulation.

## 4.1. The mechanism on the emergency response process of an earthquake disaster

In order to deal with the emergency response to the earthquake disasters, the government especial organization such as emergency departments and their executers should keep reconnaissance on the disaster zone, and make actions according to the detail disaster conditions.

How to describe such rules of the emergency process is a big problem. Considering the OODA (Observe, Orient, Decide, Act) loop which is John. Boyd, the famous U.S. captain, first put forth. It tells us the combat rule which runs as a loop, seeing as Figure 10.

In fact, the rapid response to the earthquake disaster is similar to a combat. And therefore, we should build the emergency response framework based on OODA loop as follows, seeing the Figure 11.

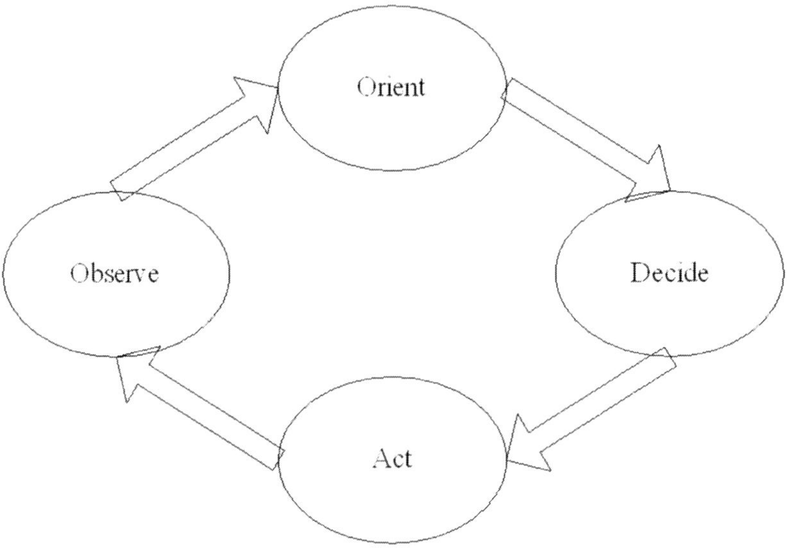


**Figure 10.** OODA loop diagram

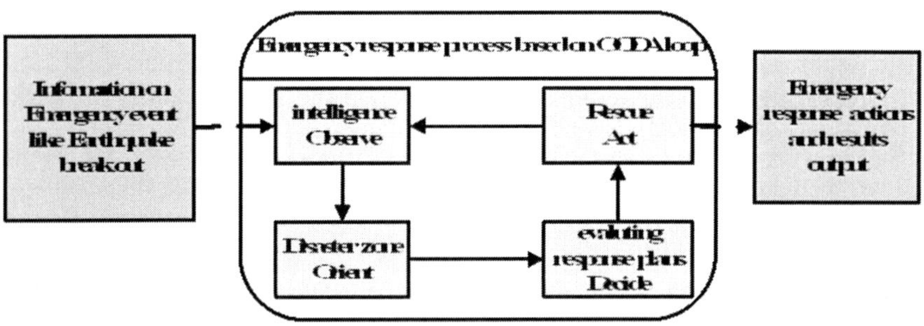

**Figure 11.** The response framework is based on OODA loop for the earthquake disaster

## 4.2. Emergency response modeling theory based on OODA

With the emergency response framework for the earthquake disaster, we need build a simulation model in scientific language. DEVS (Discrete Event System Specification) is a formalism language for modeling and analysis of discrete event systems (DESs). The emergency response system just is such a system, so that it is reasonable to using DEVS to describe the simulation process.

### 4.2.1. The OODA loop and the DEVS theory

The DEVS formalism was invented by Bernard P. Zeigler, who is emeritus professor at the University of Arizona. It is a modular and hierarchical formalism for modeling and analyzing general systems. There are two key models: atomic DEVS and coupled DEVS. *Atomic DEVS* captures the system behavior, while *Coupled DEVS* describes the structure of system.

The atomic DEVS is fit for describing the emergency response system behavior. Combining with the emergency response process framework based on OODA, we can build up the control structure in the DEVS factors, seeing in Figure 12.

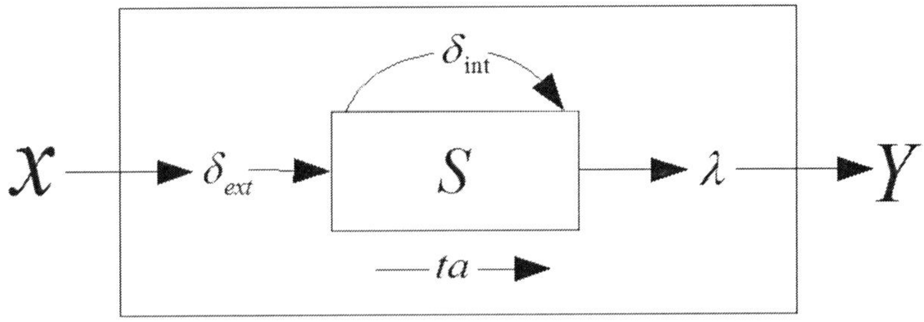


**Figure 12.** DEVS atomic model diagram

An atomic DEVS model is defined as a 7-tuple

AtomicDEVS= $<S, ta, \delta_{int}, X, \delta_{ext}, Y, \lambda>$

Where:

S is the set of sequential states (or also called the set of partial states);

X is the set of input events;

Y is the set of output events;

$Q=\{(s, t_e) \mid s \in S, 0 \leq t_e \leq ta(s)\}$ is the set of total states, and

*Te* is the elapsed time since the last event;

$\delta int: S \rightarrow S$ is the internal transition function which defines how a state of the system changes internally (when the elapsed time reaches the lifetime of the state);

$\delta ext: Q \times X \rightarrow S$ is the external transition function which defines how an input event changes a state of the system.

$\lambda{:}S\rightarrow Y$ is the output function. This function defines how a state of the system generates an output event (when the elapsed time reaches the lifetime of the state);

The output function $\lambda(s)$ is decided by the state s before transition.

$ta{:}S\rightarrow R{+}0,\infty$ is the time advance function which is used to determine the lifespan of a state, $R{+}0,\infty$ means Non negative real numbers; when no input events arrive, the state s will maintain the lifetime. When $ta(s)=0$ , we call the state s is executing state; while $ta(s)=+\infty$ , the state s is called dead state, and the system always waits for the input event.

### 4.2.2. Describe the emergency response process in OODA-DEVS

The kernel of a simulation system is to the simulation engine. The simulation engine of the emergency response system is required to design based on the DEVS simulation theory. Considering of the emergency response process is an OODA loop, the simulation engine of the emergency response system for disaster is designed as follows, seeing the Figure 13.

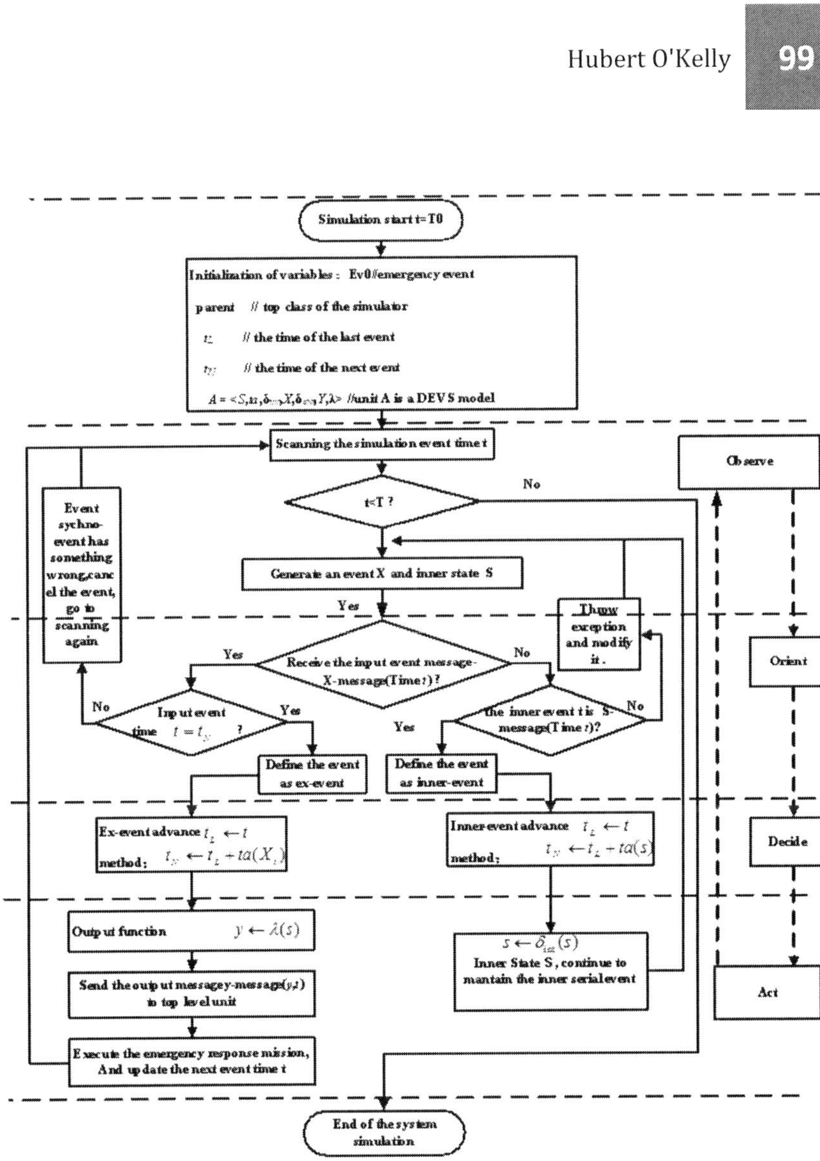


**Figure 13.** OODA-DEVS simulation engine

In the Figure 13, it t ells us the workflow of the system simulation engine. When simulation start at $T_0$, each variable in the simulation system is initiated. Then scan the current simulation time t, and judge whether the t is more than the total simulation time T. If t>T, then the emergency simulation process is terminated, else it requires to judge the emergency events. Before dealing with the messages of the emergency system, it is needed to judge the kinds of the messages. 1) If the received message is the input event message-X-message, what's more, the simulation time t of the input event must meet the requirement of the next event time, we can define such an event as the ex-message; if the simulation time of the input event is wrong, show that the synchronization time is invalid, and the event must be cancelled and to be scanned again.

After confirm the ex-message, it is required to update the simulation time of last event $t_L$ and the time of the next event $t_N$, and then output the simulation result information by means of the output function$\lambda(S)$. While the response units like rescuer teams send the simulation information to the top level simulation unit (decision making departments), the response units themselves will continue to finish their to update the simulation time t again, and return to scan the simulation event time t. so far, an external input event is finished dealing with.

2) If the message received is an inner message, then we define it as an inner serial event. Other wise, confirm it as an exceptional message, and it needs to modify in the time management (a simulation service rule). For the inner serial event, the inner state transits, and update the time of the last event, and the next event time t. continue to these steps till the ex-message is coming, and transfer to the ex state loop. From above flow chart analysis, it is easy to find out, the simulation engine process just like the OODA four parts loop.

## 4.3. Simulation system and simulation data list acquisition

### 4.3.1. The simulation architecture of the emergency response management

As we all know that, the earthquake emergency response system is very complex. The simulation system includes many parts: the emergency scenario, the behaviors of response units, the interaction communications between the rescuer and the refugees. All the parts are based on the OODA loop in the response process.

Considering the emergency response plans need the many interaction operations, so that we build up a simulation system architecture based on the HLA/RTI (High Level Architecture/Run-time Infrastructure) technology. It is suitable for the modeling developers to use HLA and RTI to describe emergency system with many uncertainty events.

The simulation architecture includes many software nodes: response plan and rescue units, the emergency information like weather and location, situation description, the environment of disaster area, emergency events generator, and rescue command system, and so on.

The simulation architecture of the emergency response for earthquake disaster is shown in Figure 14.

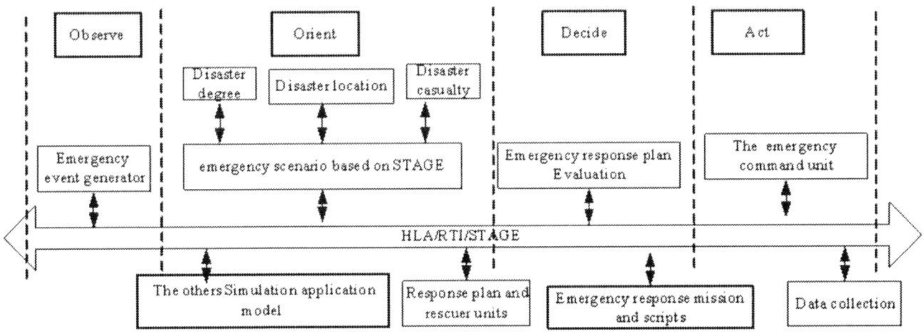


**Figure 14.** Simulation architecture of the emergency response

### 4.3.2. Response units behaviors simulation scripts

After we build a simulation system, we need analyze the simulation behaviors. In the simulation toolkit of STAGE (Scenario Toolkit and Generation Environment), we can create the behavior simulation scripts based on the above simulation theory and the simulation engine.

Emergency response unit is the force to execute the rescue mission like troop, emergency medical services, police, and firefighters and so on. For each of the emergency response unit, there is a script program to describe the unit behavior. The script codes are written in STAGE as follows.

# 5. EARTHQUAKE DISASTER SCENARIO AND EMERGENCY RESPONSE SIMULATION

In order to analyze the emergency response plans of earthquake disasters, a typical earthquake disaster is designed as an application scenario. The next emergency response plans are evaluated based on such a scenario.

## 5.1. Earthquake disaster scenario

Suppose a terrible catastrophe like earthquake breaks out in some uncertainty place, seeing the Figure 15, we should carry out the emergency response plans to quickly rescue the victims. Usually the routes to the disaster zone are far away from city and the disaster conditions are very complicated and uncertainty, so that we are required a decision-making support system to get the best emergency plan from many available plans.

(1) The script codes for detection of the emergency response units

```
    int detect_platforms[500];                    //the numbers of detect platforms
    track.cycle_on(detected_platforms);         //start scanning the detect platforms
      while((track.next()>0)and(track.indentify=danger)and(track.type=land)) do
      detect_platforms[index]=1;      //detected the emergency response units, its value is 1.
    endwhile
```

(2) The orient script code of the emergency response unit

```
    Boolean detected[500]; int i=l;boolen identify; //initiating the variables
    enum {best, better, good, worse, worst} Degree;// the enum type about disaster degree
    while(i<500)do
     if (identify)
        detected[i]=1;
      else detected[i]=0;
    while(detected[i]!=0)do
      object.orient(i,Degree);
      data_link.assign_int(i);
      i=i+l;
    endwhile;
```

(3) The decision and action of the emergency response unit:

```
    int act_index; //the index which need to rescue
    track.cycle_on(act_index);                  //scan the act_index
    while(track.next()>0)do                     //there are objects required to rescue
    if(boolen rescue==1 and the orient[i]==worse)then
     entity.request_ground_track=1; // require ground tracks to rescue.
     entity.request_speed=entity.actual_speed+entity.max_acceleration*10; //require the speed of the
rescue truck
    endif;
      if(rescue==0)then
        entity.request--speed=0; //the emergency response unit is stopping
      endif;
    endwhile;
```

In this earthquake disaster scenario, the advantages and disadvantages of the emergency response plans can be shown in Table 1.

**Table 1.** he advantages and disadvantages of some emergency response plans

| Id | response plans | Advantage | Disadvantage |
|---|---|---|---|
| 1 | Truck Vehicle | Less relied on weather, fast and land easily. | More relied on highway traffic, less safety. |
| 2 | Air- plane | Less relied on highway, high speed and large loading capacity | More relied on weather, hard to land on disaster area and high cost. |
| 3 | Helicopter | Less relied on highway, high speed, good flexibility, take off and land easily. | More relied on weather, less safety and high cost. |
| 4 | March | Less relied on highway, good flexibility, flexible route | Carry less and move slowly. |
| m | Plan M | …… | …… |

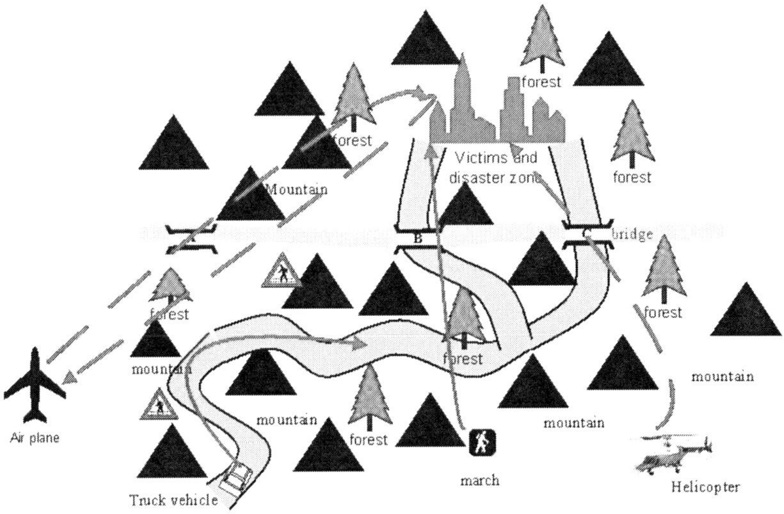


**Figure 15.** The Earthquake disaster scenario and some emergency response plans

## 5.2. The acquisition of the simulation data

For example, we select the emergency response plan by truck vehicle to simulate. In this plan simulation, first to arrange vehicles direction to the disaster zone, and then to design the routes with the random disasters conditions based on the discrete event simulation method (see Figure.16). During the simulation time, it is possible that the rescue trucks are hit by roll rock or mudslide from the secondary disasters, and it will spend much time to deal with the breakdown. We can simulate the trucks rescue action in different routes in the STAGE (see Figure.17).

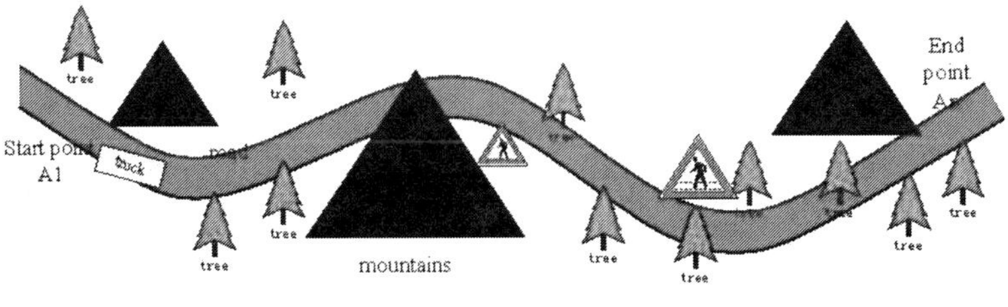


**Figure 16.** Simulation of the breakdown of the truck plan

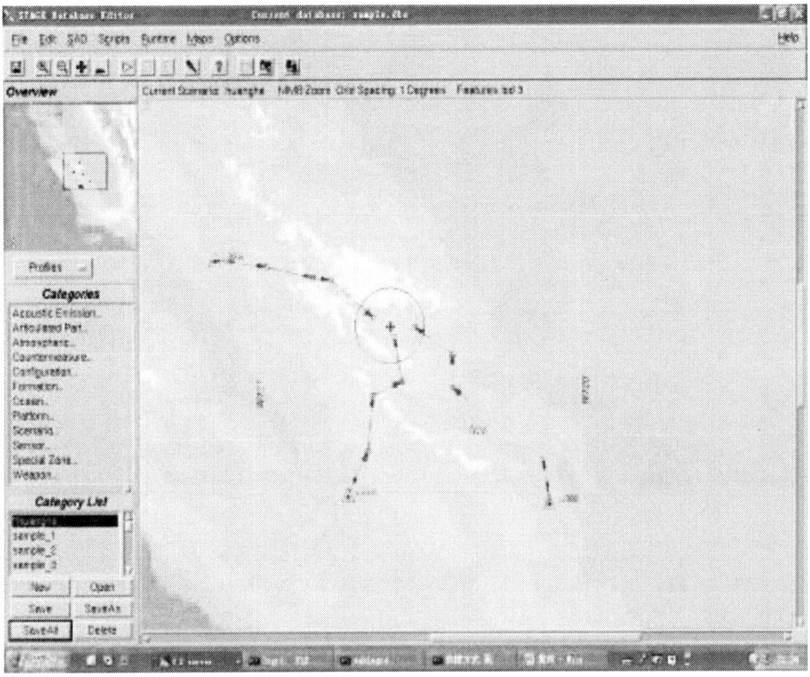


**Figure 17.** Simulation on the process of truck plan


```
<?xml version="1.0" encoding="UTF-8" ?>
< Evaluation-indices-system-of-emergency-response-plan >
    <Timeliness >
        < Rescuing-speed start-time=" 0" arrival time="4" distance=" 300.0"/>
        < Disaster–realtime-report frequency report –times-about-the-disaster="15"/>
    </ Timeliness >
    < Rescue-capability >
        < Load-volume available-weight-to-carry -and-fetch="12"/>
        < Bring-and -fetch equipment-capability-to-carry-and-fetch="0.6" />
    </ Rescue-capability >
        < Action-safety >
            < Beyond-the-bad-weather Level-of-last-bad-weather="6"/>
            < Dependability-of-rescue plan Dependability-probability="0.7"/>
    </ Action-safety >
    < Effect-and-cost >
            < Rescue-the-victims number Number-being-rescued="10"/>
    < Action-cost Cost-of–the-rescue-plan="10"/>
    </ Effect-and-cost >
</ Evaluation-indices-system-of-emergency-response-plan >
```


**Figure 18.** Simulation information in Xml file

Through the simulation system, we can collect the needed simulation information and save it as Xml file. The simulation information file (Truck.xml) is written in the format of the menu of simulation data as follows, seeing Figure 18:

The Truck.xml file is the simulation results information about adopting the truck plan in the emergency response. The simulation and data collection of the other response plans are similar to the plan of the truck. Here it is not necessary to repeat the other emergency plans.

# 6. THE INTEGRATED EVALUATION AND DECISION MAKING FOR THE EMERGENCY PLANS

As a decision-making system for the emergency response plans, it is necessary to integrate the above simulation data and the weights information of the evaluation indices system from the experts.

As for the evaluation method, because the evaluation indices are the isomer information which has different attributes. Therefore, in this paper, a good general evaluation model is adopted to build the decision-making method.

The detail evaluation method is researched as follows.

## 6.1. Build the evaluation indices system

The effectiveness evaluation indices system is built in Figure 19. $A_i$ is the index i, which describes the system attributes of the alternatives. Among the indices, there are three different types of indices: one type is the effective index, which value is the more, the better; while the second type is the cost index, which value is the more, the worse; and the third type is the proper index, which characteristic is between the two indices.

## 6.2. Effectiveness evaluation model based on TOPSIS

TOPSIS is for short of the Technique for Order Preference by Similarity to Ideal Solution. Its main idea is to evaluate the alternatives by means of the measurement scale named Euclidean distance. It is fit for evaluating many alternatives, which have different kinds of attributes.

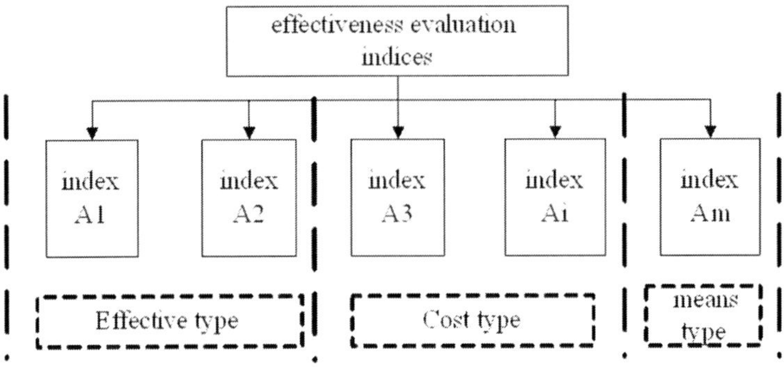


**Figure 19.** The general evaluation indices system on the emergency effectiveness

The evaluation method based on the TOPSIS is very suitable for evaluating the effect of the alternatives, which have many heterogeneous indices. In fact, this method is built on a decision-making matrix, which different rows in the same column index have the same attribute. This is the reason that the heterogeneous data from different plans can also be compared in this evaluation model.

First all, it is necessary to collect the related evaluation data. Combining with the evaluation indices system, get the data or attribute values from the different evaluation indices. Using such data or attributes values, we can build the evaluation matrix C, seeing the Table 2. In this table, the row denotes the alternatives, and the column name stands for the attributes values of the indices. Here the attributes data is from simulation information.

**Table 2.** Alternative-attribute information

| alternative | attribute 1 | attribute 2 | ... | attribute n |
|---|---|---|---|---|
| alternative1 | $C_{11}$ | $C_{21}$ | ... | $C_{1n}$ |
| ... | ... | ... | ... | ... |
| alternative m | $C_{m1}$ | $C_{22}$ | ... | $C_{mn}$ |

$C_{ij}$ means the j-th attribute of the i-th alternative or emergency plan. Where i=1,2,...,m; j=1,2,...,n.

The evaluation method based on the TOPSIS is describes as follows.

Firstly, convert the attributes in table 2 into the decision matrix C.

$$C = \begin{pmatrix} c_{11} & c_{12} & \cdots & c_{1n} \\ c_{21} & c_{22} & \cdots & c_{2n} \\ \cdots & \cdots & \cdots & \cdots \\ c_{m1} & c_{m2} & \cdots & c_{mn} \end{pmatrix}$$

$$r_{ij} = w_j c_{ij} \Big/ \sqrt{\sum_{i=1}^{n} c_{ij}^2} \quad (1)$$

For the data from the matrix C, the attributes from the same column can be compared with different rows. In order to be convenient to evaluate the different alternatives, we need to standardize the matrix C into matrix R with the unit $r_{ij}$ in formula (1). For $w_j$, it is the j-th weight of the indices. The matrix R becomes a normalization matrix which also adds the related weights from the evaluation indices, and R will be the foundation of the next evaluation steps.

Secondly, analyze the ideal point and negative-ideal point from the alternatives.

To get the ideal alternative and negative-ideal alternative, it depends on the type of the indices. Different type indices have different modes of process. The ideal alternative vector point is x* in the formula (2). At the same time, the negative-ideal alternative vector point is x-in the formula (3):

$$\{(\max_i r_{ij} \mid j \in J), (\min_i r_{ij} \mid j \in J') \mid i \in M\} = [r_1^*, r_2^*, \ldots, r_n^*] \quad (2)$$

$$\{(\min_i r_{ij} \mid j \in J), (\max_i r_{ij} \mid j \in J') \mid i \in M\} = [r_1^-, r_2^-, \ldots, r_n^-] \quad (3)$$

In the formula (2) (3), M is the set of the alternatives, J and J' is the effective type and cost type respectively.

Thirdly, calculate Euclidean distance of each alternative to the ideal alternative or the negative-ideal alternative.

The Euclidean distance of the i-th alternative to the ideal vector point.

$$S_i^* = \sqrt{\sum_{j=1}^{n} (r_{ij} - r_j^*)^2} \quad (i \in M) \quad (4)$$

The Euclidean distance of the i-th alternative to the negative-ideal vector point.

$$S_i^- = \sqrt{\sum_{j=1}^{n} (r_{ij} - r_j^{-*})^2} \quad (i \in M) \tag{5}$$

Fourthly, calculate the degree of each alternative approaching the ideal alternative $C_i$ in formula (6).

$$C_i = S_i^- / (S_i^- + S_i^*) \tag{6}$$

According to the formula (6), if the degree $C_i$ is larger, means that i-th alternative is better.

## 6.3. Response effectiveness indices system and evaluation information acquisition

In this application example, the effectiveness of several emergency response alternatives for the earthquake is evaluated based on the above evaluation model. The evaluation process is carried out as follows:

First, set up the evaluation indices system for the earthquake emergency response effectiveness, seeing the Figure 20.

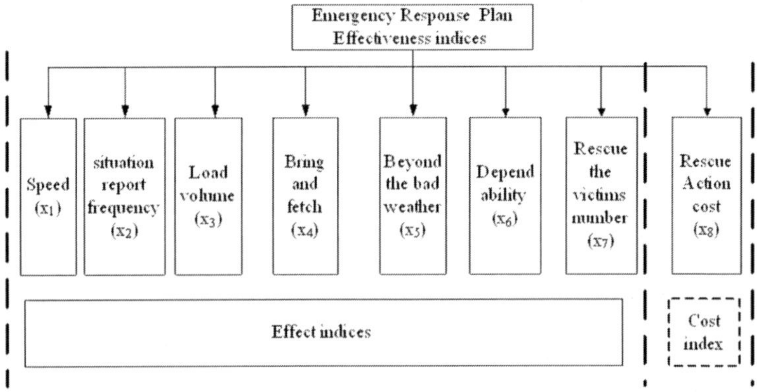


**Figure 20.** The effectiveness evaluation indices system

Second, there are two kinds of important information to be ready: indices weights and the indices simulation data.

According to the weights acquisition system, get the weights values from $x_1$ to $x_8$ as follows:

W=0.142, 0.115, 0.091, 0.10, 0.092, 0.123, 0.159, 0.178.

The simulation values of the indices from $x_1$ to $x_8$ are filled in the table 3, where, $x_8$ is the cost type index, means the less the better. The indices except $x_8$ are the effect indices.

**Table 3.** The indices simulation data of emergency response plan

| Emergency response plan A | Rescuing speed x1 | Disaster real time report frequency x2 | Load volume x3 | Bring and fetch capability x4 | Beyond the bad weather x5 | Dependability of rescue plan x6 | Rescue the victims number x7 | Action cost x8 |
|---|---|---|---|---|---|---|---|---|
| Truck | 75 | 15 | 12 | 0.6 | 6 | 0.7 | 10 | 10 |
| Air plane | 500 | 10 | 40 | 0.4 | 3 | 0.5 | 0.1* | 40 |
| Helicopter | 300 | 25 | 20 | 0.9 | 8 | 0.7 | 20 | 20 |
| march | 6 | 20 | 0.5 | 0.6 | 9 | 0.8 | 1 | 2 |

Notice *: For the airplane can not carry anything back from disaster zone, here use 0.1 stands for carrying few.

## 6.4. Process of the evaluation

With the method, we can get the original matrix C from the information in Table 3.

$$C=\begin{bmatrix} 75.0 & 15.0 & 12.0 & 0.6 & 6.0 & 0.7 & 10.0 & 10.0 \\ 500.0 & 10.0 & 40.0 & 0.4 & 3.0 & 0.5 & 0.1 & 40.0 \\ 300.0 & 25.0 & 20.0 & 0.9 & 8.0 & 0.7 & 20.0 & 20.0 \\ 6.0 & 20.0 & 0.5 & 0.6 & 9.0 & 0.8 & 1.0 & 2.0 \end{bmatrix}$$

Due to each of the index has different weight, it is necessary to think about the indices weights for the original matrix. In this paper, the weights values for the indices from $x_1$ to $x_8$ is W=0.142, 0.115, 0.091, 0.10, 0.092, 0.123, 0.159, 0.178. Using the formula (1), we can get the standard matrix P.

$$P=\begin{bmatrix} 0.018 & 0.047 & 0.0236 & 0.0462 & 0.04 & 0.063 & 0.071 & 0.0388 \\ 0.121 & 0.0313 & 0.0786 & 0.031 & 0.02 & 0.045 & 0.0007 & 0.155 \\ 0.025 & 0.078 & 0.0393 & 0.0692 & 0.0534 & 0.063 & 0.1421 & 0.0776 \\ 0.0015 & 0.0625 & 0.001 & 0.0462 & 0.06 & 0.072 & 0.0071 & 0.0078 \end{bmatrix}$$

From the matrix P, the Ideal alternative vector point can be selected as IdealPt=(0.121, 0.078, 0.0786,0.069,0.0601, 0.072, 0.142,0.0078). Notice that $X_8$ is the cost index, the less the better.

At the same time, the most negative alternative point can be chosen as NegativePt=(0.0015, 0.0313, 0.0010, 0.031, 0.020, 0.045,0.00071, 0.155)

By means of the formula (4), get the Euclidean distance of each alternative to the ideal alternative vector point as follows: IdealDis=(18.93,432.2, 232.1, 67.83).

Get the Euclidean distance of each alternative to the most negative alternative vector point as follows: anti-idealDis=(77.87, 502.41, 302.18, 21.53).

Combined with the idealDis and anti-dealDis, we can get the approaching degree F which stands for any alternative approaching the ideal alternative vector point, By the meantime, far way from the most negative alternative vector point, F=(0.4932 0.3946 0.6726 0.4451).

According to the approaching degree F, rank all the response plans in descending order as follows:

Plan 3-helicopter> Plan 1-truck > Plan 4-march > Plan 2-air plane. The diagram is described in Figure 21.

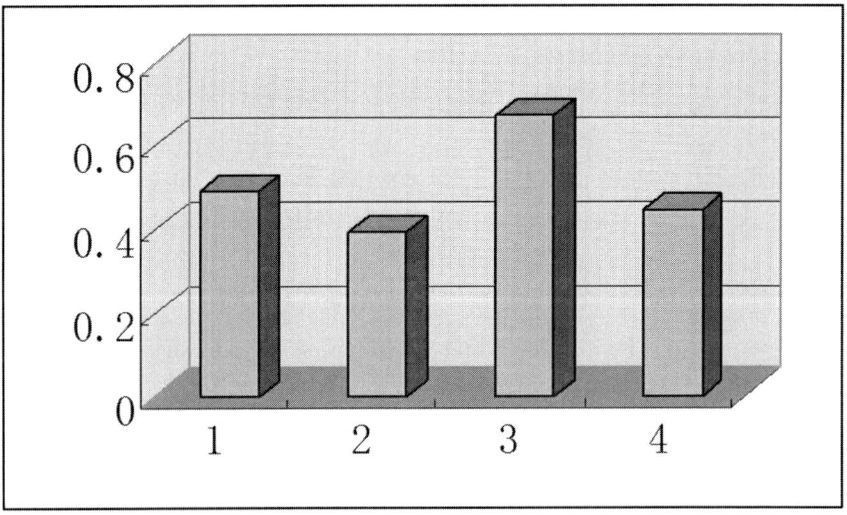


**Figure 21.** The effectiveness rank of the four Response plans

Based on the evaluation results, the plan 3(Helicopter rescue plan) is the first rank of all the plans, and the truck vehicle plan is the second. However, the last is the airplane response plan because it could not make its function to the maximum for example the plane can not land on the damage airport and pick up nothing back.

## 6.5. Future work

In this chapter, we develop a robust decision making support system, what's more, make valid application for evaluation and simulation the emergency response to the earthquake. As far as the future work concerned, there are several points to be done:

Firstly, improve the visual system of decision making system. A good visual system is welcome because we prefer the diagram to text or formulas. The future work will enforce the visual modular in the decision making system.

Secondly, improve the simulation technology for analyzing the emergency response plan, especially for the interaction simulation. There are many interactions in the simulation of the emergency response plans, and thus, it is necessary to enforce the Distributed interaction simulation technology like HLA/RTI.

Thirdly, improve the evaluation method. As we know, the evaluation method is the core of the decision-making support system. in this chapter, we adopt the experts judge method and the multi-attributes decision making method –TOPSIS. In the future work, we should adopt more evaluation methods to support the evaluation of the emergency response plans for the catastrophes.

# 7. HIGHLIGHTS

In this paper, there are some highlights as follows:

a. Put forth a set of systematic framework to simulate and evaluate the emergency response plans.

b. Define the concept of the emergency response effectiveness, and build up a set of the evaluation indices systems for emergency response plans.

c. Build the emergency response simulation model based on OODA, and describe it in DEVS theory.

d. Build up simulation system architecture based on HLA/RTI; collect simulation information by means of the simulation data list in XML format.

e. Build an evaluation method of emergency response effectiveness based on TOPSIS.

# 8. CONCLUSIONS

In this paper, put forth the concept of the emergency response effectiveness, and center on the methods of response effectiveness simulation and evaluation. A decision-making support system is built to evaluate the emergency response plans. The system is made up of several parts: indices weights acquirement subsystem, simulation mechanism of emergency response process, simulation data acquirement and the integrated evaluation method based on TOPSIS.

By means of the decision-making support system, several response plans in some earthquake disasters are evaluated for their response effectiveness. The evaluation results demonstrate that the simulation and evaluation system is available and reasonable.

# 9. SUMMARY

It is difficult to judge the most suitable preparedness plan from many emergency response plans. And therefore, it is necessary to build a decision-making support system to evaluate the response plans and to select the most suitable preparedness plan.

In fact, we face with the challenge to test and evaluate the response plans under the **real conditions** of a strong earthquake disaster. Modeling and simulation methods are adopted to evaluate and demonstrate the preparedness plans for their advantages such as low cost, safety research, time-space easily converting, and so on.

In this chapter, pointing to the strong earthquake disasters, a decision-making support system framework is designed to manage the emergency response plans, and this framework includes the simulation and evaluation methods. In detail, a whole research route map is designed and developed to simulate and evaluate the response plans for the untraditional emergency disaster like earthquake. As a result, a serial of technical methods and theory models are included in the systematic framework. There are 5 main parts (subsystems) in this framework. They are including: (A) emergency response effectiveness concept and the evaluation indices, (B) indices weights acquisition,(C)emergency response simulation theory based on OODA-DEVS, (D) simulation system and simulation data acquirement,(E) the integrated evaluation process and decision making support system.

In this chapter, the concept of the emergency response effectiveness is defined, and it can use as a comprehensive index to measure the degree of the emergency response plan. What's more, the evaluation indices of the emergency response plans are built up in systems engineering method.

In order to describe the evaluation indices clearly and be easy to design for programming, we adopt the XML (Extensible Markup Language) file format to describe the response effectiveness indices of the earthquake disaster.

To get the weights of the evaluation indices, the weights evaluation system based on MVC (Model View and Controller) mode is built.

Referring as the similar process of combat, the emergency response process framework based on OODA loop is built, and put forth a simulation model in DEVS theory. By means of the simulation model, we design the flowchart of the system simulation engines.

For many interaction operations in the emergency response simulation, we build up a simulation system architecture based on the HLA/RTI (High Level Architecture/Run-time Infrastructure) technology. It is suitable for the developers to use HLA and RTI to describe emergency response systems full of interactions and interoperability.

A typical earthquake disaster is designed as an application scenario. The following emergency response plans are evaluated based on such a scenario.

It is necessary to integrate the above simulation data and the weights information of the evaluation indices system from the experts.

As for the evaluation method, because the evaluation indices are the isomer information which has different attributes. A general evaluation model based on TOPSIS is adopted to build the decision-making method.

Taking the emergency simulation scenario of the strong earthquake disaster as example, apply with the simulation and evaluation methods in this chapter, and get the evaluation results of emergency response plans. The evaluation results show feasible. In this example, the Helicopter rescue plan is evaluated as the best plan among the response plans, and the truck vehicle plan is the second. However, the last is the airplane response plan for it can not land on the damage airport and pick up nothing back.

## NAN. ACKNOWLEDGEMENTS

This work is supported by National Natural Science Foundation (NNSF) of China under Grant No. 61374186. This work also gets some support from the 28th Research Institute of China Electronic Technology Group Corporation.

## REFERENCES

1. Kleindorfer, P.R. and Germaine, H.S.Managing Disruption Risks in Supply Chains, Production and Operations Management 2005; (14)1:53-68.
2. Rosenthal, U., A. Boin, et al. Managing Crises: Threats, Dilemmas, and Opportunities. Springfield: Charles C Thomas, 2001.
3. Yuan Hongyong, Huang Quanyi, Suguo feng,et al, theory and practice of key technologies of emergency platform system. Beijing, China: Tsinghua University,2012.
4. Takahashi, T. Agent-Based Disaster Simulation Evaluation and its Probability Model Interpretation. The 4th International Conference on Information Systems for Crisis Response (ISCRAM2007),2007.
5. Song Y, Hu, Y.H.,(2009),Group Decision-Making Method in the Field of Coal Mine Safety Management Based on AHP with Clustering,2009 IEEE International Conference on Service Operations, Logistics and Informatics, Chicago, USA, July 22–24, 2009
6. Huang,Y.Y., Wang,J.Y., An Evaluation Method of Operational Effectiveness for the Emergency Plans, 2009 IEEE International Conference on Service Operations, Logistics and Informatics, Chicago, USA, July 22–24, 2009

7. Hart, P., Rosenthal, U., Kouzmin, A., Crisis Decision Making, the centralization thesis revisited, in:Administration & Society, 1993;25(1):12-45.

8. Van Santen willem, Jonker Catholijn. Crisis Decision Making Through a Shared Integrative Negotiation Mental Model, Proceedings of the 6th International ISCRAM Conference – Gothenburg, Sweden, May 2009

9. WANG Fei-Yue. Web Social Media in Disaster Reduction and Emergency Management,Science & Technology Review;2008

10. Fan Weicheng; Advisement And Suggestion To Scientific Problems Of Emergency Management For Public Incidents; Bulletin of National Natural Science Foundation of China;2007

# 4

# CHAPTER

# CORRELATION BETWEEN EARTHQUAKES AND AE MONITORING OF HISTORICAL BUILDINGS IN SEISMIC AREAS

**Giuseppe Lacidogna *, Patrizia Cutugno, Gianni Niccolini, Stefano Invernizzi and Alberto Carpinteri**

*Department of Structural, Geotechnical and Building Engineering, Politecnico di Torino, Corso Duca degli Abruzzi 24, 10129 Torino, Italy*

## ABSTRACT

In this contribution a new method for evaluating seismic risk in regional areas based on the acoustic emission (AE) technique is proposed. Most earthquakes have precursors, i.e., phenomena of changes in the Earth's physical-chemical properties that take place prior to an earthquake. Acoustic emissions in materials and earthquakes in the Earth's crust, despite the fact that they take place on very different scales, are very similar phenomena; both are caused by a release of elastic energy from a source located in a medium. For the AE monitoring, two important constructions of Italian cultural heritage are considered: the chapel of the "Sacred Mountain of Varallo" and the "Asinelli Tower" of Bologna. They were monitored during earthquake sequences in their relative areas. By using the Grassberger-Procaccia algorithm, a statistical method of analysis was developed that detects AEs as earthquake precursors or aftershocks. Under certain conditions it was observed that AEs precede earthquakes. These considerations reinforce the idea that the AE monitoring can be considered an effective tool for earthquake risk evaluation.

**Keywords:** acoustic emission; structural monitoring; cultural heritage; earthquakes; seismic precursors

# 1. INTRODUCTION

A new method to evaluate seismic risk in regional areas is proposed as an attempt to preserve Italian historical and architectural cultural heritage. To this purpose, the spatial and temporal correlations between the acoustic emission (AE) data obtained from the monitoring sites and the earthquakes that have occurred in specific ranges of time and space are examined.

The predictive power and the non-invasive procedure of the AE technique (no external excitation is provided, since the source of energy is the damage process itself) can be exploited to preserve Italian cultural heritage, as historic buildings and monuments are exposed to seismic risk and, in general, to severe, long-term, cyclic loading conditions or harsh environmental conditions.

Buildings and structures exposed to the action of earthquakes with moderate magnitude, which are rather frequent in the central and southern regions of Italy, may undergo accelerated aging and deterioration. Such damage processes, often inaccessible to visual inspection, eventually lead to an increased vulnerability to the actions of major earthquakes, with catastrophic results [1].

The two regions considered in this paper surround the Italian cities of Varallo and Bologna, where the authors have recently carried out AE monitoring of important historical monuments for structural stability assessment. In particular, in the Italian Renaissance Architectural Complex of "The Sacred Mountain of Varallo" the structure of the Chapel XVII was analyzed, while in the City of Bologna the stability of the "Asinelli Tower", known as the highest leaning tower in Italy, was evaluated [2,3].

According to seismic analysis, several studies investigated spatial and temporal correlations of earthquake epicenters, involving, for example, the concepts of Omori's Law and fractal dimensions [4,5,6]. Other authors tried to study the complex phenomenon of seismicity using an approach that is able to analyze the spatial location and time occurrence in a combined way, without subjective a priori choices [7]. The present approach leads to a self-consistent analysis and visualization of both spatial and temporal correlations based on the definition of correlation integral [8].

Based on these considerations, we have tried not only to analyze the seismic activity in the considered regions, but also to correlate it with the AE activity detected during the structural monitoring. By adopting a modified Grassberger-Procaccia algorithm, we give the cumulative probability of the events' occurrence in a specific area, considering the AE records and the seismic events during the same period of time. In the modified integral the cumulative probabilities $C(r, \tau)$ are the function of the regional radius of interest, $r$, and of the time interval, $\tau$, both considering the peak of AEs as earthquake "precursors" or "aftershocks" [9].

## 2. DESCRIPTION OF CHAPEL XVII AND ASINELLI TOWER

In Varallo, placed in Piedmont in the province of Vercelli, there is the relevant Italian Renaissance Architectural Complex named "The Sacred Mountain of Varallo". It was built in the 15th century on a cliff above the town of Varallo and is composed of a basilica and 45 chapels.

Because of its high level of damage due to regional earthquakes that have occurred [2], we chose to consider Chapel XVII, known as the chapel of the "Transfiguration of the Christ on the Mount Tabor", for the AE analysis. This structure, having a circular plan, was built with stone masonry and mortar. The interior walls of the chapel are also equipped with some valuable frescoes (Figure 1).

In Bologna, Emilia-Romagna, there is another important masterpiece of Italian architectural and historical heritage, the "Asinelli Tower", which is also the symbol of the city. To study the damage evolution of the structure, we have recently analyzed the effect of repetitive and impulsive natural and anthropic events, such as earthquakes wind, or vehicle traffic, with the AE technique [3].

The Asinelli Tower was built in the early 12th century, and it rises to a height of 97.30 m above the ground. From the ground level, up to a height of 8.00 m, the tower is surrounded by an arcade built at the end of the 15th century. Studies conducted in the early 20th century revealed that the Asinelli Tower leaned westward by 2.25 m, and other recent studies have confirmed that its leaning is of 2.38 m, and it remains the tallest leaning tower in Italy (Figure 2).

**Figure 1.** Sacred Mountain of Varallo, Mount Tabor Chapel XVII.

# 3. ACOUSTIC EMISSION MONITORING

The AE monitoring is performed by analyzing the signals received from the transducers through a threshold detection device that counts the burst signals, which are greater than a certain voltage.

The adopted USAM (Unit for Synchronous Multichannel Acquisition) acquisition system consists of six pre-amplified AE sensors, equipped with six units of data storage, a central unit for synchronization, and a trigger threshold. The output voltage signals of the piezoelectric transducers (PZT) were filtered with a pass-band from 50 kHz to 800 kHz and a detection threshold of 100 μV. The obtained data from this device are the cumulative counting of each mechanical wave, the acquisition time, the measured amplitude in volts, the duration, and the number of oscillations across the threshold value for each AE signal [2,3]. Then, at the post-processing stage, we discarded signals with a duration shorter than 3 μs and containing less than three oscillations across the detection threshold in order to filter out electrical noise spikes.

**Figure 2.** The Asinelli and the adjacent Garisenda Towers in the city center of Bologna. The Asinelli tower is the tallest one on the right.

## 3.1. Chapel XVII

Firstly, we consider the analysis carried out in Chapel XVII of Varallo. The AE monitoring was led on the frescoed masonry on the north wall of the Chapel. The wall showed some damage: a vertical crack of about 3.00 m in length and fresco detachment. Four AE sensors were located around the vertical crack, while two were placed near the fresco detachment (Figure 3). The AE monitoring was conducted in two phases for a total duration of about 14 weeks. The first phase started on 9 May 2011 and finished on 16 June 2011; the second one was from 5 July 2011 to 5 September 2011. The monitoring results related to the chapel's structural integrity are reported in [2]. We interpreted the AE data considering the amplitude and time distribution of AE signals during the cracking phenomena. From this analysis, we found that the vertical crack, monitored on the north wall of the chapel, evolved during the acquisition period, while the process of detachment of the frescos was mainly related to the diffusion of moisture in the mortar substrate [2].

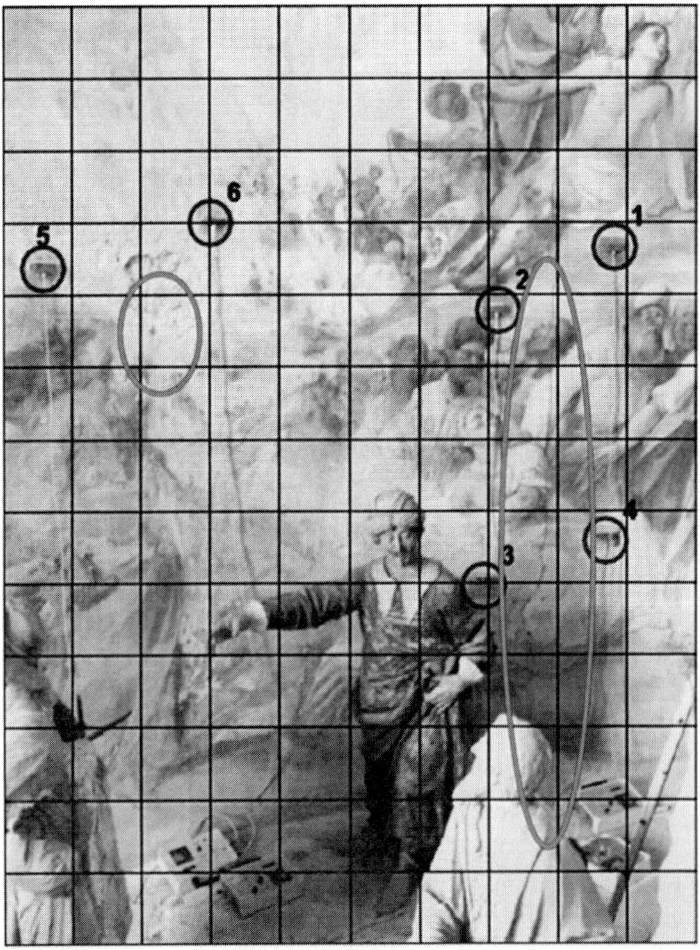


**Figure 3.** Chapel XVII. View of the monitored areas. Left side: sensors 5, 6, and the fresco detachment. Right side: sensors 1–4 and the vertical crack.

During the monitoring period, among all regional seismic events we considered 21 earthquakes with Richter magnitudes ≥1.2, having an epicenter within a radius from 60 to 100 km from Varallo. The strongest earthquake was a 4.3 magnitude event that occurred on 25 July 2011 at 12:31 p.m. in the Giaveno area (epicenter about 80 km from Varallo). Figure 4 displays the AE event rate, which counts simultaneous signal detection by multiple sensors as one event and the sequence of the earthquakes as functions of time, obtained during the monitoring period.

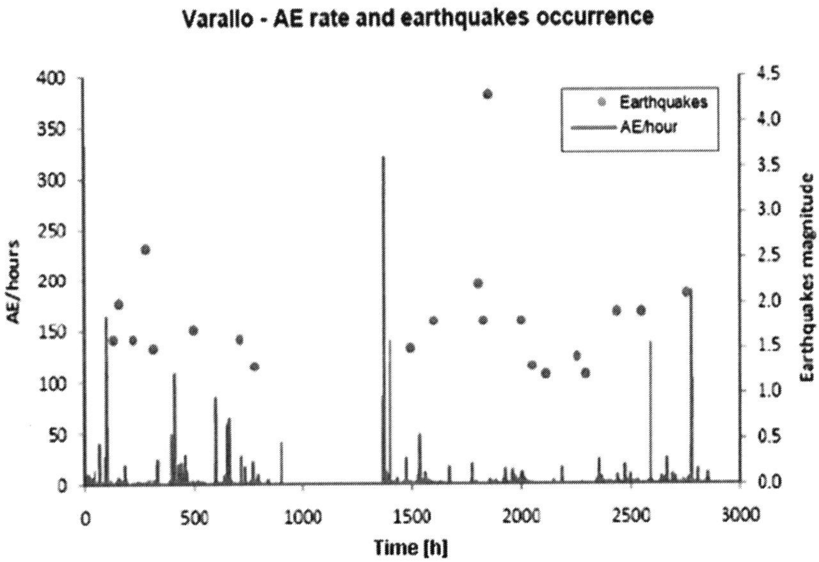


**Figure 4.** Chapel XVII: AE rate (blue chart) and nearby earthquake (red dots) occurrences as functions of time. The AE rate chart illustrates the number of AE events (averaged over 1 h), while seismic events are marked by points indicating occurrence time and Richter magnitude.

## 3.2. Asinelli Tower

The AE activity on Asinelli Tower was examined in a significant zone for monitoring purposes. This was developed by attaching six piezoelectric sensors to the northeast corner of the tower at an average height of ca. 9.00 m above ground level, immediately above the terrace atop the arcade. In this area, the double-wall masonry has an average thickness of ca. 2.45 m (Figure 5). AE monitoring began on 23 September 2010 and ended on 10 January 2011, for a total duration of about 16 weeks. By monitoring a significant part of this tower, the incidence of vehicle traffic, seismic activity and wind action on the damage evolution within the structure were assessed [3].

In this case, among all regional seismic events, we considered 43 earthquakes with a Richter magnitude ≥1.2, having an epicenter within a radius from 25 to 100 km from Bologna, as the most likely to affect the tower's stability. The strongest earthquakes were the 4.1 magnitude event that occurred on 13 October 2010 at 11:43 p.m. in the Rimini area (epicenter about 100 km from Bologna) and the 3.4 magnitude event on 21 November 2010 at 4:10 p.m. in Modena Apennines (epicenter about 50 km from Bologna). The AE instantaneous rate (averaged over 1 h) and the sequence of the earthquakes obtained during the monitoring are displayed in Figure 6. Also, in this case a correlation between peaks of AE activity in the structure and regional seismicity can be observed (Figure 6).

The tower, in fact, as in another case investigated by the authors regarding the Medieval Towers of Alba in Italy, is very sensitive to earthquake motions [9,10].

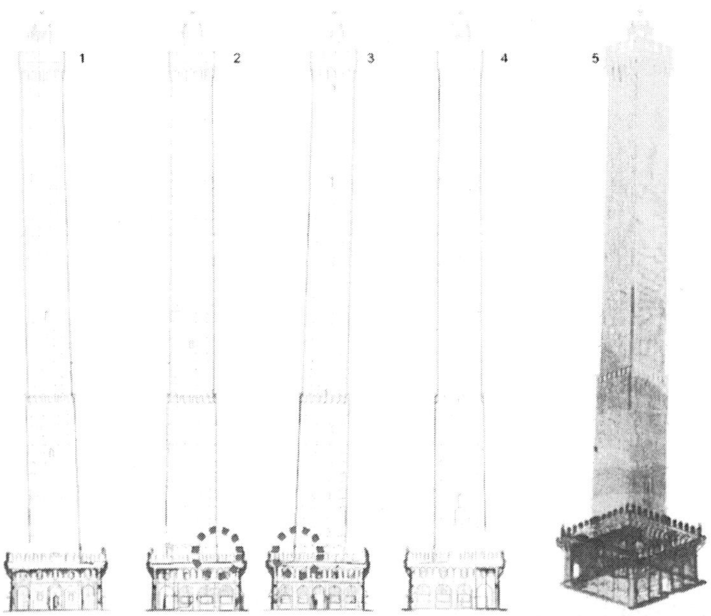


**Figure 5.** Front views and axonometric view of Asinelli Tower. Faces (**1**) South; (**2**) East; (**3**) North; (**4**) West; (**5**) Axonometric view. The AE transducers were applied to the northeast corner of the tower, in the zones marked with a circle.

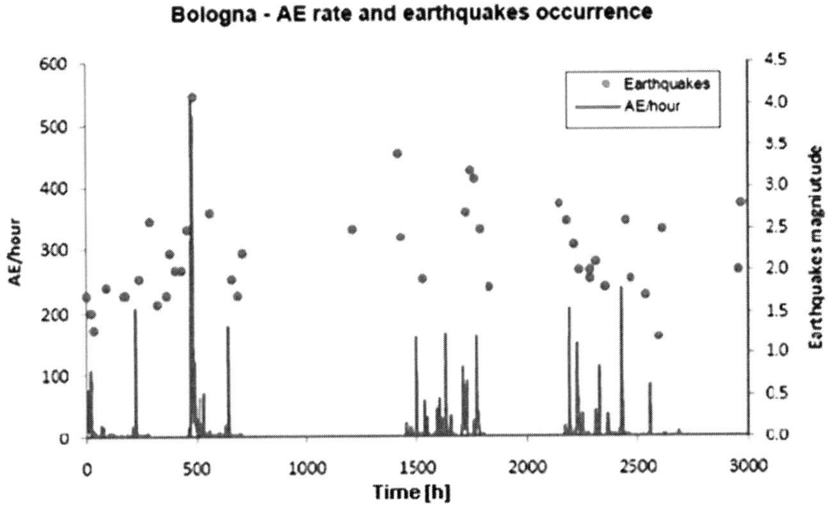


**Figure 6.** Asinelli Tower: AE rate (blue chart) and nearby earthquake (red dots) occurrences as functions of time. The AE rate chart illustrates the number of AE events (averaged over 1 h), while seismic events are marked by points indicating occurrence time and Richter magnitude.

# 4. GRASSBERGER-PROCACCIA ALGORITHM

Acoustic emissions in materials and earthquakes in the crust are very similar phenomena though on very different scales. They both involve a sudden release of elastic energy from a source located in the medium: respectively, the tip of opening micro-cracks and the seismic hypocenter [11]. The aging and the damage of historical buildings due to the action of small and intermediate earthquakes, a situation very common in central and southern Italy, implies triggering of AE activity on these structures.

However, there appear to be some seismic events which follow AE bursts and, then, do not trigger structural AE activity. In fact, intense crises of crustal stress apparently cross large areas, as revealed by increased AE activity in several case histories from Italy [12], and precede the eventual occurrence of some earthquake within them.

This experimental evidence suggests that part of the AE activity from the structures might derive from precursive microseismic activity propagating across the ground-building foundation interface.

In this sense, AE monitoring provides twofold information: one concerning the structural damage and the other the amount of stress affecting the crust, in which the building foundation represents a sort of extended underground probe.

With this spirit we introduce a statistical approach to analyze spatial and temporal correlations between AE bursts from the structures and surrounding earthquakes, along the lines of similar investigations of the spatial and temporal correlations of earthquakes [2,3,9].

Seismicity levels of the Varallo and Bologna areas can be consistently compared by investigating the earthquakes' spatial distribution over equally sized regions (i.e., circular regions with a radius of 100 km) centered at the two monitoring sites, during analogous monitoring periods (3000 h, as shown in Figure 4 and Figure 6).

Quantification of this feature is possible through the notion of the fractal dimension of hypocenters, which is frequently estimated by the correlation integral because of its great reliability and sensitivity to small changes in clustering properties [13,14,15]. The correlation integral for hypocenter points embedded into a three-dimensional space is defined as follows:

$$C(r) = \frac{2}{N(N-1)} \sum_{k=1}^{N-1} \sum_{j=k+1}^{N} \theta\left(r - \left|x_k - x_j\right|\right) \tag{1}$$

where N is the total number of seismic events, x is the hypocentral coordinate vector, and Θ is the Heaviside step function ($\Theta(x) = 0$ if $x \leq 0$, $\Theta(x) = 1$ if $x > 0$). The correlation integral C(r) gives the probability of finding another point in the sphere of radius r centered at an arbitrary point. For a fractal population of points, C(r) scales with r as a power law, D being the correlation dimension:

$$C(r) \sim r^D \tag{2}$$

We apply the correlation integral to the earthquake sequences under consideration, i.e., the earthquakes with epicentral distance in the range of 60–100 km from the Varallo Chapel and 25–100 km from the Asinelli Tower, which occurred during the AE monitoring periods. The choice of the minimum magnitude threshold, $m_L \geq 1.2$, was driven by completeness criteria.

The seismic events were taken from the Italian Seismological Instrumental and Parametric Data-Base ISIDE [16].

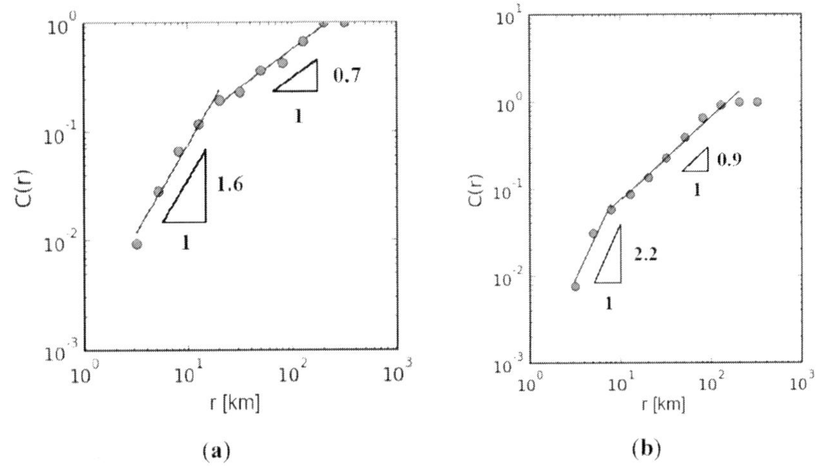


**Figure 7.** Correlation integral C(r) vs. r (km) and its slope D in bi-logarithmic scale for Varallo regional seismicity (**a**) and Bologna regional seismicity (**b**).

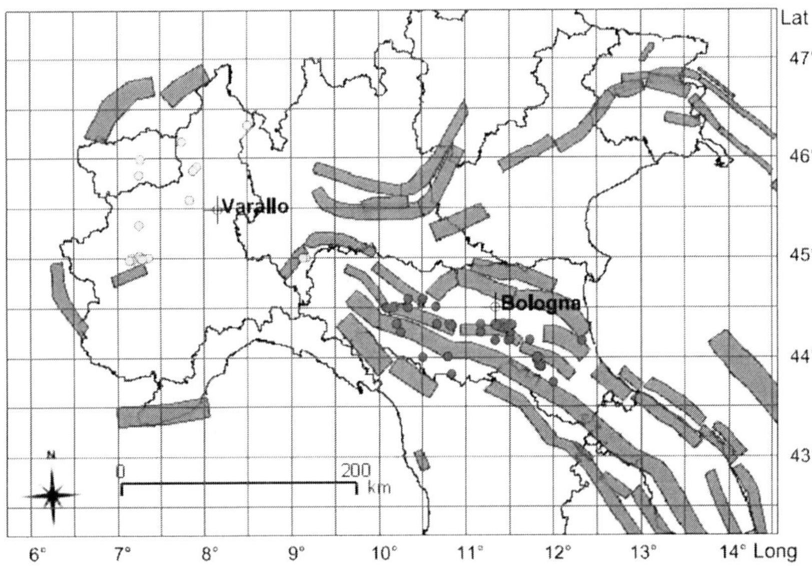


**Figure 8.** Map of Northern Italy depicting faults and epicenters.

Two patterns are evidenced in Figure 7a,b: shorter ranges, 3 km < r < 10–20 km, are defined by a unique slope giving a relatively high value of the correlation dimension, D = 1.6 (Varallo) and D = 2.2 (Bologna); for distances over 10–20 km, the slope gives a value of D = 0.7 (Varallo) and D = 0.9 (Bologna). For short ranges, D ≈ 2 corresponds to a distribution of events over a fault plane, such as the active faults in the region surrounding Bologna; the highest ranges, corresponding to a value of D < 1, suggest a dust-like setting of seismicity as in the region surrounding Varallo (see Figure 8).The higher dimension for Bologna apparently reflects the recent and active tectonics of the Emilia-Romagna in the northeastern Italy, with respect to the seismically quiet zone of Varallo [17].

# 5. SPACE-TIME CORRELATION BETWEEN AE AND SEISMIC EVENTS

Along the lines of studies on the space-time correlation among earthquakes based on a generalization of the Grassberger-Procaccia correlation integral [8], we calculate the degree of correlation between the AE time series and the sequence of nearby earthquakes recorded during the same period. No a priori assumptions on causal relationships between AE and seismic activities drive this approach, which is merely statistical. Only the choice of the time window and the region size is arbitrary.

The space-time combined correlation integral is defined as follows [7,9]:

$$C(r,\tau) = \frac{1}{N_{EQ}N_{AE}} \sum_{k=1}^{N_{EQ}} \sum_{j=1}^{N_{AE}} \theta(r - |x_k - x_0|) \cdot \theta(\tau - |t_k - t_j|) \tag{3}$$

where the index j runs over all the $N_{AE}$ event bursts $\{x_0, t_j\}$, with $x_0$ being the coordinate vector of the monitoring site, while k runs over all the $N_{EQ}$ seismic events $\{x_k, t_k\}$, with $x_k$ being the epicentral coordinate vector.

Since the double-sum in Equation (3) counts pairs formed by an AE event $(x_0, t_j)$ and a seismic event $(x_k, t_k)$ with mutual epicentral distance $|x_0 - x_k| \leq r$ and time intervals $|t_j - t_k| \leq \tau$ among all $N_{AE}$ and $N_{EQ}$ possible pairs, $C(r,\tau)$ can easily be interpreted as the probability of an AE burst and an earthquake occurring with an inter-distance $\leq r$ and a time interval $\leq \tau$.

If $C(r, \tau,)$ exhibits power-law behavior both in space and time variables, time and space fractal dimensions $D_t$ and $D_s$ can be defined similarly to Equation (2):

$$D_t(r,\tau) = \frac{\partial \log C(r,\tau)}{\partial \log \tau} \tag{4}$$

and

$$D_s(r,\tau) = \frac{\partial \log C(r,\tau)}{\partial \log r} \tag{5}$$

The time correlation dimension $D_t(r, \tau)$ characterizes the time-coupling of AE bursts to the earthquakes occurring up to a given distance r from the AE monitoring site, whereas the space correlation dimension $D_s(r, \tau)$ characterizes the spatial distribution of nearby earthquakes with separation in time from AE bursts not exceeding a given $\tau$.

It is worth noting that Equation (5) identifies with Equation (2) if $\tau$ equals the whole time data span, as all earthquakes would be taken into account in the Grassberger-Procaccia integral $C(r, \tau)$.

Actually, the imposed condition $|t_j - t_k| \leq \tau$ does not specify the chronological order between the two types of event. On the other hand, the AE time series and the sequence of nearby earthquakes are two very closely interrelated sets in the time domain. Therefore, a given AE burst might be either due to structural damage triggered by a seismic event or due to precursive microseismic activity. In spite of intrinsic difficulties in high-frequency propagation across disjointed media (in particular at the ground-building foundation interface), AE bursts apparently indicate widespread crustal stress crises during the preparation of a seismic event [9,12], when part of the related deformation energy stored in the Earth's crust might be transferred to the building foundations.

It is interesting to carry out a probabilistic analysis of the available data considering AE events once as preceding an earthquake, i.e., as potential seismic precursors, and next as following an earthquake, i.e., as structural aftershocks. In this way, the obtained conditioned probability distributions can be compared in order to discover the prevailing trend. This analysis is performed applying the modified correlation integral [9]:

$$C^{\pm}(r,\tau) = \frac{1}{N_{EQ}N_{AE}} \sum_{k=1}^{N_{EQ}} \sum_{j=1}^{N_{AE}} \theta(r - |x_k - x_0|) \cdot \theta(\tau - |t_k - t_j|) \tag{6}$$

where "+" and "–" are used to account for AE events, respectively, as seismic precursors and as aftershocks. $C^{+}(r, \tau)$, for example, gives the occurrence

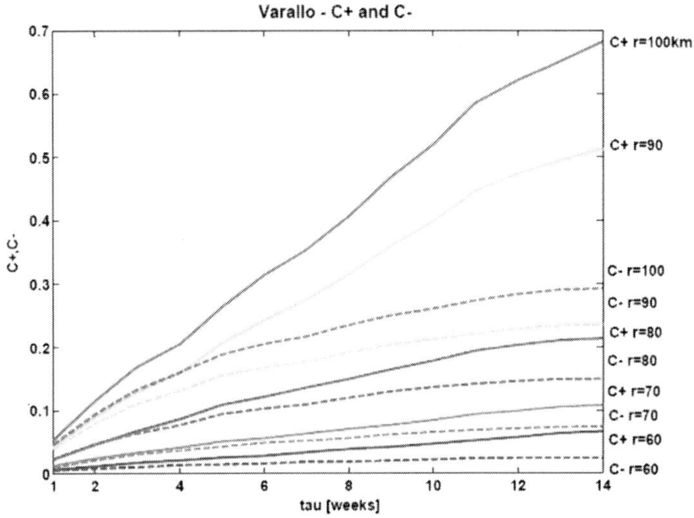


(a)

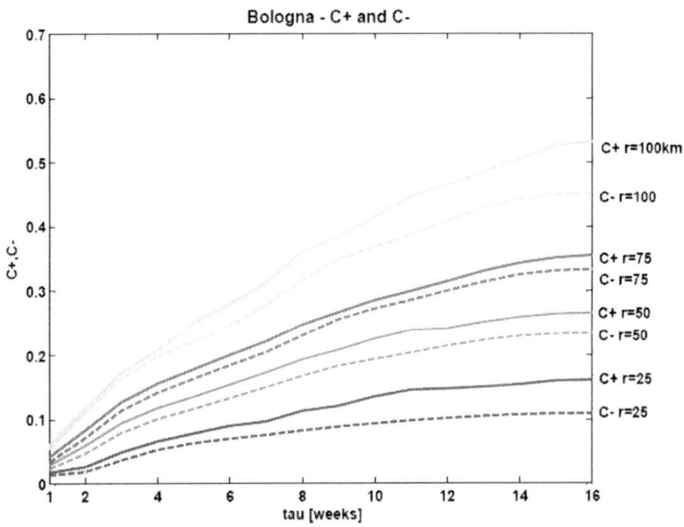


(b)

**Figure 9.** Modified correlation integrals $C^{\pm}_r(\tau)$, both considering AEs as earthquake "precursors" (+) and as "damage aftershocks" (−), plotted as functions of the time separation τ for different values of the spatial range r for: (**a**) Chapel XVII and area around the city of Varallo; (**b**) Asinelli Tower and area around the city of Bologna.

probability of an AE burst followed by an earthquake in the next interval $\tau$ and within a radius r of the monitoring site.

We applied the modified integral (Equation 6) to investigate the space-time correlation between the two above-mentioned regional seismic sequences and the AE time series from the Asinelli Tower and the Varallo Chapel (see Figure 4 and Figure 6).

The AE signals originating from damage sources definitely localized in the building's masonry are filtered out. In this way, we can preliminarily distinguish between environmental contributions due to crustal trembling (external source) and structural damage contributions (inner sources localized by means of triangulation techniques [18]). Then, by applying the modified Grassberger-Procaccia correlation integral to the remaining AE data series and to the earthquake sequence, we obtained the cumulative probabilities $C^{\pm}(r, \tau)$ as functions of timc $C^{\pm}_{r}(\tau)$ for different values of range r. This provides two-dimensional plots, which are easier to read than a three-dimensional representation of $C^{\pm}(r, \tau)$.

Figure 9a,b show that $C^{+}_{r} > C^{-}_{r}$ for all considered values of range r, suggesting that AE bursts are more likely to precede earthquakes than to follow them. The interpretation of this evidence is that the monitored structures behave as receptors of microseismic precursive activity during the preparation of a seismic event, i.e., as sensitive earthquake receptors.

# 6. AE CLUSTERING IN TIME AS A SEISMIC PRECURSOR OR AN AFTERSHOCK

Since the cumulative probabilities $C^{\pm}_{r}(\tau)$ are represented in a time domain ranging from 1 to 14–16 weeks, shorter values of $\tau$ are further investigated.

In particular, time coupling of regional seismicity to AE activity from Chapel XVII and the Asinelli Tower is analyzed in terms of $D_t$ for $\tau$ ranging from 3 to 24 h after an earthquake occurrence, and from 10 min to 24 h before an earthquake occurrence.

The varying local slope $D_t$ of Log $C^{-}$ vs. Log $\tau$ reveals that the AE activity following seismic events is not equally probable over the time (see Figure 10a,b). For short time delays, $\tau$ = 3–10 min, high values of the time fractal dimension, $D_t$ = 1.38 and 1.24, indicate tight coupling of the AE activity to the earthquakes. In other words, AE bursts following an earthquake are more likely to occur within $\tau$ = 3–10 min after the seismic event. Then, such short time delays suggest a triggering action exerted by nearby earthquakes on damage processes of Chapel XVII and the Asinelli Tower. Contrarily, for $\tau$

ranging from 1.0 × $10^3$ to 1.440 × $10^3$ min, the obtained low values of $D_t$, 0.48 and 0.45, suggest that the effects of nearby earthquakes on the structural damage evolution disappear after 24 h.

In regards to the AE events preceding earthquakes, i.e., considering AE as "seismic precursors", the linear bi-logarithmic plots of $C^+$ ($D_t$ = 1.06 and 0.94) shown in Figure 11 describe a uniform probability density of finding AE events prior to an earthquake for a wide range of time intervals, up to τ = 1.440×$10^3$ min. In other words, we observe a constant AE activity in the 24 h preceding an earthquake. The absence of accelerating precursive activity in the presented case studies confirms the need for further investigation, possibly with the aid of electromagnetic seismic precursors.

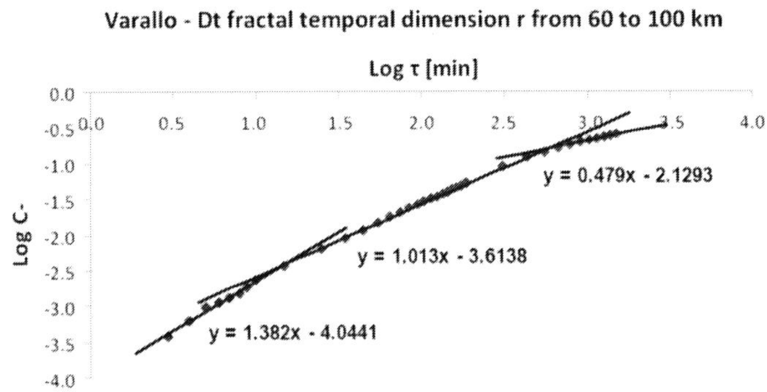


(a)

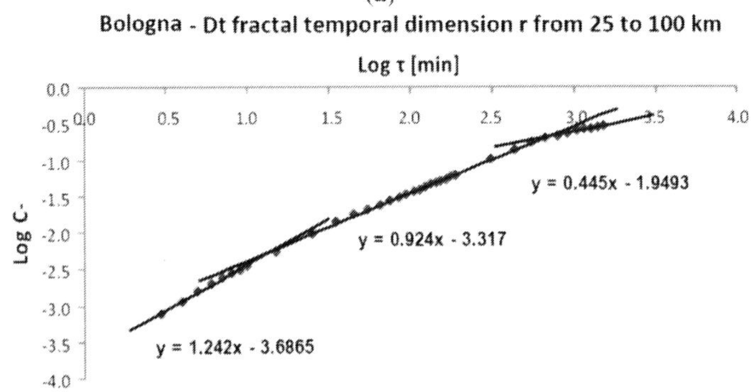


(b)

**Figure 10.** Analysis of time-clustering features of AE events considered as "seismic aftershocks" by the time correlation dimension $D_t$, as a local slope of Log $C^-$ vs. Log τ. The time range for τ is 3 min to 24 h. The selected seismic events are all those considered during the defined monitoring periods. (**a**) Chapel XVII (Varallo); (**b**) Asinelli Tower (Bologna).

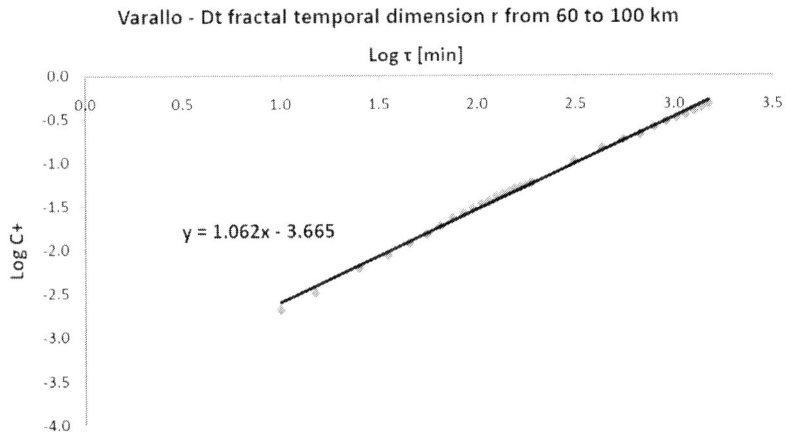


(a)

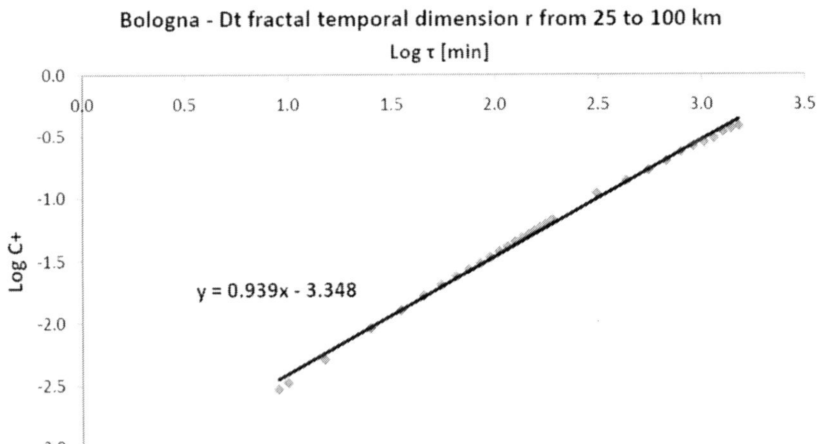


(b)

**Figure 11.** Analysis of time-clustering features of AE events considered "seismic precursors" by the time correlation dimension $D_t$, as the local slope of Log $C^+$ vs. Log $\tau$. The time range for $\tau$ is 10 min to 24 h. The selected seismic events are all those considered during the defined monitoring periods. (**a**) Chapel XVII (Varallo); (**b**) Asinelli Tower (Bologna).

# 7. CONCLUSIONS

In order to assess the seismic hazard in different areas, it is suggested to observe the spatial and temporal correlation of seismic events and monitored AEs. It emerges that AEs behave as earthquake precursors. A statistical method of analysis is proposed, based on the Grassberger-Procaccia integral. It gives the cumulative probability of the events' occurrence in a specific area, considering the AE records and the seismic events during the same period of time. Two important constructions of Italian cultural heritage were considered: the "Sacred Mountain of Varallo" and "Asinelli Tower". Given the interesting premises that emerged from this study, the Grassberger-Procaccia integral could also be applied more extensively in several monitoring sites belonging to different seismic regions, not only for defining the seismic hazard, but also to handle it as an effective tool for earthquake risk mitigation.

However, the authors are aware that a rigorous investigation to ascertain the existence of crustal stress crises crossing areas of a few hundred kilometers' radius during the preparation of a seismic event would require simultaneous and numerous operations of AE monitoring sites, adequately placed in the territory.

# ACKNOWLEDGMENTS

E. de Filippis, Director of the Piedmont Sacred Mountains Institute, is gratefully acknowledged. The authors thank the Municipality of Bologna and Eng. R. Pisani for having allowed the study on the Asinelli Tower.

# AUTHOR CONTRIBUTIONS

A.C. supervised the research; P.C. and S.I. performed the numerical simulations; G.N. carried out AE and NDT tests; G.L. wrote the paper with discussions from all the authors.

# CONFLICTS OF INTEREST

The authors declare no conflict of interest.

# REFERENCES


1. Niccolini, G.; Carpinteri, A.; Lacidogna, G.; Manuello, A. Acoustic emission monitoring of the Syracuse Athena Temple: Scale invariance in the timing of ruptures. Phys. Rev. Lett. **2011**.
2. Carpinteri, A.; Lacidogna, G.; Invernizzi, S.; Accornero, F. The Sacred Mountain of Varallo in Italy: Seismic risk assessment by Acoustic Emission and structural numerical models. Sci. World J. **2013**, 2013, 1–10.
3. Carpinteri, A.; Lacidogna, G.; Manuello, A.; Niccolini, G. A study on the structural stability of the Asinelli Tower in Bologna. Struct. Control HLTH **2015**.
4. Bak, P.; Christensen, K.; Danon, L.; Scanlon, T. Unified scaling laws for earthquakes. Phys. Rev. Lett. **2002**, 88, 178501–178504.
5. Parson, T. Global Omori law decay of triggered earthquakes: Large aftershocks outside the classical aftershock zone. J. Geophys. Res. **2002**, 107, 2199–2218.
6. Corral, A. Long-term clustering, scaling and universality in the temporal occurrence of earthquakes. Phys. Rev. Lett. **2004**.
7. Tosi, P.; de Rubeis, V.; Loreto, V.; Pietronero, L. Space-time combined correlation integral and earthquake interactions. Ann. Geophys. **2004**, 47, 1–6.
8. Grassberger, P.; Procaccia, I. Characterization of strange attractors. Phys. Rev. Lett. **1983**, 50, 346–349.
9. Carpinteri, A.; Lacidogna, G.; Niccolini, G. Acoustic emission monitoring of medieval towers considered as sensitive earthquake receptors. Nat. Hazard Earth Syst. Sci. **2007**, 7, 251–261.
10. Carpinteri, A.; Lacidogna, G. Structural monitoring and integrity assessment of medieval towers. J. Struct. Eng. **2006**, 132, 1681–1690.
11. Scholz, C.H. The frequency-magnitude relation of microfracturing in rock and its relation to earthquakes. Bull. Seismol. Soc. Am. **1968**, 58, 399–415.
12. Gregori, G.P.; Paparo, G. Acoustic emission (AE). A diagnostic tool for environmental sciences and for non destructive tests (with a potential application to gravitational antennas). In Meteorological

and Geophysical Fluid Dynamics; Schroeder, W., Ed.; Science Edition: Bremen, Germany, 2004; pp. 166–204.

13. Kagan, Y.Y.; Knopoff, L. Spatial distribution of earthquakes: The two-point correlation function. Geoph. J. Roy. Astron. Soc. **1980**, 62, 303–320.

14. Hirata, T. A correlation between the b-value and the fractal dimension of earthquakes. J. Geophys. Res. **1986**, 94, 7507–7514.

15. Carpinteri, A.; Lacidogna, G.; Niccolini, G.; Puzzi, S. Morphological fractal dimension versus power-law exponent in the scaling of damaged media. Int. J. Damage Mech. **2009**, 18, 259–282.

16. Italian Seismological Instrumental and Parametric Data-Base (ISIDE). Available online: http://iside.rm.ingv.it/iside/standard/result.jsp?rst=1&page=EVENTS#result (accessed on 5 December 2015).

17. Burrato, P.; Vannoli, P.; Fracassi, U.; Basili, R.; Valensise, G. Is blind faulting truly invisible? Tectonic-controlled drainage evolution in the epicentral area of the May 2012, Emilia-Romagna earthquake sequence (northern Italy). Ann. Geophys. Italy **2012**, 55, 525–531.

18. Carpinteri, A.; Xu, J.; Lacidogna, G.; Manuello, A. Reliable onset time determination and source location of acoustic emissions in concrete structures. Cem. Concr. Compos. **2012**, 34, 529–537.

# 5
# CHAPTER

# ROTATIONAL COMPONENTS OF THE SEISMIC FIELDS CAUSED BY LOCAL EVENTS

**Anna Kurzych[1], Krzysztof P. Teisseyre[2], Zbigniew Krajewski[1] and Leszek R. Jaroszewicz[1]**

*[1] Institute of Applied Physics, Military University of Technology, Warsaw, Poland*

*[2] Institute of Geophysics, Polish Academy of Sciences, Warsaw, Poland*

## 1. INTRODUCTION

The existence of rotation effects at the Earth surface associated with earthquakes has been observed probably at least from the times when scientific approach to the ground motions during the quake had started. They are described in several classical monographs, such as Hobbs [1] and Davison [2], in which cited examples concern, among other things, twisting of some obelisks, tombs and segments of columns. However, early publications explain such phenomena as incidental effects of interference between linear vibrations [3, 4]. For instance, Imamura [5] explained the rotation effects of some objects at the ground surface by the impact of body/surface waves: due to such impact, an object can be inclined, partly losing contact with the ground surface, and when returning to the vertical some twist occurs with respect to its former position. Hence, from the beginning, the rotational effects have been treated as derivative effects, and it was stated that although such effects are observed, they cannot be explained as effect of any rotational waves - or rotational components of seismic waves - because existence of such waves or components would contradict the ideal elastic theory [6].

In the second half of last century, it was observed a spectacular development of continuum mechanics including defects, granular structure and other deviations from the ideal linear elasticity. Special interests were concentrated on the micropolar and micromorphic continua. In such elastic continua, the real rotation can be accompanied by other kind of axial motion – the twist-bend motion. On above base, it was theoretically proved that so-called the seismic rotation waves could propagate through grained rocks, initially by Teisseyre [7] who initially attributed the appearance of rotation components in seismic wave by coupling the seismic waves with the micromorphic response of the medium characterized by the an internal/granular structure [7, 8]. From this time, this possibility was extended to rocks with microstructure or defects [9, 10] or even without any internal structure [11 -13], due to the asymmetric stresses in the medium. On this base, various types of rotational waves have been discussed theoretically [14, 15].

It should be stressed that seismologists share different opinions about the nature of rotation waves – see the preface of a monograph on rotational seismology [16]. Perhaps, as it is underlined in preface of a book [16], still the majority believes that such rotation motions are not related to inner rotations but are directly related to rotation in the displacement field which may reach much higher magnitudes in materials with an internal structure than in homogenous layers; considering damages in the high buildings, there are many examples indicating enormous increase of rotation effect caused by consecutive impacts of seismic body and surface waves.

Nevertheless, all above aspects can be treated as elements of rotational seismology. It is an emerging field for studying all aspects of rotational ground motions induced by earthquakes, explosions, an ambient vibrations. It should be noticed that nowadays there is observed rapid growth of the rotational seismology interest in many geophysical fields of knowledge [17] which includes wide seismology disciplines, seismological apparatus, seismic-origin phenomena, physical and engineering aspects of earthquakes as well as geodesy.

However practical aspect of rotational events and phenomena investigation is connected with method of their recording, and different rotational seismology branches need different devices. For example, earthquake physics need devices operating with sensitivity below $10^{-9}$ rad/s/Hz$^{1/2}$, whereas the engineering of a strong-motion seismology devices operating with a frequency range 0.05-100 Hz with sensitivity $10^{-6}$-$10^{-1}$ rad/s/$Hz^{1/2}$ [18]. In this subject, it should be noticed that the seismic rotation waves were for the first time effectively recorded in Poland in 1976 [19]. Even though, from this time, waves or phenomena of this type have been studied in a few centers over the world, a further experimental verification of the existing rotational phenomena needs a new approach to the construction of the measuring devices, because the conventional seismometers are inertial sensors detecting only linear velocities [20]. Thus, during measurement of the rotation present in the seismic field, with the use of a special array or set of conventional seismometers (for example based on a set of two classical mechanical seismometers [21]), data are disturbed by linear movements [22]. Therefore, an innovative device is necessary to detect the rotational seismic phenomena/events. According to our knowledge we can confirm that the technical implementation of the Saganc effect [23] is the most proper way to measure rotation directly. One can find instances of such solution: a ring laser [24] as well as a fibre optic seismometer [25-27]. It gives the opportunity to carry out the measurements without any reference system.

It should be underlined that all experimental data recorded during earthquakes shows that rotational components are small in comparison to linear motions - less than 10% [14, 19] or have half of above value and exist with some delay regarding last one [25].

Based on above review in this chapter we present an analysis of a few examples of the rotational seismograms. Authors have concentrated on the local seismic events obtained at the Książ Observatory in Poland. These signals were obtained from two kinds of sensors described in section 2: the micro-array of TAPSs – Twin Antiparallel Pendulum Seismometers (also named rotation seismometers or double pendulum horizontal electromagnetic seismometers) and with the Sagnac interferometer of AFORS - Autonomous Fibre-Optic Rotational Seismometer, constructed at the Institute of Geophysics and the Military University of Technology, respectively. It should be also underlined that signals derived from micro-array include two components which, according to Asymmetric Continuum Theory, have character of rotational wave: rotation $\omega$ and shear $E$ (called also pure shear) – see [14]. On the other side, the Sagnac-type seismometers detect only rotation and are completely insensitive to translations [25, 28, 29] which may contaminate rotational measurements. Nevertheless, probably all the signals analyzed here suffer from some disturbances, this is referred to in the section 3.

# 2. INSTRUMENTATION FOR RECORDING ROTATIONAL COMPONENTS OF THE SEISMIC EVENTS

Figure 1 presents the general view of the measurement devices installed, at the beginning of July 2010, in the Książ Observatory, Poland (located at 50.84380333N, 16.291755 E). There are AFORS-1, the micro-array of seismometers consisting of two TAPSs (TAPS-1 and TAPS-2) oriented perpendicularly in the N-S and E-W directions, and other instruments such as accelerometers (parallel positioned with TAPSs), etc.

**Figure 1.** The general view of measurement devices installed in the Książ Observatory

The data detected by TAPSs (two channels for each of them) are stored by standard seismological system KST while data detected by AFORS are stored both by FORS-Telemetric Server and KST. The KST system uses sampling of the signals with frequency equals 12,8 kHz. The process of data storing by KST uses frequency of 100 Hz. Figure 2 presents an example of a diagram with data collected on March 11$^{th}$, 2011 at 6 h 58 min. (after the Honshu earthquake M=9.0 on March 11$^{th}$, 2011 at 5 h 46 min. 23 s UT, recorded in Książ, Poland on March 11$^{th}$, 2011 at 5 h 58 min. 35 s UT), used in previously presented analysis [30].

## 2.1. Design of the Twin Antiparallel Pendulum Seismometers

The micro-array of seismometers (system of two TAPSs perpendicularly oriented) is an experimental apparatus, devised in the Institute of Geophysics, and manufactured according to description presented below, on the base of short period SM-3 seismometers. This is one of the simplest micro-arrays for measuring the rotation and twist (shear) [14]. It was deployed at two Polish observatories, in Książ and Ojców (see [31]). The third identical set of sensors was used in Central Italy [32].

The idea of using the classical short period SM-3 seismometer as a new kind of mechanical rotational seismometer named TAPS is presented in Figure 3 [21]. It is a set of two SM-3 seismometers (named in Figure 3b as left – L and right – R) situated on a common axis and connected in parallel, but with opposite orientation. In the case of the ground motion containing displacements $w(t)$ and rotation $\alpha(t)$, the $u$(t) - electromotive force recorded by each simple seismometer contains a component of displacement ($\pm w$) and rotation motion ($\alpha$) multiplied by a proper length of pendulum ($l$) [33]:

$$u(t)_{L,\ R} = \pm w(t) + l \cdot \alpha(t), \quad (1)$$

where sign "+" and "-" are for R and L seismometer, respectively.

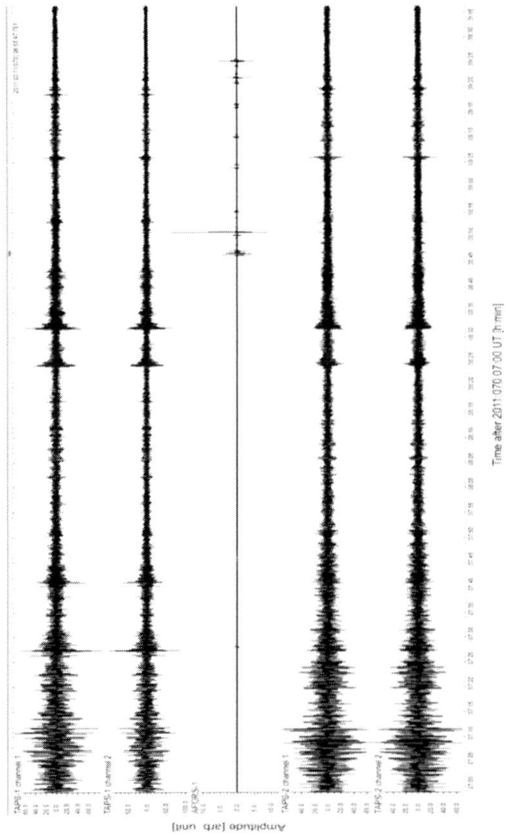


**Figure 2.** Plots of the seismic events recorded in the Książ Observatory, Poland on March 11th, 2011 starting from 06:58 UT, after the Honshu M=9.0 earthquake [30]

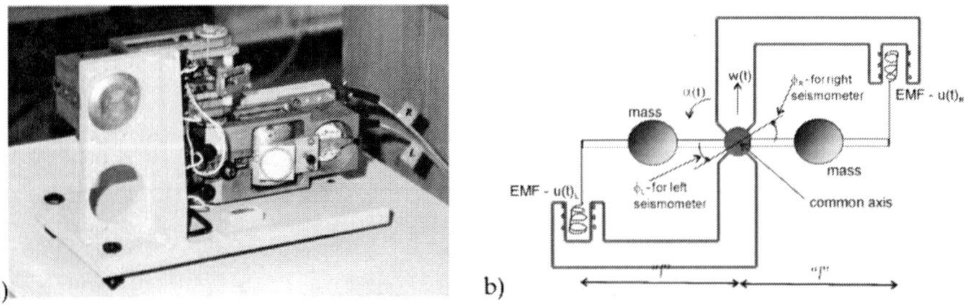


**Figure 3.** The TAPS – Twin Antiparallel Pendulum Seismometer: a) general view, b) schematic view

As one can see, in the case of identical two seismometers the rotational motions and displacement can be obtained from the sum and difference of the two recorded signals as:

$$\alpha(t) = [u(t)_R + u(t)_L]/2l, \quad \text{a}$$
$$w(t) = [u(t)_R - u(t)_L]/2l. \quad \text{b} \tag{2}$$

If the ground could be treated as a perfect rigid body, then the rotational motion recorded by sole one TAPS was identical to rotation. But rocks and the ground surface are not perfectly rigid; they transfer the mechanical waves due to slight, transient deformations which, seen along different axes, may different. Consequently, rotation $\omega$ in the plane of measurements and the given moment are calculated as a mean of rotational motions received at one and the other TAPS, while the twist $E$ (pure shear) is obtained as half of their difference:

$$\omega(t) = [\alpha_1(t) + \alpha_2(t)]/2, \quad \text{a}$$
$$E(t) = [\alpha_1(t) - \alpha_2(t)]/2. \quad \text{b} \tag{3}$$

Relations (2a) and (2b) remain valid, however, only when both seismometers forming the system have exactly the same response characteristics. Because, as a matter of fact, the pendulum seismometers are, inevitably, slightly different, the special TAPS channels calibration algorithm is used. In this system, for the aim of comparing both sensors, it is possible to rotate the position of one seismometer in such a way that both the pendulum seismometers, suspended on the common axis become oriented in the same directions, one just above the other – this is the test position. The working and test position for the case of the horizontal seismometers are schematically shown in Figure 4. The records obtained in the test positions can differ mainly due to differences in their response characteristics, and to minimize these errors, the following left channel signal calibration procedure is usually applied:

$$u_L' = u_L \sqrt{\sum u_R \cdot u_R / \sum u_L \cdot u_L}, \tag{4}$$

where: $u_L$ and $u_R$ are electromotive forces recorded by left and right seismometers in test position.

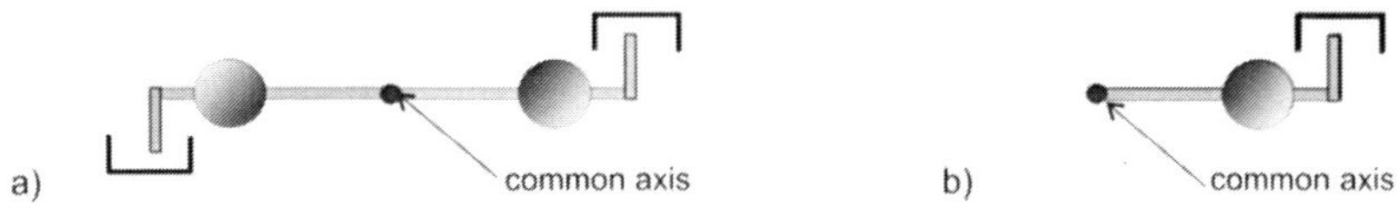


**Figure 4.** The TAPS: working a) and calibration b) positions of the pendulum seismometers

Nevertheless, there is some discrepancies due to the recording procedure in the case of the present of the difference of TAPS' pendulums attenuation characteristics. Figure 5 presents the simulation which indicates that there can be some errors in the data caused by the recording proceeding. The considered simulation was made for the attenuation difference equals $|\beta_L-\beta_R|=0.05$ between left and right seismometer attenuation. It is easy to see that the major error of the signal is present when the simulated rotation is characterized by smaller value of amplitude compared to the simulated translation component. However, this is the region where the seismic-origin rotation is expected. For above reason the process of TASP calibration seems to be an essential complication of the system work's correctness. Moreover, the extremely high sensitivity to the translational motions of the seismometers (preferred for the component of displacement detection) taken into account in their construction can limit the accuracy of such devices, too.

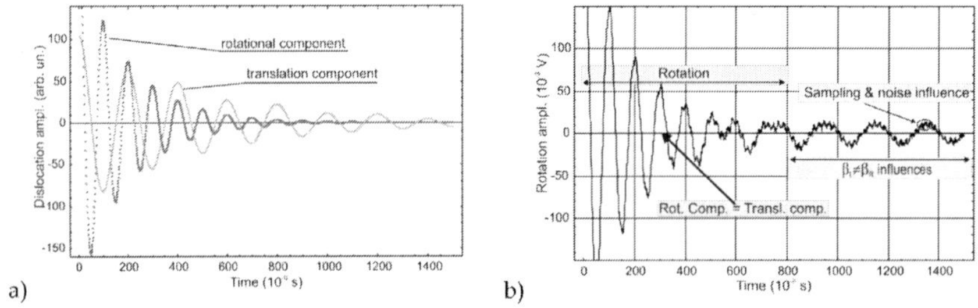


**Figure 5.** The simulated displacement and rotational components of the seismic event - a) and rotational signal detection by the TAPS - b) [22]

From above mentioned reasons, the additional numerical procedures for improving the TAPS performance may be applied, based on filtering in frequency [34] or in time [35] domains. The respective filters can then be applied to records in the normal working position to reduce the influences of non-equal operation of pendulum seismometers, presented above.

However, these methods use test position of the TAPS, that generally changes the condition of the TAPS operation. For this reason, another procedure of the recorded data processing, based on smoothing by the spline functions has been also proposed [36]. It should be noticed, that the main disadvantage of all listed methods is that they operate on recorded data, which can limit TAPS usefulness for some applications. In the research presented in this paper, the mentioned methods of signal correction were not used.

## 2.2. Design of the Autonomous Fibre-Optic Rotational Seismograph

The AFORS-1, used in our research, is one of three such devices existing in Poland and manufactured on the base of fibre-optic gyroscope, all dedicated for direct measurement of rotational components existing in seismic events and having accuracy below $5 \cdot 10^{-9}$ rad/s for 1 Hz detection band.

The physical principle for these devices is the Sagnac effect [23] which is a result of difference between two beams propagating around closed optical path, in opposite direction. Figure 6a presents the basic principle of the Sagnac's experiment. The input light beam is splitted by a beam splitter into a beam circulating in the loop in a clockwise - cw direction (Figure 6a - beam T) and a beam circulating in the same loop in a counterclockwise - ccw direction (Figure 6a – beam R). One can observe the interference pattern, in the output light, caused by the interference phenomenon of the two waves. In the case of the present of rotation with an angular rate represented by vector $\vec{\Omega}$

rad/s then a fringe shift ΔZ is observed in the output of the interferometer. The fringe shift is given by the formula [23]:

$$\Delta Z = \vec{\Omega} \bullet \vec{S} / (\lambda_0 \cdot c), \quad (5)$$

in which $\vec{s}$ is a vector of sensor loop area, $\lambda_0$ is the wavelength of the used light in vacuum, c is velocity of light in vacuum. According to the above formula a fringe shift is proportional to the cosine of angle between the axis of rotation and the normal to the sensor plane. There have been used the indigo mercury wavelength as well as the sensor area S = 866cm$^2$ in the Sagnac's experiment [23] which gave the fringe shift equals 0.07 fringes for the rotational speed equal to 2 rps. Nevertheless other papers [37] indicate that the possible detectable fringe shift was of the order of 0.01 fringe in those time. Thus, perhaps Sagnac has been carried out the research with a maximal accuracy. Sagnac also established that the effect does not depend on the shape of surface area S or on the location of the centre of rotation, whereas future investigation shown that this effect it does not depend on the presence of a commoving refracting medium in the path of the beam [37].

Figure 6b presents the Sagnac interferometer in the optical fibre solution which uses optical waveguide of the long length L wound on sensor loop of the diameter D which was shown firstly in 1976 [38]. In this approach, a phase shift $\Delta\phi$ is produced between cw and ccw propagating light, given by:

$$\Delta\phi = \frac{2\pi \cdot L \cdot D}{\lambda_0 \cdot c}\Omega, \qquad (6)$$

where $\Omega$ is the rotational component perpendicular to the sensor loop. It is clearly to indicate that the sensitivity can be change by physical dimension of the sensor loop as well as by the length of the applied waveguide. It should be noticed that application of three such systems, which loops are jointly perpendicular, provides data about space vector of the rotation. One can obtain the position change in space by integrating the data in time domain. The above procedure is used in the configuration of the fibre optical gyroscope - FOG which now, nearly 40-years from 1976, is the best recognized interferometric sensor performed in the fibre-optic technology.

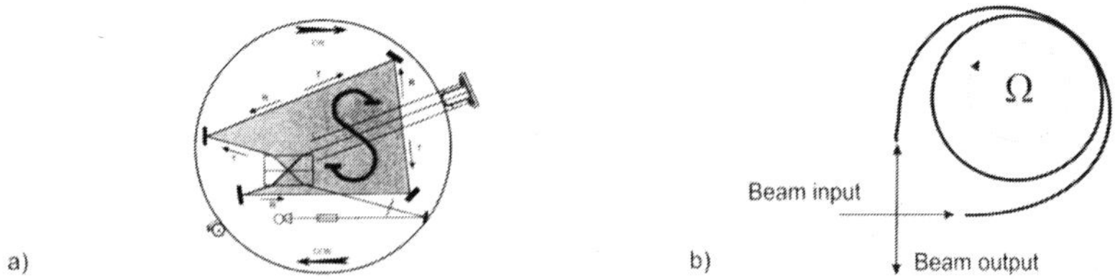


**Figure 6.** Schematic of the original Sagnac's experiment a) and its implementation in fibre optic technique b)

However, for a desired rotation rate in the range of $10^{-6}$ – $10^{-9}$ rad/s, the Sagnac effect generates a very small phase shift, so it is needful to separate and protect this effect from other disturbances so that the Sagnac effect is the unique nonreciprocal effect in the device. For this reason all FOGs use, shown in Figure 7, the reciprocal configuration [39] which is also called minimum gyro-configuration [40]. This configuration guarantees an ideal equilibrium of two counter-propagating beams in the interferometer by obtaining true single mode operation at the common input-output port of the system. It is not disturbed even by non-single mode operation in the another part of the interferometer.

It is well known that each interferometric devices yield a cosine response. For above reason the detected signal practically do not change during the small changes of the rotation due to slow changes of the cosine function at the zero. In order to obtain higher sensitivity the operation point of the interferometer is shifted by applied additional phase shift modulation. The FOG utilizes the reciprocal phase modulator which is placed in the end of the sensor loop. It caused the modulation of the phase shift by propagation delay without any residual zero offset [41]. In this way one obtain the odd response instead of even one. An ultimate performance is, however, obtained only if the unbiased response is perfectly even and the biasing modulation has only odd frequencies. Therefore, the applied phase modulator is also a delay line filter operating at the eigen-frequency [42] – the delay in the loop is equal to a half of modulation period which suppresses the residual even harmonic signals. Nowadays the FOG utilizes broad-band light source for eliminating the Kerr effect which produces the phase shift in the optical fiber Sagnac interferometer [43]. Such a broadband source is also needed to remove coherence related with noise and drift due to backscattering and backreflection as well as lack of rejection of the polarizer [44–46]. Finally, for achieving the high scale factor linearization, FOG uses a digital phase step feedback [47] by the same reciprocal phase modulator as the biasing modulator and all-digital processing procedures where demodulation is carried by a digital subtracting and sampling of the modulated signal is obtained by using analogue-digital converter [48, 49].

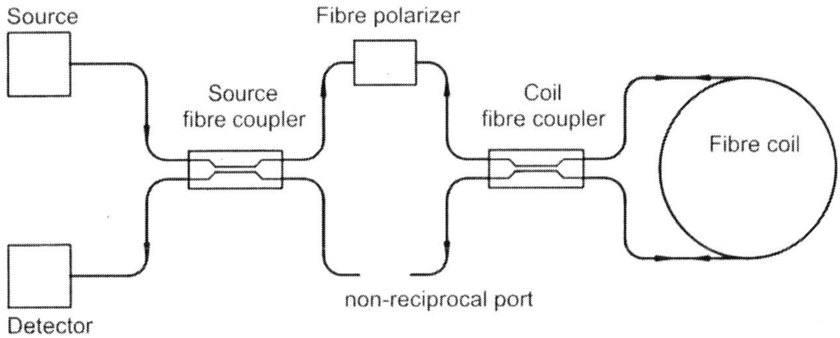


**Figure 7.** The minimum configuration of the FOG [39]

Currently, a digital processing of FOG systems is designed to record angular changes instead of rotation rates, thus, the optimization of such system to register the interesting phenomena from the rotational seismology point of view is problematic. Therefore it should be emphasized that the AFORS construction based on experiences according to the FOG development described above, but with system optimization for a direct measurement of the rotation rate only [22]. Such an approach gives a system which through a direct use of the Sagnac effect can limit drift influence on a device operation.

A detailed description of the AFORS system was published previously [29, 30, 50], hence here we summarized the above data regarding AFORS-1 construction, calibration and management. The second device - AFORS-2 is located in Warsaw (Poland) for initial works connected to the investigation of the irregular engineering construction torsional response and the interstory drift [51]. We anticipate that the new device, based on AFORS-1 and -2, AFORS-3 will be construed in the 2014 which gives us the opportunity to mount new innovative system instead of FORS-II assembled in seismological observatory Ojców, Poland [52].

The AFORS uses the minimum configuration of the FOG, however opposite of it, AFORS operates in open-loop architecture with digital data processing [53]. This technical solution is motivated by the fact that rotation events (Ω) are registered as sudden changes of rotational rate which amplitude is determined in a direct way from the Sagnac phase shift (Δφ) by following equation [37]:

$$\Omega = S_o \cdot \Delta\phi = \frac{\lambda \cdot c}{2\pi \cdot D \cdot L} \cdot \Delta\phi \qquad (7)$$

where $S_0$ is the optical constant of interferometer which depends on the fundamental parameters of fibre coil.

Upper part of Figure 8 presents the block diagram of the AFORS-1 optical part configuration. The AFORS-1 construction, designed according to the minimum-gyro configuration, contains of the: SLED diode (ΔB=31.2 nm, $\lambda_0$ = 1305.7 nm, $P_{out}$= 9.43 mW; *Exalos*, Switzerland), isolator (α=0.34 dB, 39 dB isolation; *FCA*, Poland), depolarizator (DOP<5%, α=0.20 dB; *Phoenix Photonics*, UK), two mounted-in-line polarizators (ε=43 dB, α=0.45 dB each; *Phoenix Photonics*, UK), two X-couplers (α=0.20 dB; *Phoenix Photonics*, UK), sensor loop, phase modulator (*Piezomechanik GmbH*, Germany) and detector (S=0.9 A/W; *Optoway Technology Inc.*, Taiwan). The SLED diode has been chosen for two reasons. Firstly, SLED diode is the broad-band light source, which minimalizes the disadvantageous polarization as well as coherence effects [54, 55]. Additionally, this diode gives an opportunity to obtain a high optical power, which has a direct influence on the system sensitivity. In order to isolate the diode from the backscattering we applied the optical fibre isolator. To ensure truly depolarized light the depolarizer is used before the polarizers which guarantee the single mode operation in the entirely system as well as fulfil function of the filter. Two couplers ensure that both propagating waves have the same optical path. The detector's system consists of a PIN diode and a preamplifier. The sensor loop has been made by winding 15 000 m of the SMF-28e+ length on a special composite material which includes permalloy particles for shielding the system from external magnetic field. The double-quadrupole method of winding [56] was used in order to stabilize the work system during the temperature fluctuation. The technical optimization of AFORS-1 construction (optical fibre of 15 000 m with attenuation equals 0.451 dB/km in 0.63 m sensor loop) allows to obtain a theoretical sensitivity equal to $1.97 \cdot 10^{-9}$ rad/s/Hz$^{1/2}$ in quantum noise limitation.

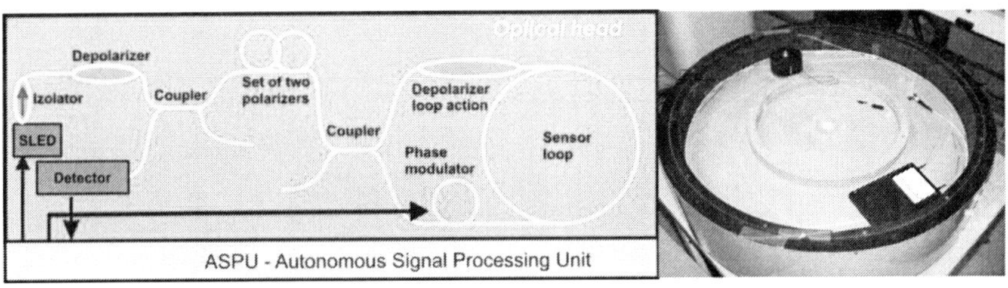


**Figure 8.** General schema and view of the AFORS optical head (generation of the Sagnac phase shift proportional to the measured rotation rate)

It should be emphasised that in the AFORS construction we have applied the special processing unit ASPU (Figure 9), which enables to obtain the detected rate of rotation in a direct way from the measured Sagnac phase shift (7). The ASPU detects the rotation rate (Ω) by selection and conversion of the first ($A_{1\omega}$) and second ($A_{2\omega}$) amplitude of the harmonic output signal [u(t)] using the following formula [50]:

$$\Omega = S_o \cdot \arctan[S_e \cdot u(t)] = S_o \cdot \arctan[S_e \cdot \frac{A_{1\omega}}{A_{2\omega}}], \tag{8}$$

in which $S_e$ is the electrical constant connected to specification of the used instrumentations and received by calibrating the sensor, while $S_o$ is the optical constant determined also during the calibration process which is described below. In order to eliminate a discontinuity the formula (8) uses the arc tangent which is extended to the four quarters (−π, π) after the Fourier transform. The ASPU utilizes the synchronous detection in a digital form. The 32 bit digital signal processor realizes all necessary processing processes including calculation the rate of rotation in a real time as well as data recording in the SD card. The electronic part sends also the stored data via fibre patch cord to the GSM/GPS modem which is connected FORS-Telemetric Server [30, 50]. The fundamental time constant equals 4.7104 ms and its value resulted from the used quartz oscillator. The multiplication (maximum multiplication is equal to $2^7$) of the time constant allows to adjust the sampling time. This approach gives detection frequency bandpass from 0.83 Hz to 106.15 Hz. This detection band is required to detect the rotational seismic-origin events. Furthermore the electronic part is equipped with digital-to-analogue converter which enable to store the data in the analogue form by the standard recording system KST.

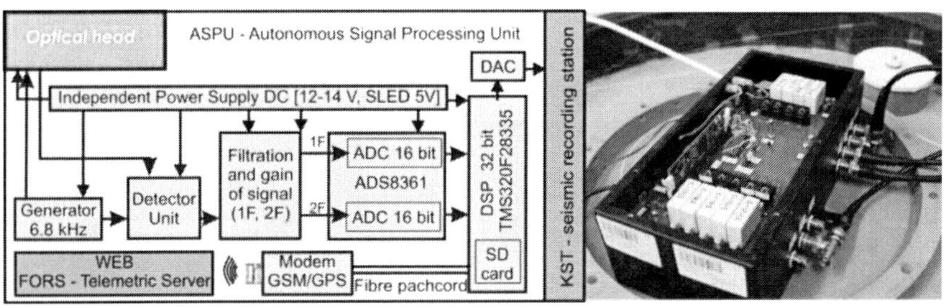


**Figure 9.** General schema and view of the AFORS's Autonomous Signal Processing Unit (rotation calculation and recording)

The calibration process was realized basing on the measurement of the defined slow rotation connected to the vector of Earth rotation in Warsaw, Poland (i.e. $\Omega_E$= 9.18 deg/h ≡ $4.45 \cdot 10^{-5}$rad/s for $\phi$ = 52°20"). AFORS was mounted vertically on a special rotation table (Figure 10a) and then rotated so that the sensor loop was directed to the North, South, East and West. The detected signal was equalled to zero for East-West (sensor loop collinear with the Earth rotation axis) while the signal was maximal, equal to ± $4.45 \cdot 10^{-5}$rad/s, for North-South (the sensor loop perpendicular to the vector component of Earth rotation). During the calibration procedure in the first step were determined the position for maximal signals, North and South, then, the position of the sensor for the West and East was defined as the midpoint between those two signals because of problems with determination this position basing on searching of signals equal zero. The maximal signals were obtained with accuracy equals 0.5 deg. In order to eliminate the drift phenomenon the 10 night hours averaging signal was applied. The above procedure allows for scaling the system and determining the constant operation parameters – optical $S_o$ and electrical $S_e$. For AFORS-1 the following values of the above constants were obtained: $S_0 = 0.043\ s^{-1}$, $S_e = 0.0144$. In order to experimental determination of the accuracy for particular device [50], which have been made according to AFORS production technology, we measured the noise level for each of them. Nevertheless the measurements were carried out at Military University of Technology, Warsaw, Poland. The place of the research could provide deviations due to urban noises. The received accuracy is equal to $5.07 \cdot 10^{-9}$ rad/s and $5.51 \cdot 10^{-8}$ rad/s, respectively, for the lower and higher working detection frequency band, as shown in Figure 10b. Additional Figure 10b shows also measured parameters for second system - AFORS-2 (theoretical sensitivity in quantum noise limitation equals $2.46 \cdot 10^{-9}$ rad/s/Hz$^{1/2}$, accuracy in detection band at level of $4.81 \cdot 10^{-9}$ rad/s and $6.11 \cdot 10^{-8}$ rad/s). It should be noted that the linear dependence of AFORSs accuracy on the detection frequency band is an advantage of this system.

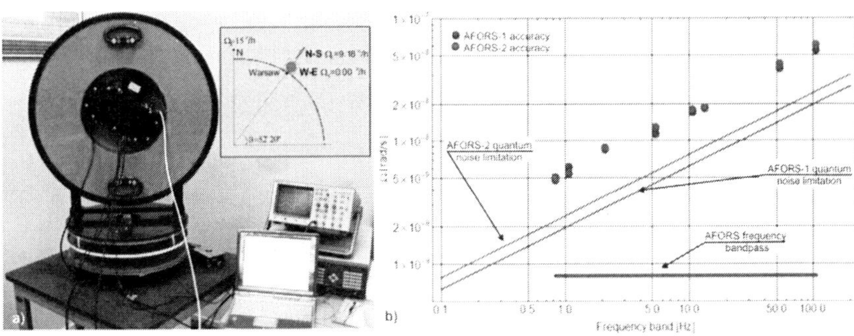


**Figure 10.** The calibration and investigation of AFORSs accuracies: a) general view of AFORS-1 during the calibration process with scheme showing its idea, b) the accuracy measured for the chosen detection frequency band for AFORS-1 and AFORS-2

# 3. ANALYSIS OF ROTATIONAL COMPONENTS OF THE SEISMIC FIELDS CAUSED BY LOCAL EVENTS

In this section we present analysis of data obtained at Książ Observatory, reveal the rotational components presence in entire seismograms (from P-wave arrival onwards), for the cases of local seismic events, of the mining (Lubin on January 20th, 2011, two events starting at 04 h 59 min. 1 s UT - shown below as Figure 11) and tectonic (Jarocin on January 6th, 2012, the event starting at 15 h 38 min. 10 s UT - shown below as Figure 12) provenience.

The results were obtained directly from KST recording system and they include five plots: channels 1 and 2 for TAPS-1; channel 3 for AFORS-1; channels 4 and 5 for TAPS-2. The TAPSs' records show linear motions that appeared during the earthquakes, and rotational components are calculated as described in the section 2.1. The channel 3 for AFORS-1 shows rotational oscillations measured in a direct way.

As one can see, the both kinds of devices (AFORS-1 as well as set of micro-array of TAPSs) recorded the events in the same time, which can confirm some correlation between devices. However the following investigation needs an additional data proceedings. To limit noise influence on recorded signals the average procedure in the beginning has been applied. From above mentioned reason the recorded seismic events have been averaged in moving windows of 100 samples (which is equal to period of 1 second).

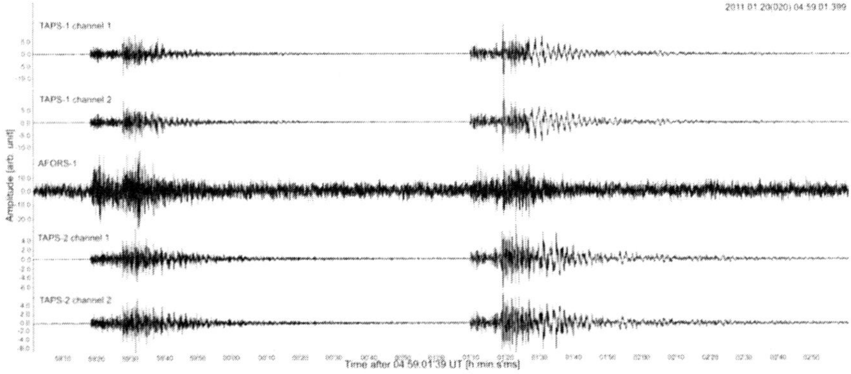


**Figure 11.** Plots of two seismic events with source in Lubin area: TAPS – ground motion velocity, AFORS – ground motion rotation rate

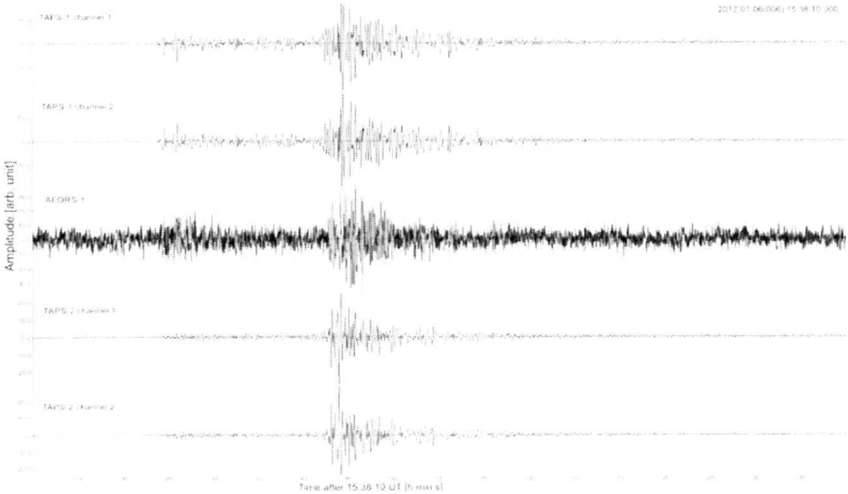


**Figure 12.** Plots of the seismic event with source in Jarocin area: TAPS – ground motion velocity, AFORS – ground motion rotation rate

The results of above operation are presented in Figures 13a-13c respectively for above three events. Consistently, we present five plots for all of them. The first plot, named TAPS-1 channel 2, presents the velocity of linear ground motion in m/s registered by second seismometer in this device. Second diagram shows channel 1 of TAPS-2. It should be noticed that second channel for any TAPS, after change of sign, has very similar plot to the first channel, so they are not presented both in our Figures. Third plot, named AFORS, presents the rotation velocity in rad/s, registered directly by AFORS-1. The result of measurement of the same component, but using the four simultaneous signals from micro-array of TAPSs (the procedure is described in section 2.1), is presented as plot four named Rotation from TAPS. Finally, last plot named Twist from TAPS, presents twist component obtained from the same micro-array in accord with, mentioned in the introduction, the Asymmetric Continuum Theory [14] – also in rad/s.

Even though the results presented in Figures 13a-13c show similar time of occurrence of the rotational components recorded by the AFORS-1 and calculated from TAPSs, their shape differs, and inevitably the correlation coefficients between them are low (though greater than zero). These coefficients are presented in Table 1, for both events recorded on January 20$^{th}$, 2011, and for their selected parts. This selection consists of: time-period when P waves arrive – 2 seconds in each case; time period of firsts S waves arrivals – again 2 seconds, and a time-period when great S-type oscillations dominate; here we choose twenty seconds in each case (such unusually high amplitudes of low frequency oscillations, dominating in the late stage of the tremor, characterize the seismic field generated by mining seismic events in the Lubin area and received at Książ observatory).

Additionally, correlation between both rotational components calculated from TAPSs recordings, that is – rotation and twist – was checked too, for the same chosen time-periods. This appeared generally high (especially in the second event), which confirms previous observations by KP. Teisseyre that in the recordings of seismic events made with a set of TAPSs, there usually occurs conformance between rotation and twist, either direct or reverse.

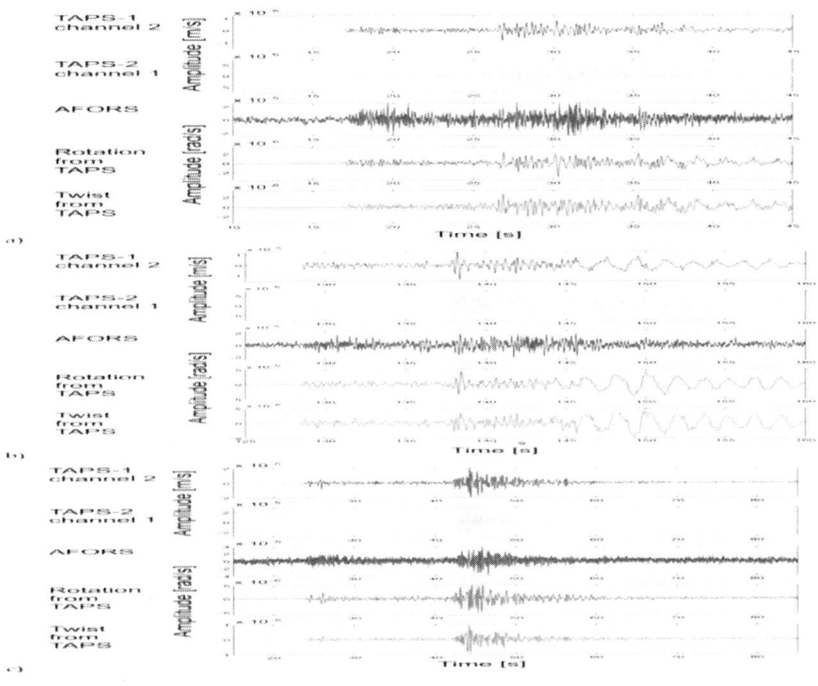


**Figure 13.** (a). Results for the first event in the Lubin area registered on January 20th, 2011; (b). Results for the second event in the Lubin area registered on January 20th, 2011; (c). Results for the Jarocin area earthquake registered on January 6th, 2012

**Table 1.** The correlation coefficients

| 20.01.2011. Lubin | First event 04 h 59 min., M=3.1 17 – 107.75 s | | | Second event 05 h 01 min., M=3.3 128.40 – 234.45 s | | |
|---|---|---|---|---|---|---|
| AFORS – Rot TAPS | 0.092 | | | 0.063 | | |
| Twist TAPS – ROT TAPS | 0.600 | | | 0.894 | | |
| Wave type time-period | P 17.00 – 19.00 s | S 26.00 – 28.00 s | great S 30.00 – 50.00 s | P 128.40 – 130.40 s | S 137.50 – 139.50 s | great S 146.00 – 166.00 s |
| AFORS – Rot TAPS | 0.200 | 0.354 | 0.073 | 0.187 | 0.050 | 0.140 |
| Twist TAPS – Rot TAPS | 0.166 | 0.483 | 0.641 | 0.534 | 0.724 | 0.958 |

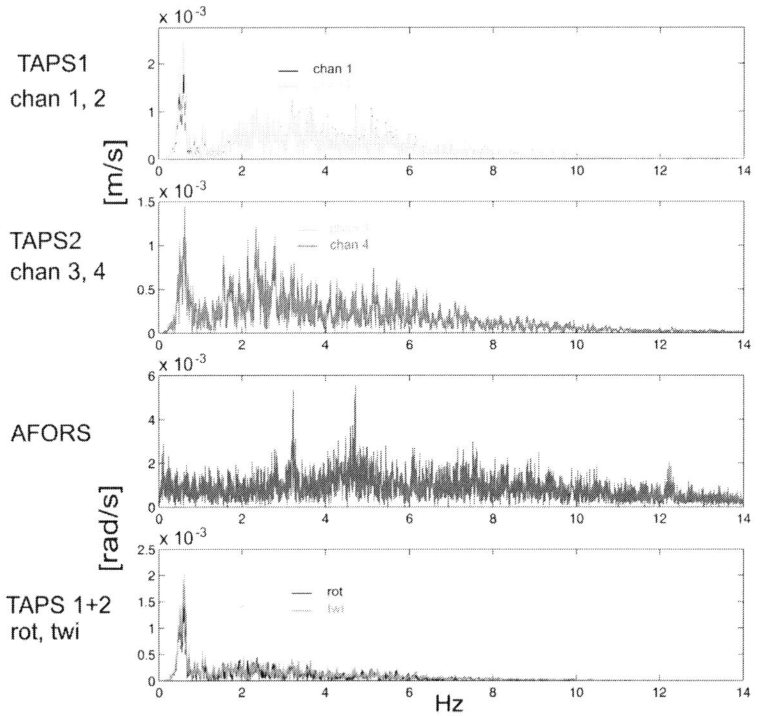


**Figure 14.** Spectra of linear and rotational signals, from ones of analysed seismic mining events, recorded on January 20th, 2011 at Książ observatory. The second one has the similar spectra

Dissimilarity between simultaneous rotation recordings obtained from AFORS and the micro-array of TAPSs may be explained by different characteristics of the used instruments or by certain errors or/and noise present in one or the other side of compared results, or in both. Here, there is dramatic difference in spectra (Figure 14), that of AFORS is always much longer. For the analyzed case of mining events, the spectrum of AFORS signal bears high amplitudes in the range of 2 – 8 Hz. Spectra of all the signals and rotational components obtained from TAPSs are short – they practically decrease to zero level at about 20 Hz – and bear sharp maximum at about 0.5 Hz, while the linear signals have also wide area of relatively high amplitudes in the range of 1 – 8 Hz. Moreover, the signals from AFORS bear high percentage of noise, probably of electronic provenience.

To find similarities in the obtained results in other way than just by sight, the following analysis has been applied. First – any of compared data chains was transformed into chain of absolute values (moduli). Then, the chains of moving averages of these moduli are created, with the window length of 100 samples, which is equivalent to 1 s. Here, absolute values of rotation (velocities) were compared – these from AFORS with those calculated from set of TAPSs. Further, analogical moving averages were investigated, but calculated from moduli of all four signals recorded by the TAPSs. If the rotation obtained from AFORS is symbolized with $\omega^o$, rotation obtained from the TAPSs with $\omega$, mean signal modulus from micro-array at the given sample with $\bar{u}$, and length of window used in moving average calculation with *w*, then formulae for described moving averages are as follows:

$$\bar{\omega}^0 = \frac{\sum_{i=1}^{i=w} |\omega_i^0|}{w}, \qquad \text{a}$$

$$\bar{\omega} = \frac{\sum_{i=1}^{i=w} |\omega_i|}{w}, \qquad \text{b}$$

$$\bar{u} = \frac{\sum_{i=1}^{i=w} \left( \frac{|u_1| + |u_2| + |u_3| + |u_4|}{4} \right)_i}{w}. \qquad \text{c} \qquad (9)$$

Results of this analysis are shown in Figures 15-17, which all have the same scheme. Diagrams a) show the plot of moving average of absolute values of the rotation rate obtained from AFORS; diagrams c) – analogical averages of rotation rate calculated from TAPSs' data; diagrams e) – analogical averages of the mean signals moduli from TAPSs – as in formula (9c). The blue plots presented in diagrams b) represent the moving correlation coefficient between moving averages of absolute rotation from AFORS and its analogue from TAPSs (compare the plots in diagrams a) and c) described mathematically by equations (9a) and (9b)). This coefficient is calculated for a window of 100 samples (1s). Additionally, in the same diagram we present reference thick lines – orange or mauve, and turquoise – which should facilitate the comparison of correlation coefficient with compared chains. The procedure for obtaining the upper reference line is following: create normalized chain (9b) to (9a) by comparing both maximum values, as (9b'). Next, make new chain as sum of (9a) with (9b'), and normalize it to 1. Thus we have obtained a doubly–normalized mean chain which joins shapes of the original two chains. The turquoise reference line is analogical to orange one, but multiplied by -1. Purpose of this line presence is to facilitate comparison of stages of high negative correlations between compared chains, again with the described doubly–normalized averages. Analogically, curves presented in diagrams d) and f) are the plots of the correlation coefficient in a moving window of 100 samples; in d) – between averages plotted in diagrams c) and e), and in f) – between averages plotted in e) and a). Reference lines are produced in analogical way as for diagrams b); the upper one is mauve in diagrams f). Figure 15 shows results of such analysis applied to recordings of first event from Lubin area; Figures 16 and 17 are made for second event from Lubin area and for Jarocin earthquake, in the same methodology.

Comparison of the moving correlation coefficients with the moving averages allows to find certain rule. In time-periods when main seismic phase arrive – as P and S (for local distances, it might be jointly for example Pg and Pb phases and analogically Sg and Sb in the S-type phases family), all investigated correlations between moving averages of the absolute signals are generally high. These are some of the time-periods in which blue line in the diagrams b), d) and f) is near 1, and the upper reference line rises. More interestingly, these time-periods starts slightly before any noticeable rise in initial moving averages, despite facts that all the moving windows had the same length. This phenomenon may have two causes: certain common order in concerned signals starts just before noticeable arrival of seismic waves, or/and this is a mere effect of filtration (which used in every contemporary seismic recording system).

The chosen time-periods of high correlations and rise of the moving averages are as follows.

For the first seismic event from Lubin - two time periods: from 16.2 to 18.5 s, this include arrival of P waves, and from 25.8 to 27 s, which include S waves arrival (see Table 1). For the second seismic event from Lubin – only one time period, from 127.9 to 129.5 s, which include arrival of P waves. For the Jarocin earthquake – one non-continuous time period, for diagrams b) and f): 21.7 – 24.1 s and for diagram d) : 20.9 – 24.1 s. This time-period, in all three diagrams, starts before arrival of P waves and comprises arrival of this seismic phase.

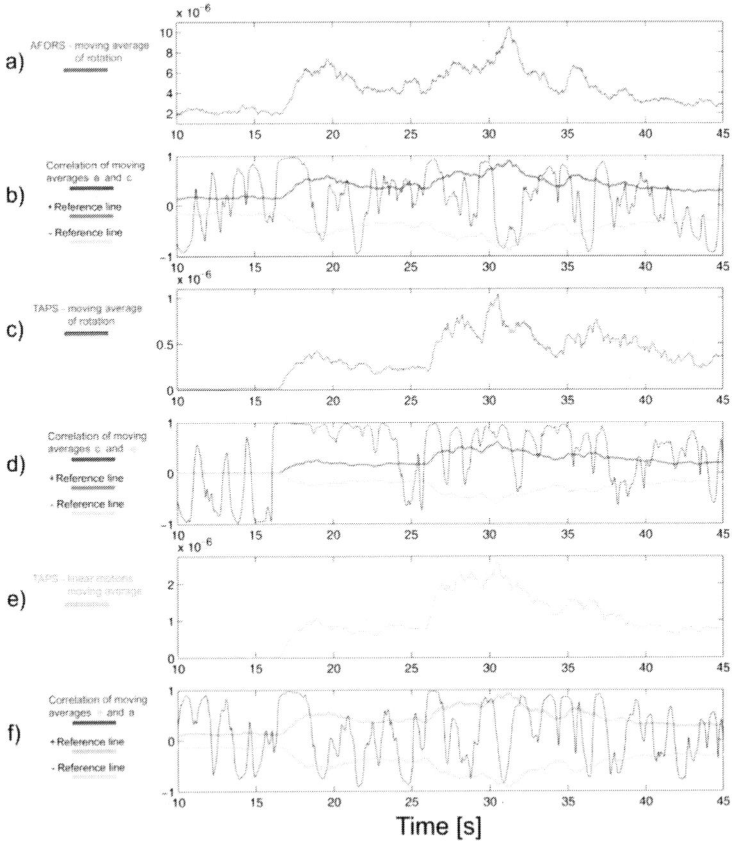


**Figure 15.** Relations of rotations and linear signals at the first event in Lubin area registered on January 20th, 2011. Diagrams: a) moving average of absolute values of AFORS – signal; b) correlation coefficient between (a) and (c) in moving window – blue line, orange reference line – doubly normalized means of (a) and (c), turquoise – the same as orange but with reversed sign; c) moving average of absolute values of rotations obtained from TAPSs; d) correlation coefficient between (c) and (e) in moving window, orange reference line – doubly normalized means of (c) and (e), turquoise – the same as orange but with reversed sign; e) moving average of absolute values of four channels of TAPSs; f) correlation coefficient between (e) and (a) in moving window, mauve reference line – doubly normalized means of (e) and (a), turquoise – the same as mauve but with reversed sign

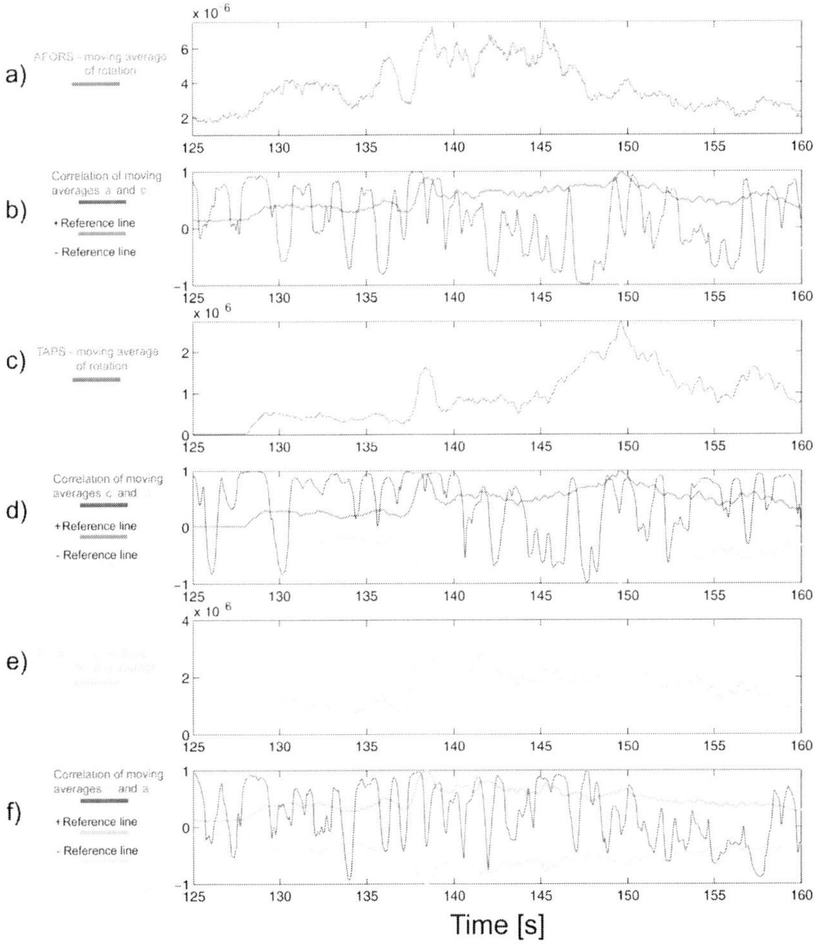


**Figure 16.** Relations of rotations and linear signals at the second event in Lubin area registered on January 20th, 2011. Diagrams descriptions as in Fig. 15

We did not find yet any other rules, especially – we did not find any relation between episodes of negative correlations (near -1) and the initial moving averages. We suppose that such comparison method may be useful in analysis of various chains, and especially their moving averages, as in this work. Crucial point is that not the original signals, but chains of their absolute values are compared.

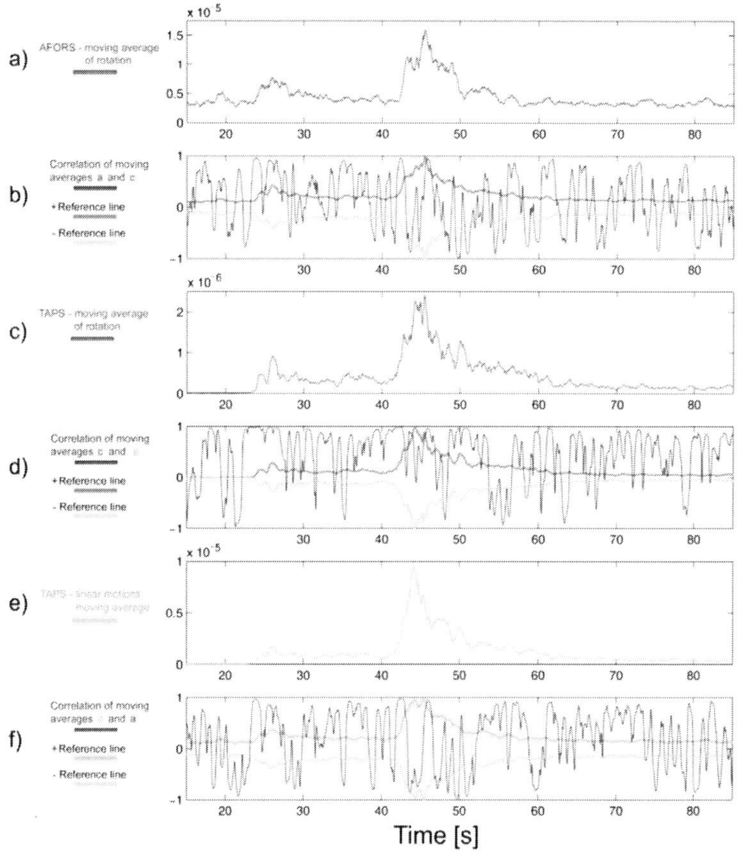


**Figure 17.** Relations of rotations and linear signals at the event in Jarocin area registered on January 6th, 2012. Diagrams descriptions as in Fig. 15

# 4. CONCLUSIONS

Simultaneous measurements of the rotations in seismic field with the use of completely different instruments – here AFORS-1 which is the Sagnac interferometer and the micro-array of TAPSs allow for comparison of the used equipment. In this work, such comparison revealed that signals differ significantly, to the degree which complicates analysis. From both kind of instruments, rotations are obtained in the same time-periods, but their plots differ. These differences are attributed partly to difference in instruments spectra and partly to disturbances in the signals, of technical provenience. Nevertheless, an analysis using the moving averages of absolute signals values, and consequently also coefficients of correlation between these averages confirmed common roots of these recorded signals, despite all their imperfections.

Research on seismic rotational effects, especially in buildings and other large constructions is widely recognized as very important and therefore these studies flourish. On the other hand, rotational components in the seismic field are also studied in various ways, but even existence of these components still evoke controversy. Presence of these components in seismograms, especially in their initial part which, according to classical theory of elasticity, contain only compressional waves, is explained in various ways. Authors believe that no one contemporary explanation is complete and proven, but this not preclude usefulness of further studies.

# 5. ACKNOWLEDGEMENTS

This work was done in 2013-2014 under the financial support of the Polish Ministry of Science and Higher Education under Key Project POIG.01.03.01-14-016/08 "New photonic materials and their advanced application", the MUT statutory activity PBS-850 and partially the the Polish National Centre for Research and Development under contract No PBS1/B3/7/2012.

# REFERENCES

1. Hobbs WH. Earthquakes. An Introduction to Seismic Geology. New York: Appleton and Co.; 1907.
2. Davison Ch. The Founders of Seismology. Cambridge: Cambridge University Press; 1927.
3. Ferrari G. Note on the Historical Rotation Seismographs. In: Teisseyre R., Takedo M., Majewski E. (eds) Earthquake Source Asymmetry, Structural Media and Rotation Effects. Berlin-Heidelberg-New York: Springer; 2006. p367-376.
4. Kozak JT. Development of Earthquake Rotation Effect Study. In: Teisseyre R., Takedo M., Majewski E. (eds) Earthquake Source Asymmetry, Structural Media and Rotation Effects. Berlin-Heidelberg-New York: Springer; 2006. p3-10.

5. Imamura A. Theoretical and Applied Seismology. Tokyo: Maruzen Co.;1937.
6. Gutenberg B. Grundlagen der Erdbebrnkunde. Frankfurt: Univ. Frankfurt a/M; 1927.
7. Teisseyre R. Earthquake Processes in a Micromorphic Continuum. Pure Applied Geophysics 1973; 102(1) 15-28.
8. Teisseyre R. Symmetric Micromorphic Tin Space Continuum: Wave Propagation, Point Source Solutions and Some Applications to Earthquake Processes. In: Thoft-Christensen P. (ed.) Continuum Mechanics Aspects of Geodynamics and Rock Fracture Mechanics. Dodrecht-Boston: D. Reidel Publ. Comp.; 1974. p201-244.
9. Eringen AC. Mirocontinuum Field Theories. Vol. 1 Foundations and Solids. New York: Springer-Verlag; 1999.
10. Teisseyre R, Boratyński W. Continuum with Self-Rotation Fields: Evolution of Defect Fields and Equations of Motion. Acta Geophysica 2002; 50(3) 223-229.
11. Teisseyre R. Asymmetric Continuum Mechanics: Deviations from Elasticity and Symmetry. Acta Geophysica 2005; 53(2) 115-126.
12. Teisseyre R, Białecki M, Górski M. Degenerated Mechanics in a Homogeneous Continuum: Potentials for Spin and Twist. Acta Geophysica 2005; 53(3) 219-223.
13. Teisseyre R, Górski M. Transport in Fracture Processes: Fragmentation of Defect Fields and Equations of Motion. Acta Geophysica 2009; 57(5) 583-599.
14. Teisseyre R, Suchcicki J, Teisseyre KP, Wiszniowski J, Palangio P. Seismic Rotation Waves: Basic Elements of Theory and Recording. Annals of Geophysics 2003; 46(4) 671-685.
15. Teisseyre R., Kozák JT. Sources of Rotation and Twist Motions. In: Teisseyre R, Takedo M., Majewski E. (eds) Earthquake Source Asymmetry, Structural Media and Rotation Effects. Berlin-Heidelberg-New York: Springer; 2006. p11-23.
16. Teisseyre R, Takeo M, Majewski E. Earthquake Source Asymmetry, Structural Media and Rotation Effects. Berlin-Heidelberg-New York: Springer; 2006.
17. Lee WHK, Cęlebi M, Todorovska MI, Igel H. (guest eds) Rotational Seismology and Engineering Applications. California: Bulletin of the Seismological Society of America 99(2B); 2009.

18. Cowsik R, Madziwa-Nussinov T, Wagoner K, Wiens D, Wysession M. Performance Characteristics of a Rotational Seismometer for Near-Field and Engineering Applications. Bulletin of the Seismological Society of America 2009; 99(2B) 1181-1189.

19. Droste Z, Teisseyre R. Rotational and Displacemental Components of Ground Motion as Deduced from Data of the Azimuth System of Seismograph. Publications od Institute of Geophysics Polish Academy of Science 1976; 97 157-167.

20. Riedesel MA, Moore RD, Orcutt JA. Limits of Sensitivity of Inertial Seismometers and Velocity Transducer and Electronic Amplifiers. Bulletin of the Seismological Society of America 1990; 80 1725-1752.

21. Teisseyre R, Nagahama H. Micro-Inertia Continuum: Rotations and Semi-Waves. Acta Geophysica Polonica 1999; 47 259–272.

22. Jaroszewicz LR, Krajewski Z, Solarz L, Marc P, Kostrzyński T. A New Area of the Fiber-Optic Sagnac Interferometer Application. In: Proceedings of the 2003 SBMO/IEEE MTT-S International Conference on Microwave and Optoelectronics, 20–23 September 2003, Rio de Janeiro, Brazil, 2003.

23. Sagnac G. L'éther lumineux démontré par l'effet du vent relative d'éther dans un interférométre en rotation uniforme. [in French] Comptes rendus de l'Académie des Sciences 1913; 95 708–710.

24. Schreiber U, Schneider M, Rowe CH, Stedmanand GE, Schluter W. Aspects of Ring Lasers as Local Earth Rotation Sensors. Surveys in Geophysics 2001; 22 603–611.

25. Jaroszewicz LR., Krajewski Z., Solarz L. Absolute Rotation Measurement Based on the Sagnac Effect. In: Teisseyre R., Takeo M., Majewski E. (eds) Earthquake Source Asymmetry, Structural Media and Rotation Effects. Berlin-Heidelberg-New York: Springer; 2006. p413–438.

26. Takeo M, Ueda H, Matzuzawa T. Development of a High-Gain Rotational-Motion Seismograph. Grant 11354004. Earthquake Research Institute University of Tokyo; 2002. p5–29.

27. Nigbor RL, Evans RJ, Hutt C. Laboratory and Field Testing of Commercial Rotational Seismometers. Bulletin of the Seismological Society of America 2009; 99, 1215–1227.

28. Schreiber KU, Stedman GE, Igel H, Flaws A. Ring Laser Gyroscopes as Rotation Sensors for Seismic Wave Studies. In: Teisseyre R.,

Takeo M., Majewski E. (eds) Earthquake Source Asymmetry, Structural Media and Rotation Effects. Berlin-Heidelberg-New York: Springer; 2006. p377-390

29. Kurzych A, Jaroszewicz LR, Krajewski Z, Teisseyre KP, Kowalski JK. Fibre Optic System for Monitoring Rotational Seismic Phenomena. Sensors 2014;14 5459-5469.

30. Jaroszewicz LR, Krajewski Z, Teisseyre KP. Usefulness of AFORS - Autonomous Fibre-Optic Rotational Seismograph for Investigation of Rotational Phenomena. Journal of Seismology 2012; 16(4) 573-586.

31. Teisseyre KP. Mining Tremors Registered at Ojców and Książ Observatories: Rotational Field Components, Publications of Institute Geophysics Polish Academy of Sciences 2006; M-29(395) 77-92.

32. Teisseyre KP. Analysis of a Group of Seismic Events Using Rotational Components. Acta Geophysica 2007; 55(4) 535-553.

33. Moriya T, Teisseyre R. Discussion on the Recording of Seismic Rotation Waves. Acta Geophysica Polonica 1999; 47 351-362.

34. Teisseyre R, Suchcicki J, Teisseyre KP. Recording the Seismic Rotation Waves: Reliability Analysis. Acta Geophysica Polonica 2002; 51 37-50.

35. Nowożyński K. Teisseyre KP. Time-Domain Filtering of Seismic Rotation Waves. Acta Geophysica Polonica 2003; 51 51-61.

36. Solarz L, Krajewski Z, Jaroszewicz LR. Analysis of Seismic Rotation Detected by Two Antiparallel Seismometers: Spine Function Approximation of Rotation and Displacement Velocities. Acta Geophysica Polonica 2004; 52 198-217.

37. Post EJ. Sagnac Effect. Review of Modern Physics 1967; 39(2) 475-493.

38. Vali V, Shorthill RW. Fiber Ring Interferometer. Applied Optics 1976; 15(5) 1099-1100.

39. Ulrich R. Fiber-Optic Rotation Sensing with Low Drift. Optics Letters 1980; 5(5) 173-175.

40. Arditty H, Lefèvre HC. (1981). Sagnac Effect in a Fiber Gyroscope, Optics Letters 1981; 6(8) 401-403.

41. Martin JM, Winkler JT. Fiber-Optic Laser Gyro Signal Detection and Processing Technique. SPIE Proceedings 1978; 139 98-102.

42. Bergh RA, Lefèvre HC, Shaw HJ. All-Single-Mode Fiber-Optic Gyroscope with Long-Term Stability. Optics Letters 1981; 6(10) 502-504.

43. Ezekiel S, Davis JL, Hellwarth RW. Intensity Dependent Nonreciprocal Phase Shift in a Fiber-Optic Gyroscope. Springer Series in Optical Sciences 1982; 32 332-336.

44. Fredricks RJ, Ulrich R. Phase-Error Bounds of Fibre Gyro with Imperfect Polariser/ Depolarizer. Electronic Letters 1984; 20(8) 330-332.

45. Lefèvre HC, Bettini JP, Vatoux S, Papuchon M. Progress in Optical Fiber Gyroscopes Using Integrated Optics. Proceedings of AGARD-NATO 1985; CPP-383 9A1-9A13.

46. Burns WK. Phase-Error Bounds of Fiber Gyro With Polarization-Holding Fiber. Journal of Lightwave Technology 1986; LT4(1) 8-14.

47. Lefèvre HC, Graindorge Ph, Arditty HJ, Vatoux S, Papuchon M. Double Closed-Loop Hybrid Fiber Gyroscope Using Digital Phase Ramp. Proceeding of OFS-3 1985, San Diego, OSA/IEEE, Postdeadline Paper 7

48. Auch W. The Fiber-Optic Gyro – a Device for Laboratory Use Only?, SPIE Proceedings 1986; 719, 28-34.

49. Arditty HJ, Graindorge Ph, Lefèvre HC, Martin Ph, Morisse J, Simonpiétri P. Fiber-Optic Gyroscope with All-Digital Processing. Proceedings of OFS- 6, Paris, Springer-Verlag Proceedings in Physics 1989; 44 131-136.

50. Jaroszewicz LR, Krajewski Z, Kowalski H, Mazur G, Zinówko P, Kowalski JK. AFORS Autonomous Fibre-Optic Rotational Seismograph: Design and Application. Acta Geophysisica 2011; 59, 578–596.

51. Jaroszewicz L.R., Krajewski Z., Teisseyre, K. P. The Possibility of a Continuous Monitoring of the Horizontal Buildings' Rotation by the Autonomous Fibre-Optic Rotational Seismograph AFORS Type. In: Lavan O., De Stefano M. (eds) Seismic Behaviour and Design of Irregular and Complex Civil Structures. Berlin-Heidelberg: Springer-Verlag; 2013. p339-351.

52. Jaroszewicz LR, Krajewski Z. Application of the Fibre-Optic Rotational Seismometer in Investigation of the Seismic Rotational Waves. Opto-Electronics Review 2008; 16(3) 314-320.

53. Böhm K, Marten P, Standigel L, Weidel E. Fiber-Optic Gyro with Digital Data Processing. 2nd International Conference on Optical Fiber Sensors, Stuttgart, Germany, 5–7 September 1984; 251–258.

54. Krajewski Z. Fiber-Optic Sagnac Interferometer as System for Rotational Phenomena Investigation Connected with Seismic Events. (In Polish) PhD thesis, Military University of Technology, Warsaw; 2005.

55. Krajewski Z, Jaroszewicz LR, Solarz L. Optimization of Fiber-Optic Sagnac Interferometer for Detection of Rotational Seismic Events. In Proceedings of SPIE 5952, Optical Fibers: Applications, Warsaw, Poland, 28 August–2 September 2005; 240–248.

56. Dai X, Zhao X, Cai B, Yang G, Zhou K, Liu C. Quantitative Analysis of the Shupe Reduction in a Fiber Optic Sagnac Interferometer. Optical Engineering 2002; 41 1155–1156.

# 6

# CHAPTER

# SELF-CENTERING SEISMIC LATERAL FORCE RESISTING SYSTEMS: HIGH PERFORMANCE STRUCTURES FOR THE CITY OF TOMORROW

**Nathan Brent Chancellor [1,*], Matthew R. Eatherton [2], David A. Roke [3] and Tuğçe Akbaş [1]**

[1] *Advanced Technology for Large Structural Systems (ATLSS) Center, Department of Civil and Environmental Engineering, Lehigh University, 117 ATLSS Drive, Bethlehem, PA 18015, USA*

[2] *Department of Civil and Environmental Engineering, Virginia Tech, 200 Patton Hall, Blacksburg, VA 24061, USA*

[3] *Department of Civil Engineering, The University of Akron, Auburn Science and Engineering Center, 244 Sumner Street, Akron, OH 44325, USA*

## ABSTRACT

Structures designed in accordance with even the most modern buildings codes are expected to sustain damage during a severe earthquake; however; these structures are expected to protect the lives of the occupants. Damage to the structure can require expensive repairs; significant business downtime; and in some cases building demolition. If damage occurs to many structures within a city or region; the regional and national economy may be severely disrupted. To address these shortcomings with current seismic lateral force resisting systems and to work towards more resilient; sustainable cities; a new class of seismic lateral force resisting systems that sustains little or no damage under severe earthquakes has been developed. These new seismic lateral force resisting systems reduce or prevent structural damage to nonreplaceable structural elements by softening the structural response elastically through gap opening mechanisms. To dissipate seismic energy; friction elements or replaceable yielding energy dissipation elements are also included. Post-tensioning is often used as a part of these systems to return the structure to a plumb; upright position (self-center) after the earthquake has passed. This paper summarizes the state-of-the art for self-centering seismic lateral force resisting systems and outlines current research challenges for these systems.

**Keywords:** self-centering; high-performance; resilient; seismic; moment frame; braced frame; shear wall

# 1. INTRODUCTION

Structures designed in accordance with even the most modern buildings codes are expected to undergo significant structural and nonstructural damage during a severe earthquake. Structures are typically designed for "life-safety" performance, in which the objective of design is to protect the lives of the occupants during the design basis earthquake (DBE), which has a return period of several hundred years [1]. Damage to the structure may include yielding, buckling or fracture of structural elements and permanent horizontal displacements after the earthquake, referred to as residual drift.

Structures experiencing large residual drifts are removed from service, and buildings requiring structural repair are often out of service for extended periods of time [2]. Repairing structural damage is expensive and time-consuming because the damage is often distributed throughout the structure in many nonreplaceable elements. If the damage is too severe, or if there is significant residual drift, the structure may require demolition. Significant damage to many structures in a large city may result in a disruption of the regional, and in some cases national, economy.

To reduce the economic disruption caused by an earthquake, and to create more resilient, sustainable cities, high-performance structural systems that can withstand strong ground shaking with little or no structural damage (outside of replaceable energy dissipating elements) are needed. To meet this need, a new class of seismic lateral force resisting systems has been developed. These damage-resistant systems reduce or prevent structural damage to nonreplaceable elements by softening the structural response through elastic gap opening mechanisms instead of yielding in primary structural elements. To dissipate seismic energy, friction elements or replaceable yielding energy dissipation elements are also included. Post-tensioning is often used as a part of these systems to return the structure to a plumb, upright position (self-center) after the earthquake has passed. These new seismic lateral force resisting systems are known as self-centering systems.

This paper summarizes the state-of-the art for self-centering seismic lateral force resisting systems for buildings and outlines current research challenges for these systems. First, a brief overview of the motivation for self-centering systems is presented in the context of how conventional seismic systems perform, including an example demonstrating the implications of the conventional design approach. Next, the mechanics and behavior of self-centering systems are described, followed by an overview of several key different types of self-centering systems that have been developed. The paper closes with a discussion of some current research challenges that are currently under investigation.

# 2. MOTIVATION FOR SELF-CENTERING SYSTEMS

## 2.1. Conventional Seismic Lateral Force Resisting Systems: Behavior and Drawbacks

Conventional seismic lateral force resisting systems are designed to dissipate seismic energy through inelastic behavior (i.e., permanent damage) of structural elements. For example, steel moment resisting frames (MRFs) are designed to dissipate energy through substantial yielding near the beam ends. For steel concentrically braced frames (CBFs), energy is dissipated through inelastic buckling and yielding of braces. Concrete shear walls dissipate energy through yielding of reinforcing steel and crushing of concrete at the base of the wall. All of these energy dissipation mechanisms, as well as mechanisms for other conventional seismic lateral force resisting systems, are associated with damage to structural elements. Figure 1 schematically shows the expected damage in these conventional structural systems.

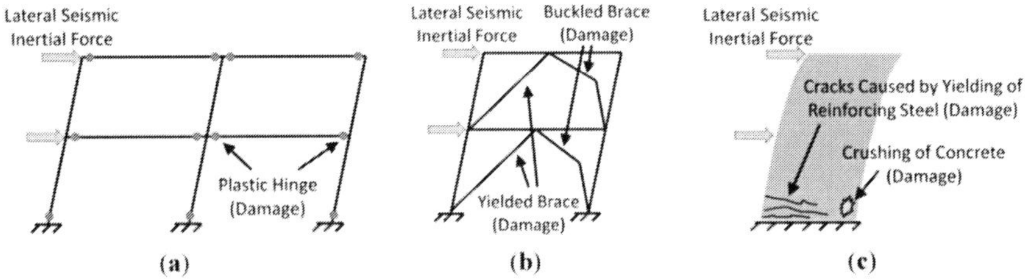


**Figure 1.** Conventional seismic lateral force resisting systems: (**a**) Steel moment resisting frame; (**b**) Steel concentrically braced frame; (**c**) Concrete shear wall.

The hysteretic behavior of a steel MRF is characterized by wide stable hysteresis loops, as shown in Figure 2a. These wide hysteresis loops are traditionally thought to be desirable due to the amount of energy dissipated per cycle of imposed displacement/rotation. However, with this type of hysteretic behavior, large residual drifts are possible when the lateral loads are removed.

The hysteretic behavior of a single brace from a computational simulation of a steel CBF is shown in Figure 2b. The brace yields in tension, but the brace buckles in compression and its load carrying capacity decreases, leading to pinched hysteresis loops. Buckling of the brace also causes very large local plastic strains near the midpoint of the brace, which can lead to low-cycle fatigue and fracture of the brace. With this type of hysteretic behavior, brace buckling and brace fracture may lead to a loss of lateral load capacity and residual drift.

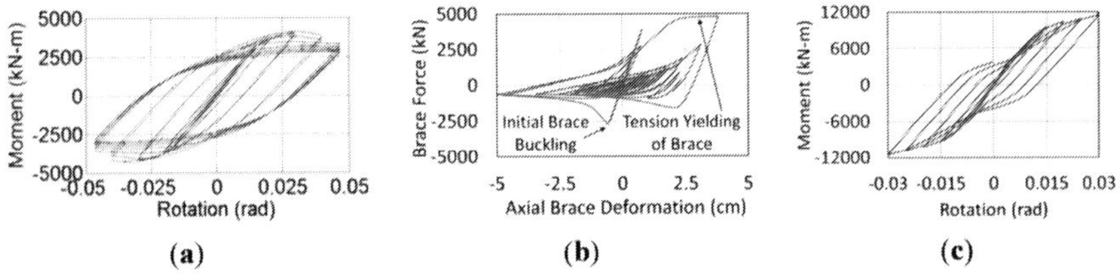


**Figure 2.** Hysteretic behavior of conventional structural systems: (**a**) Steel moment resisting frame (adapted from [3]. Copyright 2013 ASCE); (**b**) Single steel brace (adapted from [4]. Copyright 2014 Chancellor); (**c**) Concrete shear wall (from computational simulation).

The hysteretic behavior of a concrete shear wall from a computational simulation is shown in Figure 2c. The computational simulation performs a nonlinear analysis of the selected concrete shear wall section under an imposed deformation pattern to determine the moment-rotation relationship for the section. This type of hysteretic behavior is known as "stiffness-degrading" because the stiffness upon reloading the structure in the opposite direction is less than the stiffness of the previous cycle. With stiffness-degrading hysteretic behavior, less energy is dissipated per cycle than with a steel MRF, but the degrading stiffness behavior introduces some re-centering characteristics. Therefore, the residual drift at the end of an earthquake may be less than some other types of structural systems, though structural damage is still present.

## 2.2. Need for High-Performance Seismic Lateral Force Resisting Systems

The magnitude 6.3 earthquake that occurred in Christchurch, New Zealand in February 2011 is a prime example of the need for more resilient earthquake-resistant structural systems. Shortly after the earthquake, approximately 50% of the buildings in the central business district were declared unusable because they sustained significant damage or because they were adjacent to hazardous buildings [5] and nearly 1000 buildings in the central business district were subsequently demolished [6]. The estimated cost for rebuilding after this earthquake is $40 billion (New Zealand Dollars), which amounts to approximately 20% of New Zealand's GDP, without including economic losses associated with business downtime. As of May 2014, the Christchurch reconstruction is still underway [7].

Conventional seismic lateral force resisting systems depend on inelasticity in primary structural elements to resist collapse; therefore, most of these systems are inherently inefficient in limiting structural damage or residual drifts. New structural systems are therefore necessary to address performance goals associated with reducing structural repair costs and business downtime. These systems are known as "self-centering" systems due to their inherent ability to return to a plumb, upright condition after an earthquake. The remainder of this paper reviews the state-of-the-art in self-centering seismic lateral force resisting systems and outlines future research challenges for these systems.

## 3. SELF-CENTERING SEISMIC LATERAL RESISTING SYSTEMS—OVERVIEW OF BEHAVIOR

There are several key aspects of self-centering system mechanics and behavior. The first characteristic is the restoring force component that often utilizes a gap opening mechanism. The restoring force component combines with an energy dissipating element to produce flag-shaped load-deformation behavior. The shape of the hysteretic response can be tuned by the proportion of restoring force and energy dissipation components. Other key topics discussed in this section include self-centering ability in full buildings, controlling the progression of limit states, and designing for appropriate ductility demands.

### 3.1. Restoring Force Mechanism and Gap Openings

Self-centering seismic lateral force resisting systems are capable of fully re-centering when the lateral forces are removed (herein referred to as full self-centering), eliminating residual drift. In contrast to full self-centering, attempts have been made to reduce residual drifts in buckling restrained braced frames by using a linear-elastic backup moment frame as a restoring force [8]. This expected load-deformation behavior for this type of system is shown in Figure 3a, where the backup system providing a linear elastic restoring force is combined with an elastic-perfectly plastic lateral force resisting system to potentially reduce residual drifts through probabilistic self-centering. Probabilistic self-centering is described in detail later. Full self-centering, however, is usually achieved using a nonlinear elastic restoring force such as the bilinear elastic restoring force shown in Figure 3b.

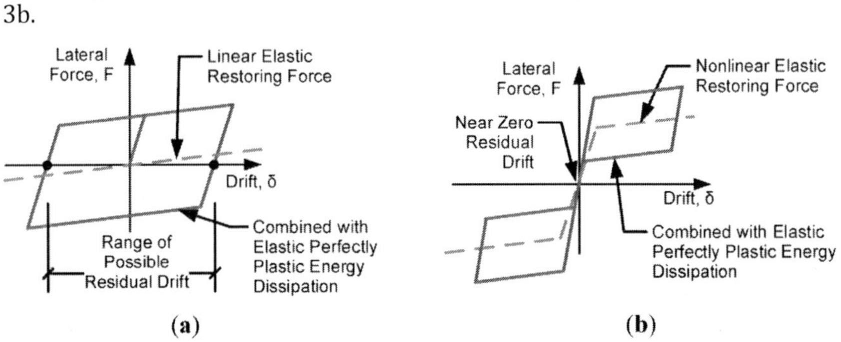


**Figure 3.** Need for nonlinear elastic restoring force to create full self-centering. (**a**) Added restoring force, but not full self-centering; (**b**) Full self-centering made possible by nonlinear elastic restoring force.

The most common approach for creating a bilinear elastic restoring force involves gaps forming between two surfaces that are initially precompressed together. Figure 4a schematically shows three categories of restoring force mechanisms that use this gap opening behavior. The first category, which includes rocking precast concrete walls (e.g., [9]), precast concrete columns (e.g., [10]), or rocking steel concentrically braced frames (e.g., [11]), forms a gap at the foundation when subjected to a prescribed overturning moment. When the lateral forces causing the overturning moment are removed, vertical unbonded post-tensioning (PT) steel pulls the structure back to a vertical, plumb condition, closing the gap. The second restoring force mechanism category includes self-centering moment resisting frames, such as precast concrete moment frames (e.g., [12]), steel moment frames (e.g., [13]), or coupled concrete shear walls (e.g., [14]), which allow gap opening between beam and column (or wall) joints when subjected to a prescribed moment. When the lateral forces causing the moment are removed, the gap is closed by compression forces provided by unbonded horizontal PT steel. The third restoring force mechanism category includes self-centering braces, in which a gap forms between telescoping concentric tubes and anchorage plates (e.g., [15]) when subjected to a prescribed axial force. Once the axial force is removed, the gap is closed and the tubes are brought back into alignment by PT steel oriented longitudinally along the length of the concentric tubes.

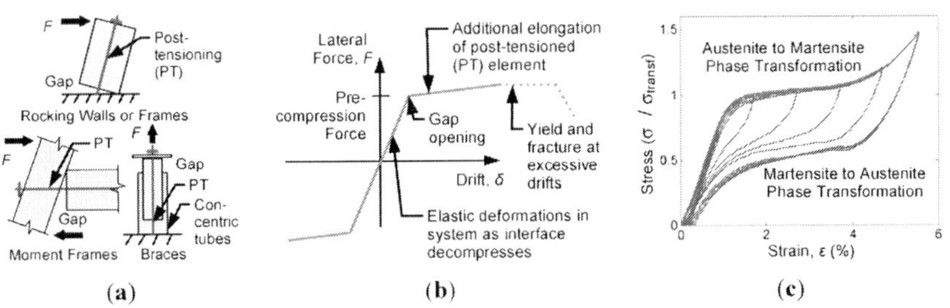


**Figure 4.** Methods for creating bilinear elastic restoring force using gap opening or shape memory alloys (SMA). (**a**) Gap opening mechanisms; (**b**) Restoring force associated with gap openings; (**c**) Nonlinear elastic behavior of SMA.

Although these systems allow gap opening at different locations, the restoring force mechanisms function identically. As shown in Figure 4b, at low levels of lateral force, the seismic lateral force resisting system behaves elastically. Once the lateral force becomes large enough to overcome the precompression force provided by the PT steel, the selected joint decompresses and a gap opens. At decompression there is a significant reduction in the stiffness of the lateral force resisting system. This reduction in stiffness is desirable because it lengthens the period of the structure and helps to limit the forces that can develop in the lateral force resisting system (i.e., softening occurs without structural damage). Because the bearing surfaces at the location of gap opening in a real structure are not all perfectly flat and the members are not exactly the same length, the actual load-deformation response may have rounded corners (e.g., [16,17]) rather than the sharp corners illustrated schematically in Figure 4b. The stiffness of the system after gap opening is primarily controlled by the axial stiffness of the PT steel and the location of the PT steel. If the PT steel is subjected to strains greater than the elastic limit, yielding and fracture may occur. The amount of inelastic strain capacity prior to fracture of the PT steel is a function of the type of PT steel system used.

Many types of PT steel systems have been used in self-centering systems. In most rocking self-centering systems and self-centering moment frames, the PT steel component consists of high strength PT strands or bars [13,14,18]. PT strands have higher ultimate stress, $F_{pu}$ = 1861 MPa, than PT bars, and have a large yield strain, which is usually taken as 1% [13]. However, the inelastic strain capacity of some PT strand systems is limited because the wedge anchorage system may cause stress concentrations in the strands [19]. This PT strand and wedge anchorage system results in mean strains at failure of 2.3% with a standard deviation of 0.5% for conventional 12.7 mm PT strands with multi-use chuck anchorages. PT bars have lower ultimate stress, $F_{pu}$ = 1030 MPa, and a smaller yield strain of 0.5%, but have significantly larger inelastic strain capacity that can be as high as 10%.

Self-centering braces typically have a short PT element length, such that it is often not practical to use steel PT because the axial deformation at yield of the PT steel is too small. Therefore, aramid fiber strands (which have an elastic strain capacity up to 4% [15]) or shape memory alloys with superelastic behavior up to 6% strain (e.g., [20]) have been used in self-centering braces. Figure 4c shows that the load-deformation response for a 23 mm round SMA rod subjected to cyclic tension (from [17]) exhibits the prerequisite nonlinear elastic behavior without requiring a gap opening mechanism. Double telescoping mechanisms have also been developed to reduce the strain demands on the PT element in self-centering braces (e.g., [21,22]).

Wherever a gap opening mechanism is allowed in a rocking or moment frame system, shear resistance across the joint interface must be considered in design. Precast concrete rocking walls [23] and precast concrete moment frames (e.g., [12]), in general, rely on friction at the joint interface to transfer shear force. As such, joint interface details (e.g., grout beds) that can develop high coefficients of friction are critical. Shear transfer across the base joint interface in steel rocking frames usually occurs through a steel curb or bumper at the column base or through an armored trough in the foundation (e.g., [11,16]) in addition to friction at the joint interface. Alternatively, yielding base plates may be used to transfer shear at the column base (e.g., [24]). Steel moment frames primarily use friction between the beam end and column face to transfer shear, but also have some redundancy for shear transfer supplied by the energy dissipation device, whether a friction damper [25], yielding angles [26], or short yielding bars [27]. Although self-centering braces are subjected primarily to axial forces, it is still necessary to restrain lateral motion of the concentric tubes using spacers or guides [15,17].

## 3.2. Energy Dissipation Options

Gap opening and the force in the PT steel provide the bilinear elastic self-centering behavior shown in Figure 4b, but do not provide energy dissipation to the structure. Computational studies on single degree of freedom structures suggest that there is a minimum amount of energy dissipation required for bilinear elastic (self-centering) systems to produce peak drifts similar to those of conventional bilinear elastic-plastic systems [28,29,30]. To produce sufficient energy dissipation, an energy dissipating element is incorporated into self-centering systems. Since self-centering seismic systems are implicitly designed to meet higher performance objectives to minimize repair costs and business downtime, energy dissipation elements should be designed to allow replacement after large earthquakes. Energy dissipating elements can be broadly categorized as hysteretic damping elements, viscous damping elements, and frictional damping elements.

A wide range of hysteretic energy dissipation devices have been used in self-centering systems. Axial acting hysteretic energy dissipating elements include: buckling restrained braces [17,31,32], short yielding elements similar to buckling restrained braces [27,33], unbonded mild steel reinforcement (e.g., [10,12]), reinforcing steel bars with reduced sections [9], and yielding anchor bolts [34]. Hysteretic energy dissipating elements that yield in a combination of shear and flexure include butterfly fuse plates [35], yielding end plates [24,36], yielding angles [26,37], web hourglass pins [38], and tapered steel cantilever plates [39].

Several types of frictional damping elements have been used in self-centering systems. These systems usually consist of two or more plates that slide relative to each other, although configurations are possible such as the Ringfeder spring which dissipates energy through friction and contributes restoring force [40]. Typically, the friction plates are clamped together with bolts or with a combination of bolts and spring washers to provide a normal force [25]. A special wearing surface is often placed between the plates to achieve a more consistent coefficient of friction. Cartridge brass [25,41] and automotive brake pads [15] have been used as wearing surfaces. It can be challenging to get consistent performance in friction damping elements. Some of the challenges include degradation of friction surfaces over time, changes in the friction coefficient during cyclic loading due to wear of the friction surfaces, and maintaining a consistent normal force. To achieve consistent performance of friction damping elements, these challenges must be addressed. In addition to hysteretic and friction damping elements, viscous damping elements have been proposed for use at the column bases of rocking frames [42].

As an alternative to clamped-plate frictional elements, a mode of frictional energy dissipation with unique dynamic properties has been developed wherein friction is encouraged at the interface connection between the lateral force resisting system and the gravity system in rocking braced frames [11]. An additional non-uplifting gravity column is placed adjacent to the uplifting rocking braced frame column. Lateral inertial forces are transferred from collector elements attached to the adjacent gravity column through lateral load bearings into the rocking braced frame. As the braced frame column uplifts, it moves relative to the non-uplifting adjacent gravity column. Energy is dissipated in the lateral load bearings through frictional forces and relative sliding. The frictional force is proportional to the applied lateral inertial force. When the earthquake ceases and lateral inertial forces are removed, the friction forces are virtually eliminated, resulting in improved self-centering ability.

## 3.3. Proportioning the Restoring Force Mechanism and Energy Dissipation Elements

The capacity of the self-centering seismic system to resist lateral loads is characterized by the sum of the resistances due to the restoring forces (PT force, gravity loads) and the force in the energy dissipation element. The sum of these two capacities must be greater than the effect of the design lateral forces calculated based on the procedures for structural system design lateral force calculations in the governing building code.

The height of the hysteresis loop (as defined by $\beta$ or $\beta_E$ in Figure 5) for a self-centering system is a function of the strength and location of the energy dissipation elements. Figure 5 assumes an energy dissipation element with a static force capacity (e.g., friction plates [25]). Incorporating viscous dampers in addition to a static energy dissipation element can reduce peak displacements and result in more consistent performance [43], but cannot be readily quantified in standard force-deformation plots like the one shown schematically in Figure 5 due to the velocity dependence of the damper forces.

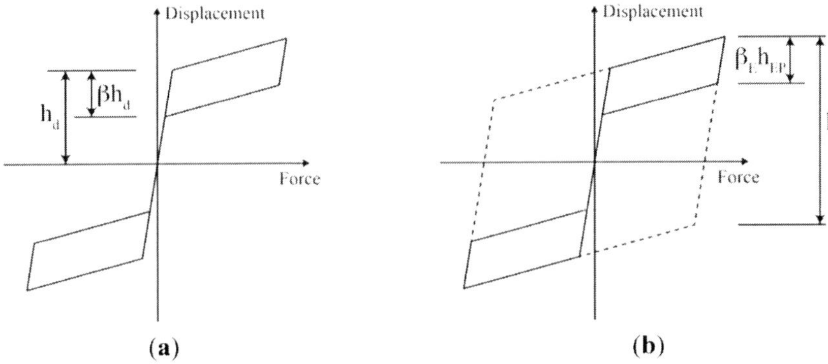


**Figure 5.** Self-centering criteria for SC systems: (**a**) $\beta \leq 1.0$ (adapted from [28]. Copyright 2002 John Wiley & Sons, Ltd.); (**b**) $\beta_E \leq 0.5$ (adapted from [29]. Copyright 2005 ACI).

The flag-shaped hysteresis demonstrates self-centering ability since the displacement returns to negligible values when the lateral forces are removed; however, the system should be proportioned such that this self-centering can occur. One method for defining self-centering capability is based on the relative height of the hysteresis loop, quantified as $\beta$ in Figure 5a [28]. For the system to fully self-center, $\beta \leq 1.0$. Alternatively, another parameter that has been used to quantify self-centering capability in the literature is defined based on the relative area contained (i.e., energy dissipated) in the self-centering hysteresis loops with respect to the area of a comparable bilinear elastoplastic hysteresis loop. If the hysteretic behavior is assumed to be a perfect flag shape, the ratio of energies is equal to the ratio of heights, $\beta_E$, as shown in Figure 5b [29]. For the system to fully self-center, $\beta_E \leq 0.5$. $\beta_E$ is a comparison of the amount of energy dissipated by a self-centering system relative to a conventional bilinear elastoplastic system with identical initial and secondary stiffness.

Regardless of the metric, self-centering capacity is adjusted by tuning the relative magnitudes of the initial PT force (which sets the capacity at decompression, when elastic behavior softens due to gap opening) and the

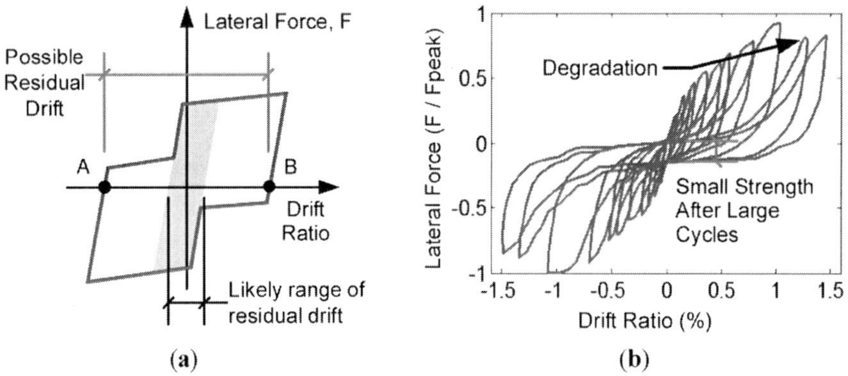


**Figure 6.** Probabilistic self-centering and the resistance of the rest of the building to self-centering. (**a**) Probabilistic self-centering concept; (**b**) Lateral resistance of typical partition wall (adapted from [49]. Copyright 1999 ASCE).

capacity of the energy dissipation elements (which controls the height of the hysteresis loop). Procedures to control the magnitude of residual displacements when the lateral forces are removed can be found in the literature for specific self-centering systems (e.g., [11,16,44,45]).

## 3.4. Probabilistic Self-Centering and the Resistance of the Rest of the Building

It has been observed that systems that do not fully prevent residual drift when the lateral force is removed (not full self-centering) can still limit residual drift if the system has a restoring force component consisting of either a nonlinear elastic restoring force (e.g., gap openings with PT steel) or an elastic restoring force that adds kinematic hardening to the system (as shown in Figure 3a). This type of partial self-centering has been referred to as dynamic stability [46] or probabilistic self-centering [30] and occurs as a result of an increased probability of yielding of the system toward plumb rather than away. Figure 6a shows a schematic diagram of the hysteretic response and potentially large residual drifts for a system without full self-centering. However, if the system is at point B in Figure 6a, and there is equal probability of further positive or negative inertial force from ground shaking, then the hysteretic behavior of the system implies that there is a higher probability of reaching the smaller magnitude yield force in the direction of zero displacement than reaching the larger magnitude yield force away from the direction of zero displacement. Many response history analyses [30,47,48] and several shake table tests [32] have demonstrated that the majority of displacement cycles are concentrated in a narrow band as shown in Figure 6a. Systems in which the restoring force was half of the yield force associated with the energy dissipation element were found to exhibit residual drifts less than the out-of-plumb tolerance for new steel buildings for 98% of the response history analyses subjected to severe ground motion (2% probability of exceedance in 50 years) [30,47].

Lateral load resistance provided by gravity framing may also have an effect on the self-centering capacity of structural systems, though the lateral load resistance of gravity framing and nonstructural components in a structure is often ignored in seismic design because the resistance can be hard to predict and neglecting it is typically assumed to be conservative. However, in the context of self-centering, components of the rest of the building can undergo inelastic actions that produce forces in opposition to the restoring force as the self-centering system attempts to bring the rest of the building back to a plumb position. A non-exclusive list of building elements that contribute to building resistance, but are typically neglected, includes: interior partitions, simple shear beam-to-column connections, stair stringers, horizontal floor diaphragms, floor coverings, ceilings, mechanical systems, and exterior cladding (e.g., precast concrete panels, stucco, brick, glass curtain walls). Figure 6b represents the load-deformation response of a typical interior partition wall made of metal studs with gypsum wall board covering both sides [49]. The hysteretic behavior of gravity framing connections and nonstructural components like partition walls can experience significant degradation, pinched hysteretic behavior, and have small residual strength after large displacement cycles. Therefore, while their consideration in response history analyses can significantly reduce peak drifts, non-structural components generally do not hinder the ability of the self-centering system to control residual drift [30].

## 3.5. Limit States and the Desired Limit State Progression

Most self-centering systems have at least four fundamental limit states [50,51,52] and while actual limit states vary depending on system configuration, it is instructive to examine these limit states in a general context. These limit states are: (1) decompression at the location where gap opening occurs coupled with yielding of hysteretic energy dissipation elements or slip on friction damper interfaces; (2) yielding of the PT elements; (3) limited damage to structural elements; and (4) severe damage to structural elements. Depending on the system, there may be other limit states to consider (e.g., shear slip in panels of precast concrete rocking walls). Concrete rocking wall systems further divide the decompression limit state into decompression of the extreme outside edge of the wall and the effective linear limit of the wall [52,53]. Limited damage of self-centering concrete rocking walls includes cracking and spalling of cover concrete at the base of the wall [52], while severe damage can include crushing of confined concrete in wall boundary elements [52,53,54]. Limited damage of self-centering rocking braced frames includes yielding of frame members, while severe damage involves frame member failure (i.e., loss of load-carrying capacity) [50]. For rocking concrete walls and rocking steel braced frames, yielding of the PT steel may be classified as limited or severe damage depending on if PT bars (limited) or PT strands (severe) are used; PT strand systems may have difficulty achieving significant inelastic strain capacity before fracture [19], while some PT bar systems can achieve significant inelastic strain capacity before fracture [18,55] and composite PT materials do not exhibit yielding prior to fracture [15]. Limited damage in self-centering steel moment resisting frames includes panel zone yield, beam bearing yield at the column face, and horizontal shear yield in the beam web [51,56], while severe damage can include beam local buckling or PT strand yielding [51,56].

The purpose of self-centering seismic lateral force resisting systems is to reduce or eliminate structural damage and residual drift under design level earthquakes so as to enable the possibility of immediately reoccupying the structure after the earthquake. In order to achieve this goal with self-centering systems, a specific limit state progression is targeted in the design of these systems [50,54,56]. Figure 7 shows a schematic of the desired limit state progression for self-centering systems using PT bars and PT strands. Decompression at the gap opening location is the first limit state to occur. At this point there is a reduction in stiffness of the lateral force resisting system, but the structure still behaves elastically. Here, yielding of hysteretic energy dissipation elements or sliding in friction energy dissipation elements is assumed to occur at decompression. The second limit state in the desired limit state progression depends on the type of PT steel used and the type of self-centering system. If PT bars are used with either rocking concrete walls or rocking steel braced frames, then the second limit state is PT steel yielding. If PT strands are used or the system is a moment resisting frame, then the second limit state is limited damage to the system. The third limit state in the progression for rocking concrete walls or rocking steel braced frames is limited damage to the structure. The final limit state for all self-centering systems is severe damage of structural members.

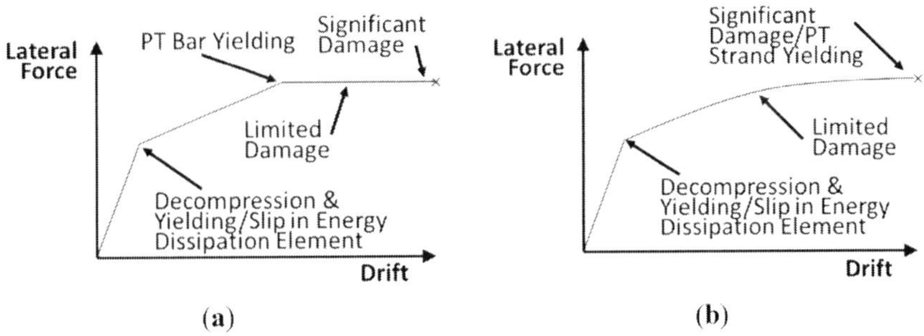


**Figure 7.** Limit state progression for self-centering systems. (**a**) Self-centering system using post-tensioning (PT) bars; (**b**) Self-centering system using PT strands.

## 3.6. Ductility Demands

The hysteretic behavior of self-centering structural systems is characterized by flag-shaped hysteresis loops and is fundamentally different from conventional structural systems. Therefore, traditional relationships between the strength, period, and ductility demand for a structure, such as the so-called "equal displacement" rule, may not be valid for self-centering systems [28,29,57]. Studies of ductility demands on self-centering single degree-of-freedom systems show that for systems with low energy dissipation, the ductility demand can be significantly higher for self-centering systems than for bilinear elastoplastic hardening systems with identical initial and secondary stiffness [28,29,57]. On the other hand, studies have shown that if the post-yield stiffness and energy dissipation ratio (e.g., $\beta$ = 0.5–1.0) are sufficient, that peak drifts can be similar to elastic perfectly plastic systems [28,29,58].

The increased ductility demand in self-centering systems can occur across nearly all periods and is more noticeable when the system strength is low. Predictive relationships for ductility demand in self-centering systems have been developed that account for the influence of the period of the structure, the strength of the structure (using a response modification factor, R), the amount of energy dissipated ($\beta_E$), the post-decompression stiffness relative to the initial elastic stiffness ($\alpha_K$), and the site soil type [57]. These predictive relationships were developed from studies of SC single degree-of-freedom structures subjected to various ground motion sets. These ground motion sets consisted of large magnitude (6.7–7.6) ground motions with a distance of less than 30 km and were organized based on the soil type at the site of the ground motion recording station. The ground motions were scaled to reduce the dispersion about the mean spectrum of the ground motion set over a period range. Figure 8 shows the ductility demand ($\mu_{DBE}$) from these predictive relationships for a site with Type D soil.

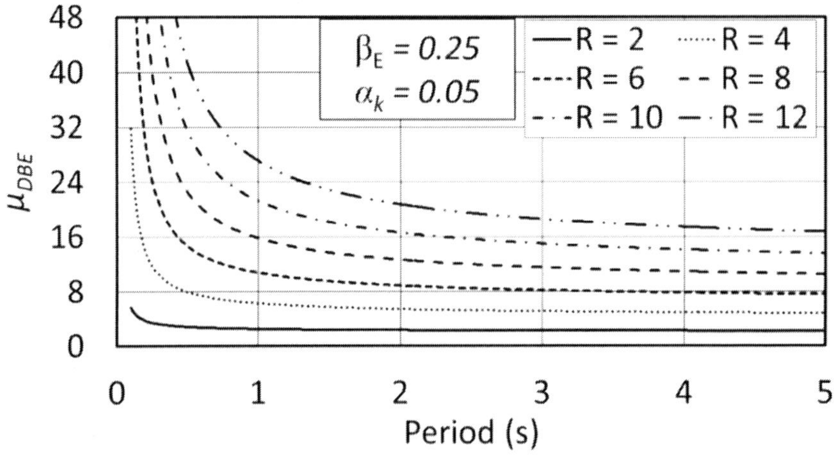


**Figure 8.** Ductility demands from predictive relationships [57] for a self-centering system on Type D soil.

# 4. SELF-CENTERING SEISMIC LATERAL RESISTING SYSTEMS

A wide variety of self-centering seismic lateral force resisting systems have been developed over the past two decades, beginning with the precast concrete moment frames studied in the PRESSS research program [59]. Self-centering was then adapted for use in concrete shear walls (e.g., [52]), steel moment frames (e.g., [13]), and steel concentrically braced frames (e.g., [11]). This section provides a brief summary of recent research into specific self-centering seismic lateral resisting systems.

## 4.1. Rocking Systems

Gap opening, typically at the base of the wall or frame (see Figure 4a), provides rocking systems' nonlinear softening behavior while still allowing the structure to remain elastic. The rocking response significantly decreases the lateral stiffness of the system relative to the initial elastic stiffness, and coupled with yielding of PT steel, effectively limits the peak base overturning moment that can develop. Main structural elements can then be designed to remain nominally elastic (undamaged) for this peak base overturning moment. Energy dissipation is achieved through the implementation of replaceable fuse elements (e.g., [60]), viscous dampers (e.g., [42,43]), or through friction (e.g., [11]). The following subsections further categorize and describe rocking systems as post-tensioned precast concrete shear walls, rocking steel frames, and vertically post-tensioned rocking steel frames. Although there is substantial literature related to free rocking systems (e.g., a stone monument rocking on its foundation) and their unique issues such as tipping over, this summary focuses on seismic systems that limit uplift to be relatively small compared to the width of the rocking element.

### 4.1.1. Post-Tensioned Concrete Shear Walls

Conventional precast concrete shear walls have high stiffness, but limited drift capacity before significant damage occurs as they are typically designed to emulate the behavior of cast-in-place concrete shear walls. Several researchers have adapted self-centering techniques, such as gap opening, to allow these systems to exhibit better seismic performance than emulative walls. The system has been implemented in practice in different forms (examples are given in [61]).

A system of unbonded post-tensioned precast concrete walls that allow gap opening at horizontal joints located at each floor level has been developed (e.g., [52]). Vertical unbonded PT bars or tendons provide a positive stiffness after gap opening, as well as a restoring force mechanism, as the PT force acts to close the gap. The lateral load behavior of this system was demonstrated through large-scale experiments [54,62,63,64]. Experimental results show that the unbonded post-tensioned precast concrete walls self-center and demonstrate better performance than conventional precast concrete shear walls (e.g., [63]).

A number of energy dissipation sources have been studied for unbonded post-tensioned precast concrete walls, including: mild reinforcing steel across the horizontal joints [62], U-shaped flexural plate connectors between two adjacent shear walls [65], dog-bone shaped energy dissipators [66], and small external energy dissipation devices that act similar to buckling restrained braces [67]. Friction damping between adjacent walls can also be used to dissipate energy [41].

A coupled rocking wall system has been developed in which gap opening may form at the end of the coupling beams as well as the base of the wall [68]. The degree of coupling of the post-tensioned precast concrete shear walls can be varied using the post-tensioned coupling beams. Experimental results indicate that the system can be designed to have high strength, ductility, and energy dissipation capacity [14].

### 4.1.2. Rocking Steel Braced Frames (Without Post-Tensioning)

Rocking braced frame systems have been studied for decades (e.g., [69]). Rocking frames have been implemented in New Zealand in bridges as early as 1981 and in a building in 2007 [61]. Recent research in damage-free structural systems has increased the visibility of such systems outside of New Zealand. These systems allow gap opening, but do not have PT across the joint and rely on gravity loads for self-centering.

Investigation of rocking steel concentrically braced frames goes back to the 1970s [69,70]. Through the years, shake table tests and computational simulations of rocking steel frames [24,31,36,42,70] have shown that allowing base uplift can reduce the peak base shear, peak accelerations, and peak member forces compared to a similar structure with elastic behavior and fixed bases. A wide range of energy dissipation elements have been used including yielding column base plates that allow uplift [24], similar yielding splice plates along the height of braced frame columns [36], torsional yielding bars [70], buckling restrained braces [31], and viscous dampers [42].

Since these systems rely on gravity loads to provide the restoring force that closes the base uplifting gap, the balance between the yield capacity of the energy dissipating element and the expected actual gravity loads is important. This design approach is effective in structures with large gravity loads such as bridge piers (e.g., [31]), but may not be appropriate for structures where little gravity load is applied directly to the rocking frame (e.g., short structures).

### 4.1.3. Post-tensioned Rocking Steel Braced Frames

Similar to systems described in the previous section, post-tensioned rocking steel braced frames have high initial stiffness (similar to that of conventional concentrically braced frames) but can withstand intense earthquakes with little or no structural damage. By introducing post-decompression stiffness through PT steel, these systems can further limit the peak drifts as compared to rocking systems without vertical PT steel.

Several configurations of post-tensioned rocking steel braced frames have been studied (e.g., [50,71,72,73]). In all of these configurations, gap opening is allowed at the base of the braced frame and vertical oriented PT steel is used to prestress the braced frame to the foundation. In some configurations (e.g., [71]), two coupled adjacent rocking concentrically braced frames were used while others consist of single rocking frames (e.g., [32,74]). In some configurations [11,74], the gravity system and lateral force resisting system were separated by introducing additional gravity columns adjacent to the rocking braced frame columns. This separation of the gravity and lateral system eliminates damage to the floor diaphragm when the braced frame rocks.

Multiple sources of energy dissipation have been studied for post-tensioned rocking steel braced frames, including yielding steel elements [60,71,72,73], friction bearings [11,74], and viscous damping elements [73].

Experimental and analytical results show that post-tensioned rocking steel braced frames can significantly reduce the probability of member yielding with respect to conventional systems and that the system self-centers after seismic excitation even after significant PT bar yielding [11]. Large-scale static [71] and shake table [72] testing indicate that post-tensioned rocking steel braced frames effectively concentrate structural damage in the replaceable fuse elements and self-center after seismic excitation. An analytical study indicated that post-tensioned rocking steel braced frames can significantly improve performance of structural and nonstructural components as shown by reduced drifts and accelerations with respect to conventional ductile systems [73]. The system has been implemented in buildings in the U.S. and New Zealand (e.g., [40,75]).

Research into post-tensioned rocking steel braced frame response indicated that higher mode effects significantly increase member force demands compared to those expected based on a pushover analysis [74]. These higher mode effects should be accounted for in design so that the braced frame members remain elastic under seismic loading. Several methods have been proposed to account for higher mode effects including modal load factors [11] and a force amplification design procedure [16].

As an alternative to designing members to remain elastic under higher mode effects, mitigation techniques have been studied (e.g., [76]) to reduce the required strength for members to remain elastic. Two higher mode response mitigation techniques that have been studied [77] are: (1) introduction of multiple rocking sections, in which gap opening is permitted at multiple locations throughout the height of the frames; and (2) a force-limiting brace member in the first story, which limits the base shear capacity of the system. Shake table testing [78] demonstrated the effectiveness of the proposed higher mode mitigation techniques.

## 4.2. Self-Centering Moment Resisting Frames

Conventional moment resisting frames are designed with so-called strong-column weak-beam criteria to dissipate seismic energy by forming plastic hinges at the beam ends and at the base of the first story columns to form a global beam-sway mechanism. Although the structure is expected to behave in a ductile manner and dissipate a significant quantity of seismic energy, there may be significant inelastic residual deformation. As an alternative to conventional moment resisting frames with rigid beam-to-column connections, self-centering moment resisting frames that allow a gap to form at the beam-column interface have been developed using both precast concrete and structural steel. This section summarizes published research on concrete and steel self-centering moment resisting frames.

### 4.2.1. Concrete Self-Centering Moment Resisting Frames

Self-centering concrete moment frames consist of concrete beams and columns horizontally post-tensioned together so that a gap can open at the beam-column interface when subjected to a specific applied moment, as shown in Figure 4a. Energy dissipation is supplied at the beam-to-column joint through a variety of mechanisms such as unbonded mild reinforcing steel [64], friction damping elements [79] and other devices.

The early development of concrete self-centering moment resisting frames came out of the PRESSS project during the 1990s [64,80,81]. The purpose of the PRESSS project was to develop recommendations for the seismic design of precast concrete seismic structural systems and develop new materials, concepts and technologies for precast concrete seismic structural systems [80]. Self-centering concrete beam-to-column connections accommodated large rotations without a loss in load carrying capacity [81]. Tests of a five story precast concrete moment resisting frame incorporating self-centering connections and designed using displacement-based design principles showed that drifts were not excessive (e.g., residual drift after DBE was 0.06%), and that the self-centering connections suffered little or no damage [64]. The damage was limited to spalling of cover concrete in the beams immediately adjacent to the columns and some crushing of the fiber grout pads at the beam-column interfaces [64]. Self-centering concrete moment frames have performed well in laboratory seismic testing, have been shown to limit damage to the structure, and have been implemented in practice (e.g., [61,82]).

### 4.2.2. Steel Self-Centering Moment Resisting Frames

Similar to concrete self-centering moment resisting frames, steel self-centering moment frames consist of steel beams and columns post-tensioned together so that a gap can open at the beam-column interface when subjected to a specific moment, as shown in Figure 4a. Energy dissipation devices for the steel SC moment resisting frames have included yielding seat angles [13,26,83], short axial yielding devices similar in behavior to buckling-restrained braces [27], yielding web hourglass pins [38], friction devices at the bottom flange of the beam [84] or at both flanges [85], friction devices at the web of the beam [41,56,86], and yielding thin steel plate infill walls [87,88,89].

Numerous tests of steel self-centering moment resisting frame beam-to-column connections (e.g., [26,27,82,84,85,90]) show reliable self-centering behavior of the connection. However, it was noted for one test of a beam-to-column connection with PT strands that individual wires in some of the PT strands fractured in the anchorage grips before reaching the ultimate strength of the strand [26]. Fracture of wires in strands was also noted for rocking braced frame tests using PT strands at strains of about 1%; although the fracture of one wire did not propagate to other wires in the strand [71]. To increase the ductility of self-centering moment resisting frame beam-to-column connections at ultimate states, design equations and detailing requirements were proposed in accordance with finite element analyses and experimental results for possible plastic hinge locations in the beams [91].

Large-scale frame tests of steel self-centering moment resisting frames [41,86,92] showed considerably better seismic performance compared to conventional moment resisting frames. Steel self-centering moment resisting frames are generally expected to remain undamaged during the DBE, allowing immediate occupancy of the structure and to prevent collapse under the maximum considered earthquake (MCE) [86]. In this case, an MCE event is defined as an intense earthquake with a return period of about 2500 years [1].

During DBE level tests, the peak drift ranged from 1.8% to 4.9%, but the frame self-centered with residual drift ratios less than 0.1% [86]. There was a small amount of yielding at a column base, but this is expected behavior when the column bases are fixed. Self-centering column bases similar to the self-centering beam-to-column connections have been developed to eliminate this source of damage [93,48]. Bottom flange beam yielding was reported during a maximum considered earthquake (MCE) test, but this is the expected response under this very rare earthquake according to the

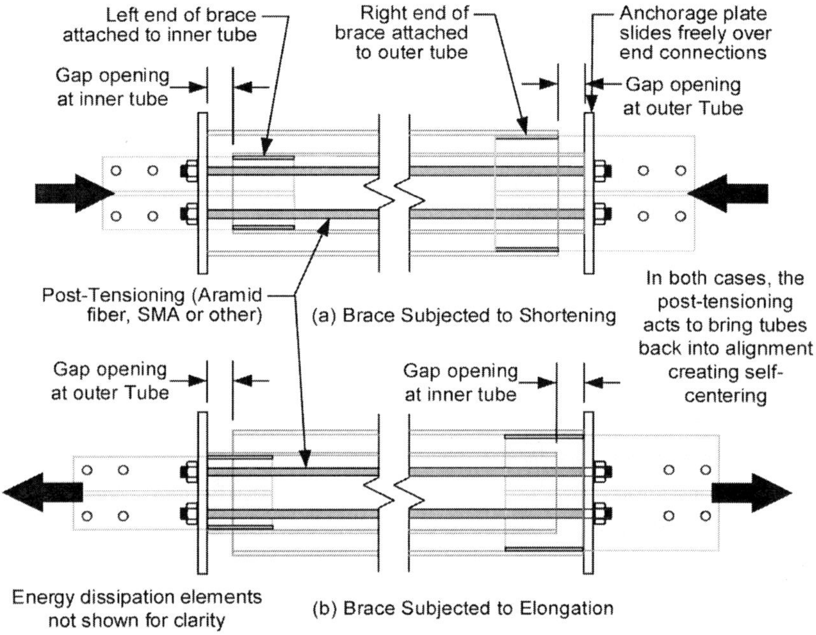


**Figure 9.** Schematic representation of concentric telescoping tubes used in self-centering braces (adapted from [47]. Copyright John 2014 Wiley & Sons, Ltd.). (**a**) Brace subjected to shortening; (**b**) Brace subjected to elongation.

design objectives [41]. There was essentially no residual drift under the MCE.

## 4.3. Self-Centering Braces for Steel Braced Frames

Self-centering braces return to their original length after undergoing axial elongation or shortening. One self-centering brace that has been developed uses a restoring force mechanism consisting of two concentric tubes precompressed by aramid fiber PT strands and an energy dissipation mechanism using friction pads [15]. Figure 9 schematically shows the configuration of telescoping concentric tubes and PT elements for the restoring force mechanism [47]. Several researchers have developed self-centering braces that use shape memory alloy (SMA) wires [20,94,95]. Some use telescoping mechanisms (e.g., [17,94,95]) similar in concept to the one shown in Figure 9, while others make use of the superelastic SMA material behavior directly to create self-centering [20]. There have been a number of generalized research programs into the use of SMA for seismic design of structures starting in the early 1990's [96] and including large research projects in Europe (e.g., MANSIDE project [97]), and in the U.S. (e.g., [98]). SMA materials are "superelastic," which means very large strains (up to 10%) can be recovered elastically upon the removal of load.

Computational studies of prototype buildings show that self-centering braces that do not have full self-centering hysteretic behavior (i.e., the load-deformation curve does not pass near zero displacement when the lateral load is removed), can still reliably limit residual drifts to negligible magnitude [47]. Computational studies [47] and shake table testing [99] show that prototype buildings constructed with self-centering braces can dissipate significant energy, exhibit peak drifts similar to conventional systems, and nearly eliminate residual drifts.

## 4.4. Self-Centering Timber Systems

To improve seismic performance and reduce structural damage of multi-story timber buildings, self-centering timber systems have been proposed [100,101,102]. Several configurations of self-centering timber systems have been investigated, including post-tensioned beam-to-column subassemblies, wall, and frame systems [100,101]. In addition to using post-tensioned bars and gap-opening mechanisms to provide self-centering behavior, internal and external energy dissipaters have also been provided [100,101,102]. Experimental tests on self-centering timber systems have shown these systems to be highly ductile, while still self-centering [100,101].

Coupled rocking timber walls with U-shaped energy dissipaters, which were also used in post-tensioned concrete rocking shear walls [64,65], were also examined [101]. Based on experimental results, coupled rocking timber walls with U-shaped energy dissipaters have better seismic performance and increased drift capacity compared to single rocking timber walls [101].

Although past studies on self-centering timber systems have been carried out in New Zealand [100,101], there is an increasing interest and ongoing research projects in other countries (e.g., the United States) as well (e.g., [103]). Promising experimental test results have encouraged the construction of structures with self-centering timber systems in New Zealand [102].

# 5. SOME CURRENT RESEARCH CHALLENGES IN SELF-CENTERING SEISMIC LATERAL FORCE RESISTING SYSTEMS

## 5.1. Isolating Deformations Related to Gap Opening

A key component of most self-centering systems is the gap opening mechanism. The gap opening mechanism allows for nonlinear elastic behavior of the structure. However, allowing gaps to form between beam-to-column connections in moment resisting frames and at the base of rocking systems presents connection detailing challenges. These challenges are outlined in this section.

### 5.1.1. Floor Diaphragm Connections in Self-Centering Moment Resisting Frames

In conventional moment resisting frames, the concrete floor diaphragm transfers seismic inertial forces from the floor diaphragm to each beam in the moment resisting frame, as shown schematically in Figure 10a. Transferring load to every beam in the moment resisting frame is problematic for self-centering moment resisting frames. When lateral load is applied to the self-centering moment resisting frame, gaps open at the interface between the end of the beam and the column face. Since there is very little axial shortening of the beam, this gap opening causes the length of the frame to grow [51], as shown in Figure 11; however, the floor diaphragm restrains the expansion of the frame. Limiting frame expansion will affect the behavior of the frame and may damage the floor diaphragm. One proposed detail [51] to address this problem is shown schematically in Figure 10b, in which the floor diaphragm is attached to only one beam in the moment resisting frame per floor. The floor diaphragm and beams in all other bays of the moment resisting frame are non-composite, permitting the beam to move relative to the floor diaphragm. In another study, the frame expansion problem is eliminated by providing the rocking joints at the opposite ends of the beam to have counteracting behavior. A type of post-tensioned rocking connection detail [75] is proposed in which the connection is permitted to rock only about beam bottom flange while the contact of the beam top flange with the column is preserved at all time. Other diaphragm and gravity beam connection details have also been proposed [85,92,104] to address the frame expansion issue.

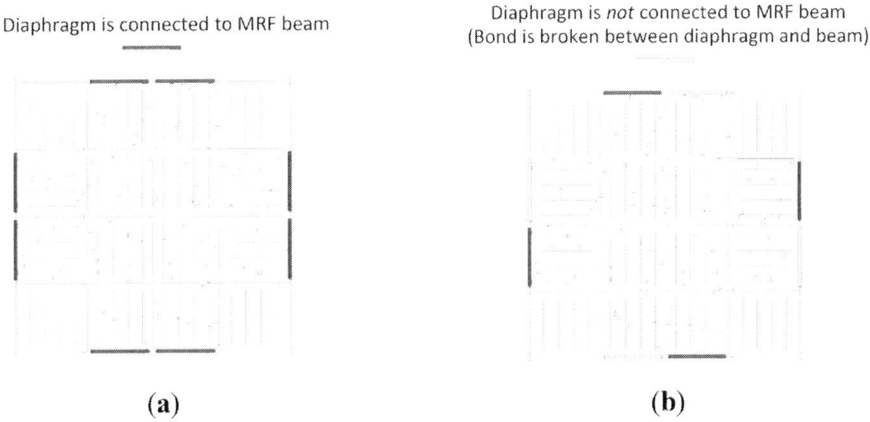


**Figure 10.** Schematic floor diaphragm connection. (**a**) Floor diaphragm connection to conventional moment resisting frame (MRF); (**b**) Floor diaphragm connection to self-centering MRF (SC-MRF).

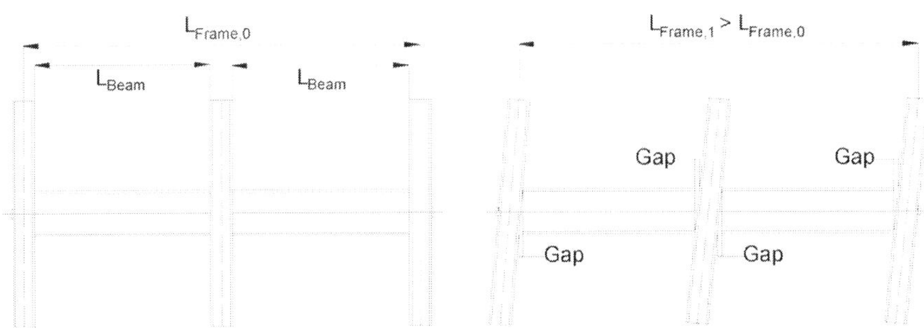


**Figure 11.** Frame expansion in self-centering moment resisting frames.

### 5.1.2. Floor Diaphragm Connections in Rocking Self-Centering Systems

Column uplift in rocking self-centering systems can cause local floor damage and may create challenges in connecting the floor system to the rocking frames. There are two general approaches for connecting the diaphragm to a rocking frame or rocking wall: (1) isolating the rocking element from the diaphragm to limit damage; and (2) connecting the diaphragm to the rocking element and accepting local damage.

Figure 12a shows one possible configuration for isolating a rocking steel braced frame from the gravity framing and floor diaphragm. The collector forces are transferred from an adjacent gravity framing column into the rocking frame column through a bearing connection. The bearing can be a low friction connection or a controlled friction interface used to dissipate seismic energy [11]. As an alternative, shear plates capable of transferring shear while allowing uplift could be connected between the rocking frame beams and adjacent gravity framing beams.

Figure 12b shows the approach wherein gravity framing and the floor diaphragms are connected directly to the rocking frame beam (e.g., [64]). Localized damage would be expected around the uplifting columns. If gravity framing beams are connected to uplifting columns, the rotational demands can be magnified as compared to those of typical gravity connections, particularly if the adjacent gravity beam span is short. A modified collector detail that can transfer axial force while allowing large rotations may be desirable. Shear plates with strategically slotted holes (some vertical, some horizontal) have been used in practice [105]. Additionally, true pins or top flange plates with horizontally slotted web holes are also possible. If rocking frames or walls are placed on the perimeter of a building, it may also be necessary to detail the building façade to accommodate the vertical motion of the rocking system.

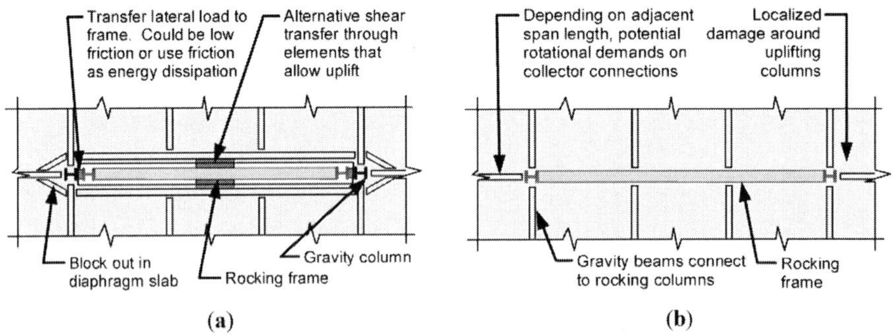


**Figure 12.** Approaches for attaching rocking frames to surrounding gravity framing. (**a**) Isolating the rocking frame from the gravity framing; (**b**) Direct attachment of the rocking frame to the gravity framing.

## 5.2. Higher Mode Effects

Researchers have recognized for some time that building code provisions do not adequately account for higher mode effects [64] and the effects are particularly pronounced for rocking self-centering systems [10,76]. As described in a previous section, the inter-story shear forces associated with the first mode are reduced because rocking is allowed, but higher modes can produce inter-story shear forces significantly larger than those predicted by the building code (e.g., [4,11]). Although design approaches have been proposed such as applying modal load factors in response spectrum analysis [4,11], more research is needed to accurately quantify and predict higher mode response in self-centering systems.

An alternative to designing for the unreduced higher mode response in rocking structures involves adding multiple softening mechanisms throughout the building height to limit the higher mode response (e.g., [76]). Placing the uplifting mechanism only at the base of the wall or frame can limit the base overturning moment, but does not limit the base shear, story shears, or story moments. Analytical and experimental studies [77,78] have shown that these other response parameters, which may be controlled by higher modes rather than the first mode, can be reduced by placing additional softening mechanisms into the structure. These additional restoring force mechanisms might be additional rocking mechanisms at other stories in the structure, shear type mechanisms using self-centering braces, or both. While this approach appears promising, further research is needed on this topic.

## 5.3. Collapse Safety Assessment of Self-Centering Systems

Self-centering seismic lateral force resisting systems are used to reduce or eliminate damage under design level earthquakes. However, under very rare and intense earthquakes, even self-centering systems may be damaged such that collapse resistance must be assessed. The Federal Emergency Management Agency (FEMA) has adopted a methodology for assessing the safety of seismic lateral force resisting systems and evaluating new systems for inclusion in future building code provisions (FEMA P695) [106]. Under this methodology, archetype structures incorporating the seismic lateral force resisting systems are modeled and subjected to a suite of ground motion records using nonlinear response-history analysis. The ground motion records are applied to the nonlinear numerical model and incrementally increased until at least 50% of the ground motion records cause the structure to collapse. The probability of collapse is then established for a hazard level with 2% probability of exceedance in 50 years. Collapse safety assessment of self-centering systems is important if these systems are to be included in building code previsions. Published research on collapse safety assessment of self-centering structures is currently very limited (e.g., [107,108]) and is an ongoing topic of research.

### 5.4. Efficiency and Strategic Use of Self-Centering Systems Considering Life-Cycle Cost

The development of self-centering systems is driven by socio-economic costs; high post-earthquake costs show the need for a damage-free alternative to conventional seismic lateral force resisting systems. Self-centering systems were developed to increase the drift capacity prior to structural damage; however, in some cases, there is additional complexity in design, fabrication, and erection of the self-centering system, which may increase construction costs. Though these increased construction costs have not been directly quantified, there is an expectation that the savings due to the damage-free performance of self-centering systems in strong earthquakes will offset the increased initial cost over the life of the structure.

For a building to be damage-free, more than just the seismic lateral force resisting system must remain undamaged. The gravity force resisting system, floor diaphragms, and nonstructural elements must also be serviceable after an earthquake to achieve damage-free building performance [109]. Therefore, life-cycle cost analysis must consider not only the structural repair costs, but all elements of the structure. Quantifying the life-cycle costs of self-centering seismic lateral force resisting systems compared to those of conventional seismic lateral force resisting systems is the subject of ongoing [110,111,112] and future research.

## 6. CONCLUSIONS

Conventional seismic lateral force resisting systems dissipate seismic energy through distributed damage to primary structural elements and residual drifts, which can result in significant socio-economic losses due to repair costs and business downtime. This review outlined the need for new seismic lateral force resisting systems that remain essentially undamaged and virtually eliminate residual drift after design level earthquakes so the building may be occupied immediately after the earthquake. A new class of seismic lateral force resisting systems, self-centering systems, has been developed to meet this need.

Self-centering systems capitalize on a nonlinear elastic gap opening mechanism and concentrate damage in replaceable energy dissipation elements without damaging the structural elements. Post-tensioning across the gap opening joint provides post-gap opening stiffness and self-centering capability as the post-tensioning force acts to close the gap. Energy dissipation devices (e.g., yielding, viscous, and friction) dissipate the seismic energy and are replaceable. The hysteretic behavior for self-centering systems is fundamentally different from that for conventional seismic lateral force resisting systems; self-centering systems dissipate less energy per displacement cycle, but exhibit no residual drift. Limit states for self-centering systems are reasonably well defined. The strength of the individual components can be proportioned such that the progression of limit states prevents damage to the seismic lateral force resisting system under design level earthquakes. Ductility demands on self-centering systems can be larger than for conventional systems due to reduced hysteretic damping, but self-centering systems can be designed and detailed to have significantly higher ductility capacity than conventional seismic lateral force resisting systems.

There are multiple types of self-centering seismic lateral force resisting systems, but these systems fall into three broad categories: rocking systems, moment resisting frames, and self-centering braces. Rocking systems have been developed and studied for unbonded post-tensioned concrete shear walls and steel braced frames (with and without post-tensioning). Laboratory tests of concrete and steel rocking self-centering systems have shown that these systems have little or no structural damage under design level earthquakes. Self-centering moment resisting frames made from steel and precast concrete have been studied. Steel self-centering moment resisting frames have been noted to be lighter than conventional steel moment resisting frames while providing damage free performance [86], which potentially makes them cost-competitive with the conventional steel moment resisting frames. Self-centering braces have been developed that can undergo large axial deformations and return to their original length without damage.

Ongoing challenges for self-centering systems include detailing to accommodate gap openings, designing rocking systems for higher mode effects, collapse assessment, and life-cycle cost analysis. Detailing to accommodate gap opening in the structure is an issue for both self-centering moment resisting frames and rocking systems. Details have been proposed and studied to accommodate the gap opening behavior. The response of higher modes is significant for rocking systems and must be appropriately accounted for in member design or mitigated to satisfy damage resistant performance goals. Strategies for both cases were described herein. Collapse safety assessment of self-centering systems is also an important issue for providing adequate structural reliability and is an ongoing topic of research. Because self-centering systems are expected to have a higher initial construction cost than conventional systems, but are expected to pose lower costs over the life of the structure, ongoing research is being conducted to quantify life cycle costs so that architects and engineers can make educated choices about using self-centering seismic systems.

Self-centering seismic lateral force resisting systems have been shown to reduce or eliminate structural damage associated with conventional lateral force resisting systems during significant earthquakes. High-performance self-centering structural systems have the potential to reduce socio-economic loss and disruption of cities after significant earthquakes. Self-centering seismic structural systems are an important tool for building the resilient, sustainable city of tomorrow.

## CONFLICTS OF INTEREST

The authors declare no conflicts of interest.

## REFERENCES


1. Building Seismic Safety Council (BSSC). NEHRP Recommended Provisions for Seismic Regulations for New Buildings and Other Structures: FEMA 450; BSSC: Washington, DC, USA, 2003.

2. Eguchi, R.T.; Goltz, J.D.; Taylor, C.E.; Chang, S.E.; Flores, P.J.; Johnson, L.A.; Blais, N.C. Direct economic losses in the Northridge earthquake: A three-year post-event perspective. Earthq. Spectra 1998, 14, 245–264.

3. Eatherton, M.R.; Toellner, B.W.; Watkins, C.E.; Abbas, E. The Effect of Powder Actuated Fasteners on the Seismic Performance of Protected Zones in Steel Moment Frames. In Proceedings of the Structures Congress, Pittsburgh, PA, USA, 2–4 May 2013.

4. Chancellor, N.B. Seismic Design and Performance of Self-Centering Concentrically-Braced Frames. Ph.D. Thesis, Department of Civil and Environmental Engineering, Lehigh University, Bethlehem, PA, USA, 2014.

5. Field Manual: Postearthquake Safety Evaluation of Buildings; Applied Technology Council (ATC): Redwood City, CA, USA, 2005.

6. New Zealand 2013 Treasury Budget Speech. Available online: http://www.treasury.govt.nz/budget/2013/speech/06.htm (accessed on 14 May 2014).

7. Canterbury Earthquake Recovery Authority. Available online: http://cera.govt.nz/news/2014/new-program-to-show-progress-of-public-projects-14-may-2014 (accessed on 14 May 2014).

8. Kiggins, S.; Uang, C.M. Reducing residual drift of buckling-restrained braced frames as a dual system. Eng. Struct. 2006, 28, 1525–1532.

9. Holden, T.; Restrepo, J.; Mander, J.B. Seismic performance of precast reinforced and prestressed concrete walls. J. Struct. Eng. 2003, 129, 286–296.

10. Palermo, A.; Pampanin, S.; Marriott, D. Design, modeling, and experimental response of seismic resistant bridge piers with posttensioned dissipating connections. J. Struct. Eng. 2007, 133, 1648–1661.

11. Roke, D.; Sause, R.; Ricles, J.M.; Chancellor, N.B. Damage-Free Seismic-Resistant Self-Centering Concentrically-Braced Frames; ATLSS Report 10–09; Lehigh University: Bethlehem, PA, USA, 2010.

12. Rahman, M.A.; Sritharan, S. Performance-based seismic evaluation of two five-story precast concrete hybrid frame buildings. J. Struct. Eng. 2007, 133, 1489–1500.

13. Ricles, J.M.; Sause, R.; Garlock, M.M.; Zhao, C. Posttensioned seismic-resistant connections for steel frames. J. Struct. Eng. 2001, 127, 113–121.

14. Kurama, Y.C.; Weldon, B.D.; Shen, Q. Experimental evaluation of posttensioned hybrid coupled wall subassemblages. J. Struct. Eng. 2006, 132, 1017–1029.

15. Christopoulos, C.; Tremblay, R.; Kim, H.J.; Lacerte, M. Self-centering energy dissipative bracing system for the seismic resistance of structures: Development and validation. J. Struct. Eng. 2008, 134, 96–107.

16. Eatherton, M.; Ma, X.; Krawinkler, H.; Deierlein, G.G.; Hajjar, J.F. Quasi-static cyclic behavior of controlled rocking steel frames. J. Struct. Eng. 2014, 2014.

17. Miller, D.J.; Fahnestock, L.A.; Eatherton, M.R. Development and experimental validation of a nickel–titanium shape memory alloy self-centering buckling-restrained brace. Eng. Struct. 2012, 40, 288–298.

18. Gonner, N.; Chancellor, N.B.; Sause, R.; Ricles, J.M.; Roke, D. Design, Experimental Setup, and Testing of Self-Centering Steel Concentrically-Braced Frame Test Structure; ATLSS Report 10-06; Lehigh University: Bethlehem, PA, USA, 2010.

19. Bruce, T. Behavior of Post-Tensioning Strand Systems Subjected to Inelastic Cyclic Loading. Master Thesis, Virginia Tech, Blacksburg, VA, USA, 2014.

20. Zhu, S.; Zhang, Y. Seismic analysis of concentrically braced frame systems with self-centering friction damping braces. J. Struct. Eng. 2008, 134, 121–131.

21. Chou, C.C.; Chen, Y.C. Development and Seismic Performance of Steel Dual-Core Self-Centering Braces. In Proceedings of the 15th World Conference on Earthquake Engineering, Lisbon, Portugal, 24–28 September 2012.

22. Erochko, J.; Christopoulos, C.; Tremblay, R. Design and Testing of an Enhanced-Elongation Telescoping Self-Centering Energy-Dissipative (T-SCED) Brace. In Proceedings of the 15th World Conference on Earthquake Engineering, Lisbon, Portugal, 24–28 September 2012.

23. Innovation Task Group, American Concrete Institute (ACI). Requirements for Design of Special Unbonded Post-Tensioned Precast Shear Wall Satisfying ACI ITG-5.1 (ACI ITG-5.2-09) and Commentary; ACI ITG-5.2-09, Reported by ACI Innovation Task Group 5; ACI: Farmington Hills, MI, USA, 2009.

24. Midorikawa, M.; Azuhata, T.; Ishihara, T.; Wada, A. Shaking table tests on seismic response of steel braced frames with column uplift. Earthq. Eng. Struct. Dyn. 2006, 35, 1767–1785.

25. Wolski, M.; Ricles, J.M.; Sause, R. Seismic Resistant Self-Centering Steel Moment Resisting Frames with Bottom Flange Friction Devices. In Behavior of Steel Structures in Seismic Areas: 5th International Conference on Behavior of Steel Structures in Seismic Areas; Taylor & Francis: London, UK, 2006.

26. Garlock, M.M.; Ricles, J.M.; Sause, R. Experimental studies of full-scale posttensioned steel connections. J. Struct. Eng. 2005, 131, 438–448.

27. Christopoulos, C.; Filiatrault, A.; Uang, C.M.; Folz, B. Posttensioned energy dissipating connections for moment-resisting steel frames. J. Struct. Eng. 2002, 128, 1111–1120.

28. Christopoulos, C.; Filiatrault, A.; Folz, B. Seismic response of self-centring hysteretic SDOF systems. Earthq. Eng. Struct. Dyn. 2002, 31, 1131–1150.

29. Seo, C.Y.; Sause, R. Ductility demands on self-centering systems under earthquake loading. ACI Struct. J. 2005, 102, 275.

30. Eatherton, M.R.; Hajjar, J.F. Residual drifts of self-centering systems including effects of ambient building resistance. Earthq. Spectra 2011, 27, 719–744.

31. Pollino, M.; Bruneau, M. Seismic testing of a bridge steel truss pier designed for controlled rocking. J. Struct. Eng. 2010, 136, 1523–1532.

32. Ma, X.; Krawinkler, H.; Deierlein, G. Seismic Design, Simulation and Shake Table Testing of Self-Centering Braced Frame with Controlled Rocking and Energy Dissipating Fuses; Stanford University: Stanford, CA, USA, 2011.

33. Maurya, A.; Eatherton, M.R. Self-Centering Beams with Resilient Seismic Performance. In Proceedings of the ASCE Structures Congress, Boston, MA, USA, 3–5 April 2014.

34. Takamatsu, T.; Tamai, H.; Yamanishi, T.; Matsuo, A. Self-Centering Performance of Non-Slip-Type Exposed Column-Base. In Behavior of Steel Structures in Seismic Areas: 5th International Conference on Behavior of Steel Structures in Seismic Areas; Taylor & Francis: London, UK, 2006.

35. Ma, X.; Borchers, E.; Peña, A.; Krawinkler, H.; Deierlein, G. Design and Behavior of Steel Shear Plates with Openings as Energy-Dissipating Fuses; Stanford University: Stanford, CA, USA, 2010.

36. Wada, A.; Yamada, S.; Fukuta, O.; Tanigawa, M. Passive Controlled Slender Structures Having Special Devises at Column Connections. In Proceedings of the 7th International Seminar on Seismic

Isolation, Passive Energy Dissipation and Active Control of Vibrations of Structures, Assisi, Italy, 2–5 October 2001.

37. Shen, Q.; Kurama, Y.C. Nonlinear behavior of posttensioned hybrid coupled wall subassemblages. J. Struct. Eng. 2002, 128, 1290–1300.

38. Vasdravellis, G.; Karavasilis, T.L.; Uy, B. Large-scale experimental validation of steel posttensioned connections with web hourglass pins. J. Struct. Eng. 2012, 139, 1033–1042.

39. Toranzo, L.A.; Restrepo, J.I.; Mander, J.B.; Carr, A.J. Shake-table tests of confined-masonry rocking walls with supplementary hysteretic damping. J. Earthq. Eng. 2009, 13, 882–898.

40. Gledhill, S.M.; Sidwell, G.K.; Bell, D.K. The Damage Avoidance Design of tall steel frame buildings—Fairlie Terrace Student Accomodation Project, Victoria University of Wellington. In Proceedings of the 2008 New Zealand Society for Earthquake Engineering, Wellington, New Zealand, 11–13 April 2008.

41. Lin, Y.C.; Sause, R.; Ricles, J. Seismic performance of a large-scale steel self-centering moment-resisting frame: MCE hybrid simulations and quasi-static pushover tests. J. Struct. Eng. 2012, 139, 1227–1236.

42. Tremblay, R.; Poirier, L.P.; Bouaanani, N.; Leclerc, M.; Rene, V.; Fronteddu, L.; Rivest, S. Innovative Viscously Damped Rocking Braced Steel Frames. In Proceedings of the 14th World Conference on Earthquake Engineering, Beijing, China, 12–17 October 2008.

43. Kam, W.Y.; Pampanin, S.; Palermo, A.; Carr, A.J. Self-centering structural systems with combination of hysteretic and viscous energy dissipations. Earthq. Eng. Struct. Dyn. 2010, 39, 1083–1108.

44. El-Sheikh, M.T.; Sause, R.; Pessiki, S.; Lu, L.W. Seismic behavior and design of unbonded post-tensioned precast concrete frames. PCI J. 1999, 44, 54–71.

45. Kurama, Y.C. Simplified seismic design approach for friction-damped unbonded post-tensioned precast concrete walls. ACI Struct. J. 2001, 98, 705–716.

46. MacRae, G.A.; Kawashima, K. Post-earthquake residual displacements of bilinear oscillators. Earthqu. Eng. Struct. Dyn. 1997, 26, 701–716.

47. Eatherton, M.R.; Fahnestock, L.A.; Miller, D. Computational study on the behavior of self-centering buckling restrained braces. Earthq. Eng. Struct. Dyn. 2014.

48. Chi, H.; Liu, J. Seismic behavior of post-tensioned column base for steel self-centering moment resisting frame. J. Constr. Steel Res. 2012, 78, 117–130.

49. Gad, E.F.; Chandler, A.M.; Duffield, C.F.; Stark, G. Lateral behavior of plasterboard-clad residential steel frames. J. Struct. Eng. 1999, 125, 32–39.

50. Roke, D.; Sause, R.; Ricles, J.M.; Seo, C.Y.; Lee, K.S. Self-Centering Seismic-Resistant Steel Concentrically-Braced Frames. In Proceedings of the 8th US National Conference on Earthquake Engineering, San Francisco, CA, USA, 18–22 April 2006.

51. Garlock, M.M.; Sause, R.; Ricles, J.M. Behavior and design of posttensioned steel frame systems. J. Struct. Eng. 2007, 133, 389–399.

52. Kurama, Y.; Pessiki, S.; Sause, R.; Lu, L.W. Seismic behavior and design of unbonded post-tensioned precast concrete walls. PCI J. 1999, 44, 72–89.

53. Perez, F.J.; Pessiki, S.; Sause, R. Seismic design of unbonded post-tensioned precast concrete walls with vertical joint connectors. PCI J. 2004, 49, 58–79.

54. Perez, F.J.; Sause, R.; Pessiki, S. Analytical and experimental lateral load behavior of unbonded posttensioned precast concrete walls. J. Struct. Eng. 2007, 133, 1531–1540.

55. Perez, F.J.; Pessiki, S.; Sause, R.; Lu, L.W. Lateral load tests of unbonded post-tensioned precast concrete walls. ACI Spec. Publ. 2003, 211, 161–182.

56. Lin, Y.C.; Ricles, J.M.; Sause, R.; Seo, C.Y. Earthquake Simulations on a Self-Centering Steel Moment Resisting Frame with Web Friction Devices. In Proceedings of the 14th World Conference on Earthquake Engineering, Beijing, China, 12–17 October 2008.

57. Seo, C.Y. Influence of Ground Motion Characteristics and Structural Parameters on Seismic Responses of SDOF Systems. Ph.D. Thesis,

Department of Civil and Environmental Engineering, Lehigh University, Bethlehem, PA, USA, 2005.

58. Tremblay, R.; Lacerte, M.; Christopoulos, C. Seismic response of multi-storey buildings with self-centering energy dissipative steel braces. J. Struct. Eng. 2008, 134, 108–120.

59. Priestley, M.J.N.; Tao, J.R. Seismic response of precast prestressed concrete frames with partially debonded tendons. PCI J. 1993, 38, 58–69.

60. Eatherton, M.; Ma, X.; Krawinkler, H.; Billington, S.; Hajjar, J.F.; Deierlein, G. Design concepts for controlled rocking of self-centering steel-braced frames. J. Struct. Eng. 2014, 2014.

61. Buchanan, A.H.; Bull, D.; Dhakal, R.; MacRae, G.; Palermo, A.; Pampanin, S. Base Isolation and Damage-Resistant Technologies for Improved Seismic Performance of Buildings; Technical Report 2011-02 for the Royal Commission of Inquiry into Building Failure Caused by the Canterbury Earthquakes; University of Canterbury: Christchurch, New Zealand, 2011.

62. Kurama, Y.C. Hybrid post-tensioned precast concrete walls for use in seismic regions. PCI J. 2002, 47, 36–59.

63. Smith, B.; Kurama, Y.; McGinnis, M. Comparison of Hybrid and Emulative Precast Concrete Shear Walls for Seismic Regions. In Proceedings of the ASCE Structures Congress, Las Vegas, NV, USA, 14–16 April 2011.

64. Priestley, N.M.J.; Sritharan, S.; Conley, J.R.; Pampanin, S. Preliminary results and conclusions from the PRESSS five-story precast concrete test building. PCI J. 1999, 44, 42–67.

65. Aaleti, S.; Sritharan, S. A simplified analysis method for characterizing unbonded post-tensioned precast wall systems. Eng. Struct. 2009, 31, 2966–2975.

66. Restrepo, J.I.; Rahman, A. Seismic performance of self-centering structural walls incorporating energy dissipators. J. Struct. Eng. 2007, 133, 1560–1570.

67. Marriott, D.; Pampanin, S.; Palermo, A. Quasi-static and pseudo-dynamic testing of unbonded post-tensioned rocking bridge piers

with external replaceable dissipaters. Earthq. Eng. Struct. Dyn. 2009, 38, 331–354.

68. Kurama, Y.C.; Shen, Q. Posttensioned hybrid coupled walls under lateral loads. J. Struct. Eng. 2004, 130, 297–309.

69. Clough, R.W.; Huckelbridge, A.A. Preliminary Experimental Study of Seismic Uplift of a Steel Frame; Earthquake Engineering Research Center (EERC) Report No. UCB/EERC-77-22; University of California: Berkeley, CA, USA, 1977.

70. Kelley, J.; Tsztoo, D. Earthquake Simulation Testing of a Stepping Frame with Energy-Absorbing Devices; Earthquake Engineering Research Center (EERC) Report No. EERC 77-17; University of California: Berkeley, CA, USA, 1977.

71. Eatherton, M.; Hajjar, J.; Ma, X.; Krawinkler, H.; Deierlein, G. Seismic Design and Behavior of Steel Frames with Controlled Rocking—Part I: Concepts and Quasi-Static Subassembly Testing. In Proceedings of the ASCE Structures Congress, Orlando, FL, USA, 12–15 May 2010.

72. Ma, X.; Eatherton, M.; Hajjar, J.; Krawinkler, H.; Deierlein, G. Seismic Design and Behavior of Steel Frames with Controlled Rocking—Part II: Large Scale Shake Table Testing and System Collapse Analysis. In Proceedings of the ASCE Structures Congress, Orlando, FL, USA, 12–15 May 2010.

73. Pollino, M. Structural and Non-Structural Seismic Demands on Controlled Rocking Steel Braced Frame Buildings. In Proceedings of the ASCE Structures Congress, Chicago, IL, USA, 29–31 March 2012.

74. Roke, D.; Sause, R.; Ricles, J.M.; Gonner, N. Design Concepts for Damage-Free Seismic Resistant Self-Centering Steel Concentrically-Braced Frames. In Proceedings of the 14th World Conference on Earthquake Engineering, Beijing, China, 12–17 October 2008.

75. Dowden, D.M.; Bruneau, M. NewZ-BREAKSS: Post-tensioned rocking connection detail free of beam growth. AISC Eng. J. 2011, 48, 153–158.

76. Wiebe, L.; Christopoulos, C. Mitigation of higher mode effects in base-rocking systems by using multiple rocking sections. J. Earthq. Eng. 2009, 13, 83–108.

77. Wiebe, L.; Christopoulos, C.; Tremblay, R.; Leclerc, M. Mechanisms to limit higher mode effects in a controlled rocking steel frame. 1: Concept, modeling, and low-amplitude shake table testing. Earthq. Eng. Struct. Dyn. 2013, 42, 1053–1068.

78. Wiebe, L.; Christopoulos, C.; Tremblay, R.; Leclerc, M. Mechanisms to limit higher mode effects in a controlled rocking steel frame. 2: Large-amplitude shake table testing. Earthq. Eng. Struct. Dyn. 2013, 42, 1069–1086.

79. Morgen, B.G.; Kurama, Y.C. Seismic response evaluation of posttensioned precast concrete frames with friction dampers. J. Struct. Eng. 2008, 134, 132–145.

80. Priestley, M.N. Overview of PRESSS research program. PCI J. 1991, 36, 50–57.

81. Priestley, M.N. The PRESSS program: Current status and proposed plans for phase III. PCI J. 1996, 41, 22–40.

82. Englekirk, R.E. Design-construction of the paramount—A 39-story precast prestressed concrete apartment building. PCI J. 2002, 47, 56–71.

83. Ricles, J.M.; Sause, R.; Peng, S.W.; Lu, L.W. Experimental evaluation of earthquake resistant posttensioned steel connections. J. Struct. Eng. 2002, 128, 850–859.

84. Wolski, M.; Ricles, J.M.; Sause, R. Experimental study of a self-centering beam—Column connection with bottom flange friction device. J. Struct. Eng. 2009, 135, 479–488.

85. Kim, H.J.; Christopoulos, C. Friction damped posttensioned self-centering steel moment-resisting frames. J. Struct. Eng. 2008, 134, 1768–1779.

86. Lin, Y.C.; Sause, R.; Ricles, J.M. Seismic performance of steel self-centering, moment-resisting frame: Hybrid simulations under design basis earthquake. J. Struct. Eng. 2013, 139, 1823–1832.

87. Clayton, P.M.; Berman, J.W.; Lowes, L.N. Seismic design and performance of self-centering steel plate shear walls. J. Struct. Eng. 2011, 138, 22–30.

88. Dowden, D.M.; Purba, R.; Bruneau, M. Behavior of self-centering steel plate shear walls and design considerations. J. Struct. Eng. 2011, 138, 11–21.

89. Clayton, P.M.; Berman, J.W.; Lowes, L.N. Subassembly testing and modeling of self-centering steel plate shear walls. Eng. Struct. 2013, 56, 1848–1857.

90. Tsai, K.C.; Chou, C.C.; Lin, C.L.; Chen, P.C.; Jhang, S.J. Seismic self-centering steel beam-to-column moment connections using bolted friction devices. Earthq. Eng. Struct. Dyn. 2008, 37, 627–645.

91. Kim, H.J.; Christopoulos, C. Numerical models and ductile ultimate deformation response of post-tensioned self-centering moment connections. Earthq. Eng. Struct. Dyn. 2009, 38, 1–21.

92. Chou, C.C.; Chen, J.H. Seismic design and shake table tests of a steel post-tensioned self-centering moment frame with a slab accommodating frame expansion. Earthq. Eng. Struct. Dyn. 2011, 40, 1241–1261.

93. Chi, H.; Liu, J.; Garlock, M. Design and Analytical Validation of Post-Tensioned Column Bases. In Proceedings of the ASCE Structures Congress, Vancouver, Canada, 24–26 April 2008.

94. Dolce, M.; Cardone, D. Theoretical and experimental studies for the application of shape memory alloys in civil engineering. J. Eng. Mater. Technol. 2006, 128, 302–311.

95. Yang, W.C.S.; DesRoches, R.; Leon, R.T. Design and analysis of braced frames with shape memory alloy and energy-absorbing hybrid devices. Eng. Struct. 2010, 32, 498–507.

96. Graesser, E.J.; Cozzarelli, F.A. Shape-memory alloys as new materials for aseismic isolation. J. Eng. Mech. 1991, 117, 2590–2608.

97. Dolce, M.; Cardone, D. Mechanical behavior of shape memory alloys for seismic applications. 1. Martensite and austenite NiTi bars subjected to torsion. Int. J. Mech. Sci. 2001, 43, 2631–2656.

98. DesRoches, R.; McCormick, J.; Delemont, M. Cyclic properties of superelastic shape memory alloy wires and bars. J. Struct. Eng. 2004, 130, 38–46.

99. Erochko, J.; Christopoulos, C.; Tremblay, R.; Kim, H.J. Shake table testing and numerical simulation of a self-centering energy dissipative braced frame. Earthq. Eng. Struct. Dyn. 2013, 42, 1617–1635.

100. Pampanin, S.; Palermo, A.; Buchanan, A.H.; Fragiacomo, M.; Deam, B.L. Code Provisions for Seismic Design of Multi-Storey Post-Tensioned Timber Buildings. In Proceedings of the 39th Meeting of the Working Commission W18-Timber Structures, CIB, International Council for Research and Innovation, Florence, Italy, 28–31 August 2006.

101. Buchanan, A.; Deam, B.; Fragiacomo, M.; Pampanin, S.; Palermo, A. Multi-storey prestressed timber buildings in New Zealand. Struct. Eng. Int. 2008, 18, 166–173.

102. Palermo, A.; Sarti, F.; Baird, A.; Bonardi, D.; Dekker, D.; Chung, S. From Theory to Practice: Design, Analysis and Construction of Dissipative Timber Rocking Post-Tensioning Wall System for Carterton Events Center, New Zealand. In Proceedings of the 15th World Conference on Earthquake Engineering, Lisbon, Portugal, 24–28 September 2012.

103. Pei, S.; Berman, J.; Dolan, D.; van de Lind, J.; Ricles, J.; Sause, R.; Blomgren, H.E.; Popovski, M.; Rammer, D. Progress on the Development of Seismic Resilient Tall CLT Buildings in the Pacific Northwest. In Proceedings of the 2014 World Conference on Timber Engineering, Quebec City, Canada, 10–14 August 2014.

104. Chou, C.C.; Chen, J.H. Development of floor slab for steel post-tensioned self-centering moment frames. J. Constr. Steel Res. 2011, 67, 1621–1635.

105. Mar, D. Design Examples Using Mode Shaping Spines For Frame and Wall Buildings. In Proceedings of the 9th US National and 10th Canadian Conference on Earthquake Engineering, Toronto, Canada, 25–29 July 2010.

106. Quantification of Building Seismic Performance Factors: FEMA P695; Federal Emergency Management Agency: Washington, DC, USA, 2009.

107. Tahmasebi, E.; Chancellor, N.B.; Sause, R.; Ricles, J.M.; Joó, A.L.; Akbas, G. Collapse Performance of Steel Self-Centering Braced

Frame Systems. In Proceedings of the 7th International Conference on Behavior of Steel Structures in Seismic Areas, Santiago, Chile, 9–11 January 2012.

108. Tahmasebi, E.; Sause, R.; Ricles, J.M.; Chancellor, N.B.; Akbas, T. Probabilistic Collapse Performance Assessment of Self-Centering Concentrically Braced Frames. In Proceedings of the 10th U.S. National Conference on Earthquake Engineering, Anchorage, AK, USA, 21–25 July 2014.

109. MacRae, G. Low Damage construction—Some Systems Issues. In Proceedings of the 10th International Conference on Urban Earthquake Engineering, Tokyo, Japan, 1–2 March 2013.

110. Roke, D.A.; Chandra, A.; Huang, Q.; Sett, K. Methodology for Life Cycle Cost Assessment of Self-Centering Concentrically Braced Frame Systems. In Proceedings of the 10th International Conference on Urban Earthquake Engineering, Tokyo, Japan, 1–2 March 2013.

111. Dyanati, M.; Huang, Q.; Roke, D.A. Seismic Performance and Economic Feasibility Evaluation of Self-Centering Concentrically-Braced-Frame System. In Proceedings of the 11th International Conference on Structural Safety and Reliability, New York, NY, USA, 16–20 June 2013.

112. Dyanati, M.; Huang, Q.; Roke, D.A. Structural and Nonstructural Performance Evaluation of Self-Centering, Concentrically Braced Frames Under Seismic Loading. In Proceedings of the ASCE Structures Congress, Boston, MA, USA, 3–5 April 2014.

# 7 CHAPTER

# KEY FACTORS INFLUENCING THE MECHANISM OF RAPID AND LONG RUNOUT LANDSLIDES TRIGGERED BY THE 2008 WENCHUAN EARTHQUAKE, CHINA

**Fawu Wang[1*], Ping Sun[2], Lynn Highland[3] and Qiangong Cheng[4]**


## ABSTRACT

### Background

The 2008 Wenchuan earthquake triggered many rapid and long runout landslides, which directly caused great loss of property and human lives and were responsible for a large percentage of total damages caused by the earthquake. It is very important for the purposes of landslide disaster prevention and mitigation to understand the earthquake triggered mechanism of initiation and motion of rapid and long runout landslides, which can potentially be the deadliest of ground failures.

### Results

In this paper, field investigations of some highly damaging landslides caused by the Wenchuan earthquake are introduced first, and followed by data from ring shear tests used to simulate the initiation and motion of one landslide in particular, the Donghekou complex landslide.

## Conclusions

It was found that groundwater and valley water played key roles in the rapid motion and long runout process of this landslide during the great earthquake.


**Keywords:** Wenchuan earthquake Rapid and long runout landslides Initiation Motion Hydrogeological condition

# BACKGROUND

The Wenchuan earthquake occurred at 14:28 local Beijing time on 12 May 2008 in Sichuan Province of China (Cui et al. [2009]; Huang and Li [2009]). The earthquake's epicenter was located at Yingxiu Town, Wenchuan County (30.40°N, 103.47°E, and its depth was 14 km). The earthquake magnitude was Mw7.9 (U.S. Geological Survey [2008]) and Ms8.0 (China Earthquake Administration [2008]). The most affected area was along the central and frontal faults of the Longmen Shan fault system in western Sichuan, in a 300 km long belt-shaped zone (Figure 1). Numerous landslides were triggered in this zone (Yin et al. [2009]; Tang et al. [2009]). According to Cui et al. ([2009]), as of 19 June 2009, the earthquake resulted in 69,200 known deaths, 18,195 missing, 374,216 injured, 5,362,500 collapsed houses, 21,426,600 badly damaged houses, and more than five million people left homeless. Among the losses and damages, about one third were due to earthquake-induced landslides (Yin et al. [2009]). The large amount of damage from landslides results from a combination of factors that include a large magnitude earthquake, topographical controls, and the population distribution of the area. This tectonically active area is characterized by steep mountains and river systems that are formed at the boundary between the Tibetan Plateau and the Sichuan Basin, making this area susceptible to erosion and landsliding (Yi et al. [2006]; Burchfiel et al. [2008]; Ouimet [2010]). The relatively gentle topography that exists in some areas of the Longmen Shan fault system attracted people, as there is a good groundwater supply as well as a mild climate. Three county capital towns, Dujiangyan City, Qushan Town (Beichuan County), and Qiaozhuang Town (Qingchuan County) were located along the fault system, making the fault belt a densely populated area. During the earthquake, all of the county towns were severely damaged by strong shaking, fault displacement and landsliding. Among them, Qushan Town that is located in the middle part of the Longmen Shan fault system was most heavily damaged. At this site, two landslides (Chengxi slide and Xinbei Mid-school slide) destroyed everything in their travel paths, and directly caused more than 2,100 deaths in Beichuan County (Yin et al. [2009]).

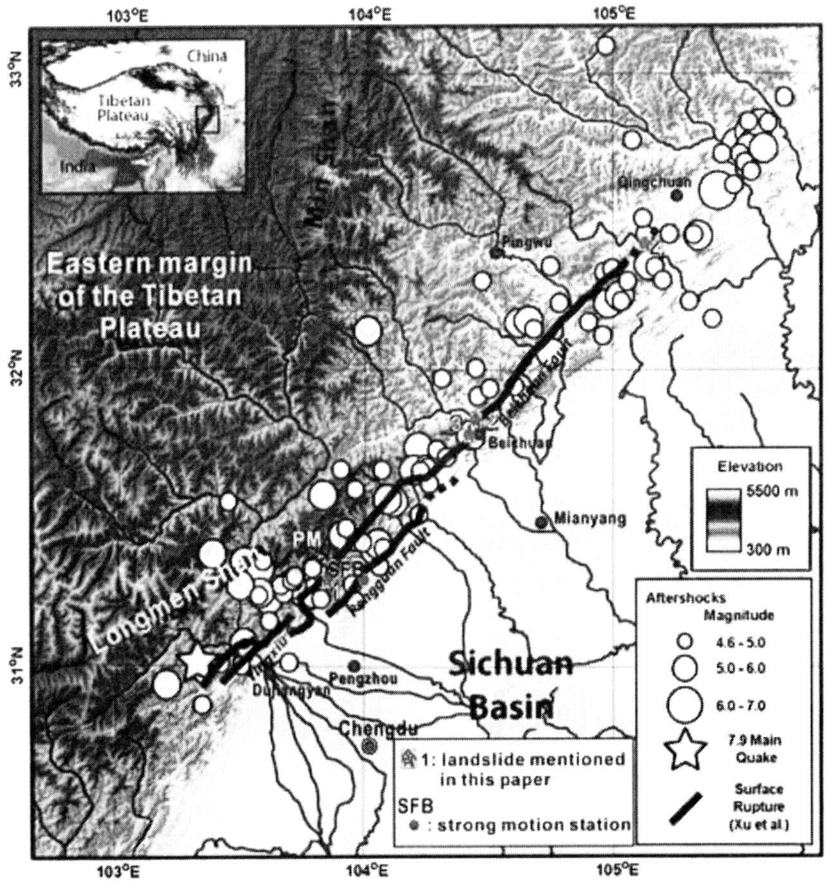


1: Xiejiadian Rockslide-rock flow in Jiufeng Village; 2 and 3: Chengxi landslide and Xinbei Mid-school rockslide; 4: Donghekou complex slide

**Figure 1. GEOGRAPHY, TOPOGRAPHY AND RIVERS OF THE EARTHQUAKE REGION (OUIMET,[2010]) (WITH PERMISSION FROM W OUIMET).** The surface rupture associated with the Wenchuan earthquake by Xu et al. ([2009]) (solid lines denote inferred rupture) and aftershocks greater than M4.5 are also included. The Yingxiu, Beichuan and Pengguan faults are labeled. Four landslides mentioned in this paper are marked, and SFB strong-motion station is shown.

Analyzing the meteorological conditions just prior to the earthquake is vital for framing the explanation as to why the landslides were so large and numerous. In the earthquake affected area, there are three meteorological monitoring stations, located in Dujiangyan City, Pingwu County, and Mianyang City. The one located in Pingwu County is near the center of the earthquake affected area, and is 50 km west of the Donghekou complex slide, which will be described. The daily rainfall data in 2008 are shown in Figure 2, indicating that the total precipitation in 2008 was 718.5 mm. Considering the rainfall intensity and frequency, the period from March to May can be classified as experiencing medium rainfall, while June to September can be classified as having heavy rainfall. The Wenchuan earthquake occurred during a medium rainfall season. The antecedent rainfall before the earthquake was 58.8 mm for one month (from 12 April to 12 May), and 33.3 mm for ten days (from 2 May to 12 May).

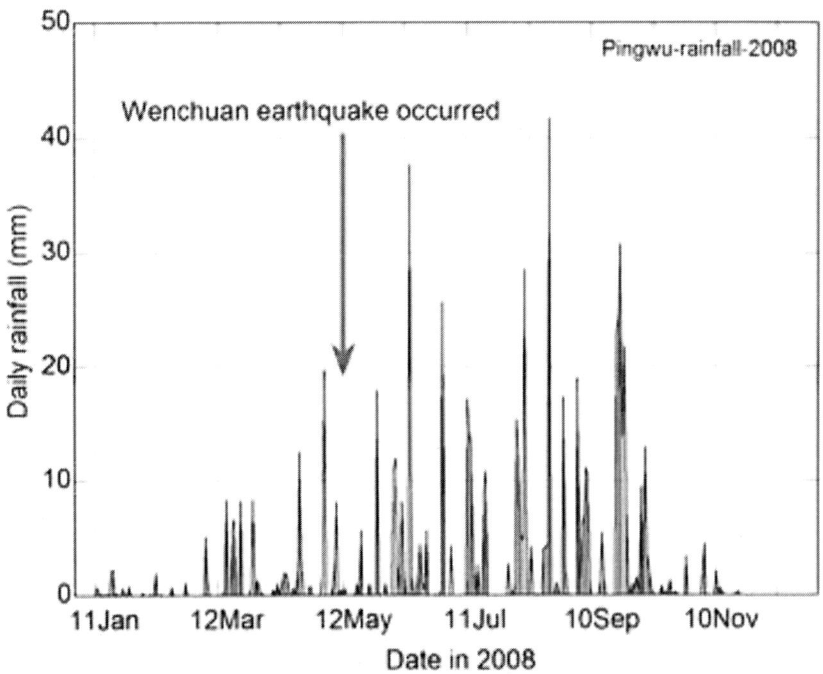


**Figure 2.** DAILY RAINFALL DATA MONITORED AT PINGWU METEOROLOGICAL STATION IN 2008.

As summarized by Bird and Bommer ([2004]), large landslides triggered by earthquakes can potentially cause massive loss of life and injuries and leave an indiscriminate trail of devastation. In this paper, some typical landslides, which moved rapidly and caused large numbers of fatalities, will be briefly introduced and compared with each other. The examples are Xiejiadian landslide at Jiufeng in Pengzhou City, the rapid and long runout Chengxi landslides and the Xinbei Mid-school rockslide in Beichuan County, and the Donghekou complex slide in Qingchuan County (Figure 1). These landslides killed all of the people and destroyed all villages in their travel paths (Wang et al. [2009]; Yin et al. [2009]; Cui et al. [2009]; Sato and Harp [2009]). Secondly, a detailed study of the Donghekou complex landslide will be presented. The detailed study includes field investigation, soil sampling, and a geo-simulation test on the soil samples from this slide, with the application of actual seismic motion from the earthquake. This study will contribute to an understanding of the mechanism of the initiation and motion of rapid and long runout landslides, especially the key factors controlling the travel distance.

# 2 METHODS, RESULTS AND DISCUSSIONS

## 2.1 Field investigations for some typical landslides

In the year following the Wenchuan earthquake, the authors investigated the earthquake affected areas for geo-hazards at four different points in time. The first time was in June 2008, one month after the earthquake occurred. At that time, most of the landslides remained in their original shape after initial occurrence and the local people had a recent memory about the topographic change around them, making the observation and investigation more closely attuned to conditions which occurred immediately after the initiation of the landslides. Investigations were conducted on Chengxi landslide, Xinbei Mid-school rockslide, and Xiejiadian landslide. The second investigation was conducted in July 2008. At that time, the first access to Donghekou complex slide was made. Two landslide dams caused by this slide were not yet excavated or modified. After that time, the landslide dams were leveled, and a memorial park was built at the lower part of the slide. The third investigation was conducted at the end of July 2008 for a duration of one week, and debris flows caused by rainfall in the earthquake-affected area were observed. During the fourth investigation conducted in February 2009 (a dry season in this area), it was found that groundwater was flowing out of the slide surface and source area of the Chengxi landslide, the Donghekou complex slide, and the

Xiejiadian landslide. During the field investigations, there was a strong indication that the hydrogeological conditions associated with the landslides controlled the initiation and motion of the landslides. In order to verify this hypothesis, characteristics of some of these typical landslides, i.e., the Xiejiadian landslide, the Chengxi landslide and the Xinbei Mid-school rockslide, and the Donghekou complex slide are presented here to show the importance of hydrogeological conditions to the landslide process.

### 2.1.1 Xiejiadian rockslide – rock flow in Jiufeng Village

In Jiufeng Village, Pengzhou City, the Xiejiadian rockslide, which occurred first, later became a flowslide since the topography of the source area made the accumulation and impounding of water possible. The head scarp of this landslide is on a steep cliff of Precambrian granite, which is located above a small valley that channeled the flow (Chigira et al. [2010]). The total travel distance is about 1.6 km, which occurred rapidly leading to more than 100 fatalities from the lack of fleeing time. In Figure 3a, the smashed houses and displaced slide mass looks very dry, and a witness did not see any water seep out from the displaced slide mass. Also, as seen in Figure 3b, which is an old map created before the earthquake, a gully with water has always existed in this area. In Figure 3c, the displaced slide mass consists of large gravels and boulders, and thus the permeability of the slide mass should be very high. The high permeability will not allow groundwater to be distributed throughout the sliding mass, but it will be most significant near the sliding surface. To find groundwater in the source area and the travel path, we investigated the Xiejiadian rockslide in February 2009 again. As shown in Figure 3d, it was found that groundwater flow existed at the bottom of the displaced slide debris. The surface gully where two persons were taking samples was formed by water erosion in the period between May 2008 after the earthquake and February 2009 when the investigation was carried out. According to Chigira et al. ([2010]), the investigators also found water-saturated clayey materials and water flow in the source area.

### 2.1.2 The Chengxi complex landslide (debris fall/slide/flow) and the Xinbei Mid-school rockslide in Qushan Town

The damages caused by landslides triggered by the Wenchuan earthquake in Qushan Town, Beichuan County are the most severe of the entire disaster (Figure 4). At the west part of Qushan Town, the Chengxi (meaning the west side of the town) complex landslide occurred. Here, debris consisting of weathered shale and sandstone moved downward at first, then slipped at the slope toe, and finally became a flow slide. It destroyed and buried all of the buildings in its travel path, which was the center of the old area of Qushan Town. Moreover, because of the landslide's extremely high speed, the buildings distributed around its deposit boundary collapsed due to the extreme force of an air-blast caused by the landslide motion. This landslide caused 1,600 fatalities.

In the fourth investigation, we confirmed that groundwater existed in the weathered shale in the source area of the landslide, and the underground soil located 30 cm beneath the surface had high water contents, while the surface was quite dry.

**Figure 3. PHOTOS AND MAP OF THE XIEJIADIAN ROCK FLOW IN JIUFENG VILLAGE. (A)** Toe of the rock flow (taken on 14 June 2008); **(B)** Old map showing a valley located along the travel path; **(C)** Travel path of the rock flow (taken on 14 June 2008); **(D)** Water flowing which is fed by a groundwater discharge from the source area (Taken on 17 February 2009).

**Figure 4. THE CHENGXI COMPLEX LANDSLIDE (LEFT) AND THE XINBEI MID-SCHOOL ROCKSLIDE (RIGHT) IN QUSHAN TOWN, BEICHUAN COUNTY.** The Chengxi landslide killed 1,600 people and the Xinbei Mid-school rockslide killed 500 people. The dotted lines show the travel path and the extent of each landslide.

The cause of the rapid sliding in the Chengxi complex slide, in addition to the steepness of the slope at the source area, was possible due to the existence of a shallow groundwater table in the basin beneath Qushan Town. As indicated by a longitudinal cross section on the Chengxi slide made by Yin et al. ([2009]), the ground level before the earthquake was just 3–5 m higher than the water level in the Jian River, a river flowing through Qushan Town. We had planned to excavate the slide mass to determine the water content in the slide surface, however, this area has been reserved for an earthquake memorial park, and we could not get permission.

The Xinbei Mid-school rockslide occurred in dolomite/limestone. Because the displaced rock mass was almost dry and of large boulder size, it slowed and stopped at about 25 degrees. The permeability of dolomite block is high because of its fractured structure. In addition, the travel path of this slide is at a much higher elevation than the Jian River. The groundwater table is estimated to be far beneath the slide surface of the rockslide. There is no

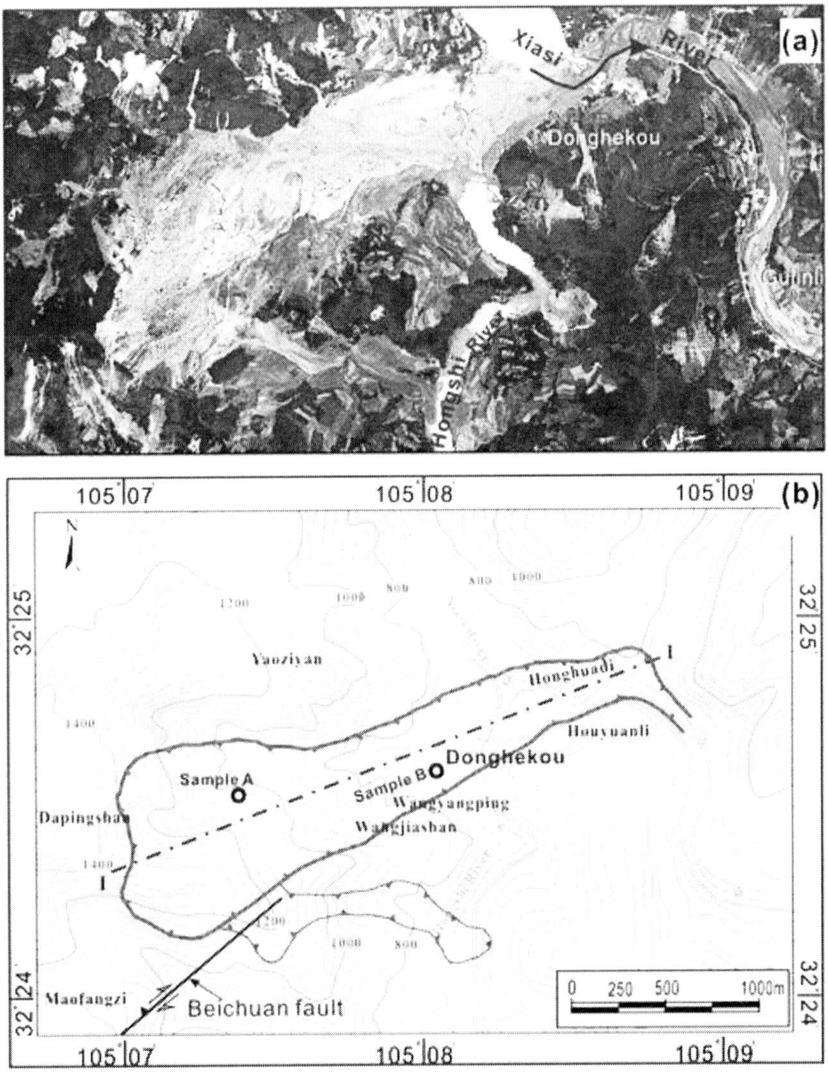


**Figure 5. AREAL PHOTO AND MAP OF THE DONGHEKOU COMPLEX SLIDE. (A)** Areal photo (from Yin et al. [2009]) and **(B)** topographic map (landslide boundaries are outlined in red).

water affecting the sliding process, so the debris could stop and deposit at 25 degrees. The Xinbei Mid-school was destroyed and about 500 students and teachers were killed, because the school was located near the foot of the slope, and in the direct path of the landslide. Additionally, the school was in very close proximity to the earthquake fault rupture, and most likely experienced a damaging level of shaking from the earthquake.

### 2.1.3 A complex landslide in Donghekou Village

A large-scale rockslide occurred on the mountain behind Donghekou Village. Movement was triggered by the earthquake, and the initial rockslide became a debris flow as it moved down the mountain. Figure 5a and Figure 5b are an areal photo and topographic map of the Donghekou complex slide. Donghekou Village was located at the convergence of the Hongshi River and the Xiasi River. The debris flow dammed the Xiasi River and its tributary Hongshi River, and formed two landslide lakes. Part of the displaced slide mass entered the Xiasi River and flowed down the stream as a debris flow. In the source area, the average length is about 1 km, the average width is about 920 m, and the total length of the rockslide and debris flow reached about 3 km.

From Figure 6a, it can be seen that many houses were built along the Hongshi River. There were about 1,500 people living in this village. At this location, the Hongshi River formed a fan with alluvial and flood deposits. At the two sides of the Hongshi River near the converging point, the ground level is less than 1 m above the water level in the river. Connecting to the water level in the Hongshi River, the groundwater table in the alluvial and flood deposit is estimated to be very shallow, and the valley deposits in fully saturated conditions.

As can be seen from Figure 6b, the landslide destroyed all of the houses in this area, and dammed the rivers. Four sub-villages and an elementary school were destroyed, and about 700 people were killed. The landslide dam in the main stream of the Xiasi River is not very thick. Because of the long travel distance, the debris flow deposits spread wide and thin.

The slickenside faces on the mountain at the source of the landslide and the adjacent mountain, which also experienced the rockslide motion, indicate intense energy and high velocity of movement of the sliding mass. Local eyewitness said three buses full of local travelers were buried by the rapidly-moving debris because the buses had no time to escape. The flat deposits of the complex slide showed characteristic of a flowslide. There are almost no large blocks in the travel path, showing the highly fractured property of the sliding mass before it formed the landslide.

**Figure 6. A COMPLEX SLIDE (ROCKSLIDE–DEBRIS FLOW) OCCURRED IN DONGHEKOU VILLAGE, QINGCHUAN COUNTY DURING THE WENCHUAN EARTHQUAKE. (A)** Photo of the region before the earthquake, taken by the local people and **(B)** Photo 2 months (July 2008) after the earthquake by the first author. The flow-path was considered to be fully saturated.

Figure 7 shows the geological components of the Donghekou rockslide–debris flow and the surrounding area. The source area of the rockslide is located in dolomitic limestone and sericitic limestone of the Yuanji formation of the Sinian period (Pre-Cambrian Era). The strata dipped to the Northwest with an average dip angle of 10 degrees. The dip direction is about 90 degrees to the sliding direction of the landslide. Within the travel path, the displaced sliding mass moved in weathered tuff sandstone, phyllite and carbon-siliceous slate of Cambrian period. These materials have a high potential for grain crushing when sheared and a relatively low permeability. According to Yuan et al. ([2010]), the position of a rupture of the Beichuan active fault was added in Figure 7.

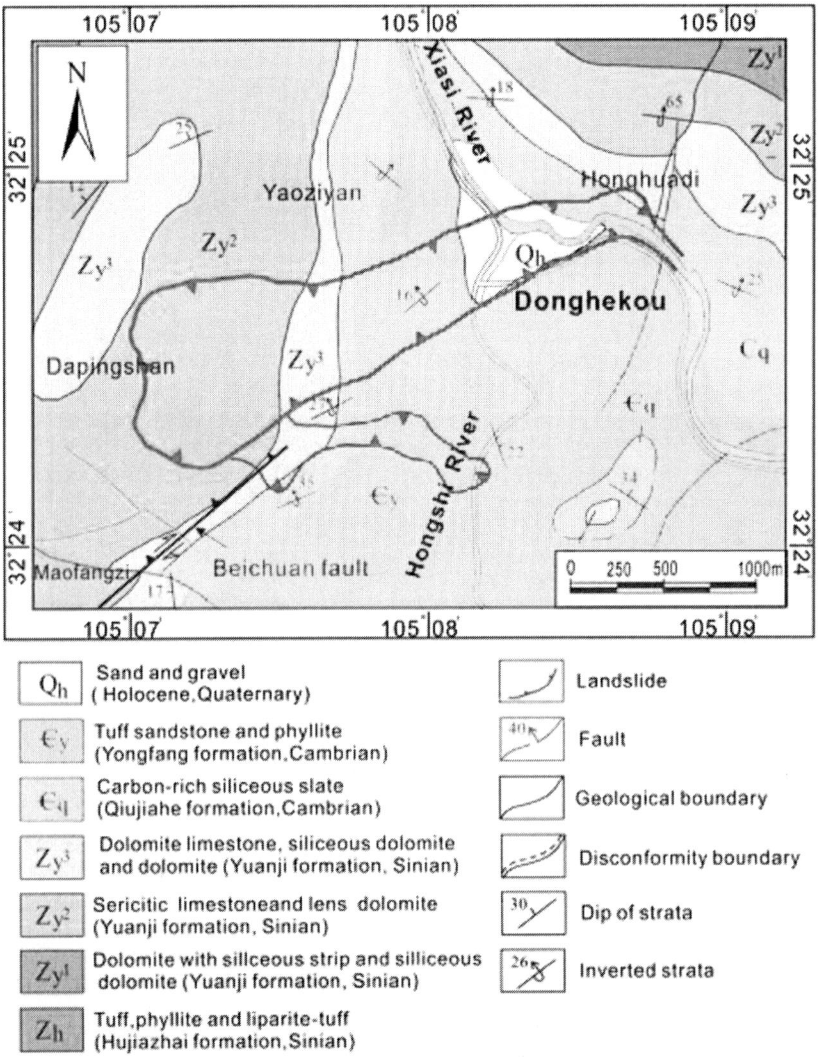


**Figure 7.** GEOLOGICAL MAP OF THE DONGHEKOU COMPLEX SLIDE AND SURROUNDING AREA.

Figure 8 shows the central longitudinal section (section 1–1 in Figure 5b) of the Donghekou complex slide. The average thickness of the slide mass in the source area is about 75 m, with the maximum thickness of 110 m, and the total volume of the Donghekou rockslide is about 69 million $m^3$. In Figure 8, by connecting the top point of the rockslide and the toe of the debris flow, the apparent friction angle of the rockslide-debris flow was determined to be 11 degrees. This indicates a high mobility for the rockslide-debris flow. In the source area, there was much slide debris that remained, covering the sliding surface of the landslide, making it difficult to determine the material in the slide zone. Observation of the slide mass distributed in the lower part (traveling path and deposit area), it is indicated that the sliding zone existed in the carbon-siliceous slate of the Qiujiahe formation of the Cambrian period. The structure of the slide mass during the earthquake consisted of two layers, i.e., fragmented limestone of the Sinian period overlaying weathered carbon-rich siliceous slate of the Cambrian period. In the investigation conducted in July 2008, it was observed that the weathered slate was in a soft-plastic state and fully saturated in a natural slope near the Donghekou slide.

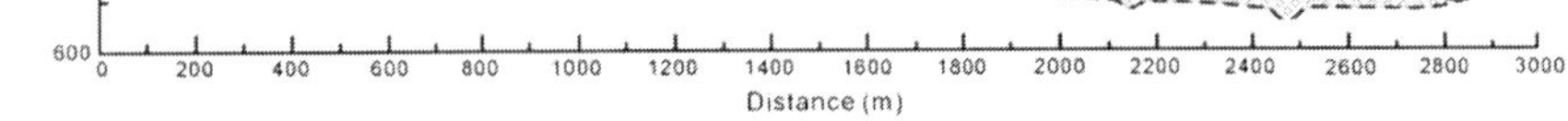


**Figure 8.** LONGITUDINAL SECTION (SECTION I-I IN FIGURE **5** B) OF THE DONGHEKOU COMPLEX SLIDE SHOWING LANDSLIDE TOPOGRAPHICAL CHARACTERISTICS, AND LOCATIONS FROM WHICH SAMPLES A AND B WERE TAKEN.

For the field investigation on Donghekou slide conducted in February 2009, sampling from the source area (sample A) and valley deposit (sample B) was carried out. At the source area (Figure 9), weathered carbon-siliceous slate was labeled as sample A. Although the debris around the mountain was dry, a small amount of water was flowing down the valley. Lineaments representing shearing, indicated by the layered structure of slate with an intrusion of soft yellow clay were observed at the sampling point. Considering the size limitation of the ring shear apparatus, during sampling, boulders and gravels were sieved out, and only the grains smaller than 4.75 mm were collected, as a disturbed sample A. This kind of treatment in sampling may affect the peak value of shear resistance, while have no effect on the residual shear strength or that in steady state (Sassa et al. [2004]), which is an important parameter for rapid and long runout slides. An old valley deposit was found beneath the displaced debris that moved in this event, and it became sample B, which was taken (Figure 10), by the same method as sample A.

**Figure 9.** LOCATION OF SAMPLE A IN A GULLY IN THE SOURCE AREA OF THE DONGHEKOU COMPLEX SLIDE.

**Figure 10.** LOCATION OF SAMPLE B IN THE OLD VALLEY DEPOSIT BENEATH THE DISPLACED SLIDING DEBRIS FROM THE LANDSLIDE.

## 2.2 Simulation tests for the initiation and motion mechanism of the Donghekou complex slide for use as a case study

We selected Donghekou complex slide as a prototype model for laboratory tests to study the initiation and motion mechanisms of landslides triggered by the Wenchuan earthquake, because the Donghekou complex slide moved for a long distance and showed a high rate of speed. Long travel distance implies a very low shear resistance during the landsliding. We try to find the mechanism to make the low shear resistance possible during the landslide motion. After the earthquake, many analyses and discussions have been conducted for this landslide, and the most comprehensive comments were given by Qi et al. ([2011]). They mentioned that besides the influence of the earthquake, the high relief and steep inclination in the source slope and plentiful saturated loose deposits formed by a river or gully in the travel path had a great contribution to the rapid and long runout landslides. In this paper, we explain the initiation and motion mechanisms using the concept of undrained shearing, although other mechanisms may also exist.

Using the undrained ring shear apparatus developed at Kyoto University, Japan (Sassa [1997]; Sassa et al. [2004]), two types of simulation tests were conducted. They are: (a) Simulation test to clarify the initiation mechanism of the Donghekou complex slide triggered by the Wenchuan earthquake and (b) Test to simulate the movement when the failed sliding mass loaded onto the valley deposits, and clarify the motion mechanism for long runout and in high speeds.

### 2.2.1 The geotechnical simulation test for the Donghekou slide with the actual seismic motion

In a geotechnical simulation test, the initial stress level, i.e., the initial normal stress, shear stress, and pore-water pressure existing in the slope before earthquake should be reproduced first to simulate the initial stress condition in the slope, and then the cyclic loading caused by a seismic motion can be applied to observe the effect on the slope caused by the earthquake. Because the Donghekou complex slide is so large and the earthquake was so strong, it is impossible to simulate both the initial stress condition of the slope and the cyclic loading from the earthquake at the same time. As commonly understood in slope stability evaluations, the slope angle is much more important than the thickness of sliding mass. That is, the relationship between the initial normal stress and shear stress is more important than the absolute value of the normal stress and shear stress. In this test, with consideration of the stress capacity of the ring shear apparatus, the initial slope angle of the Donghekou complex slide was maintained, while the thickness of the sliding mass was reduced significantly.

For the initial stress condition of the Donghekou rockslide, the slope angle of 25 degrees was used at the source area. The initial normal stress and shear stress were obtained from the slope angle, thickness of the sliding mass, and the unit weight of the soil/rock. With the consideration of the capacity of the apparatus, a thickness of 20 m was used to replace the actual average thickness of 75 m. The unit weight of the sliding mass was assumed to be 18 $kN/m^3$. Statistically, under the same slope condition, larger scale landslides always show a higher mobility (Okuda [1984]). Theoretically, for an infinite long slope consisting of non-cohesive soils, the slope stability has no relationship with gravitational acceleration. Considering that the stress level in the simulation test is about 27% of the actual, the initiation mechanism and motion mechanism of the Donghekou complex slide can be qualitatively interpreted. The actual failure should have a higher mobility than that shown in the laboratory test.

According to Wang et al. ([2010]) and Lu et al. ([2010]), there are three strong motion stations, i.e., Wolong, SFB, and MZQ, located around the Longmen Shan fault system. The SFB station experienced a similar seismic intensity as that of the Donghekou complex slide. According to a detailed analysis by Li et al. ([2008]), the Donghekou slide and the SFB station are located in the same seismic intensity zone of level-VIII in the Chinese Seismic Intensity Scale. For this reason, the SFB record was used in the geo-simulation test on Donghekou slide. The SFB strong motion station was located about 1.2 km from the secondary fault rupture and it recorded a large vertical motion in Bajiao Town, Shifang County (Lu et al. [2010]). According to Kurahashi and Irikura ([2010]) and Wang et al. ([2010]), the SFB monitoring station is near an asperity and in the forward rupture direction. The site is primarily alluvium, and the observation point is on ground (free field station). The seismometer was set 20 m below the ground surface. During the earthquake, the SFB station recorded three seismic wave components (EW and NS in the horizontal plane, and UD in the vertical plane) (Figure 11). The PGAs (peak ground acceleration) at EW, NS, and vertical directions are 557.2 gal, 633.7 gal, and 581.9 gal, respectively.

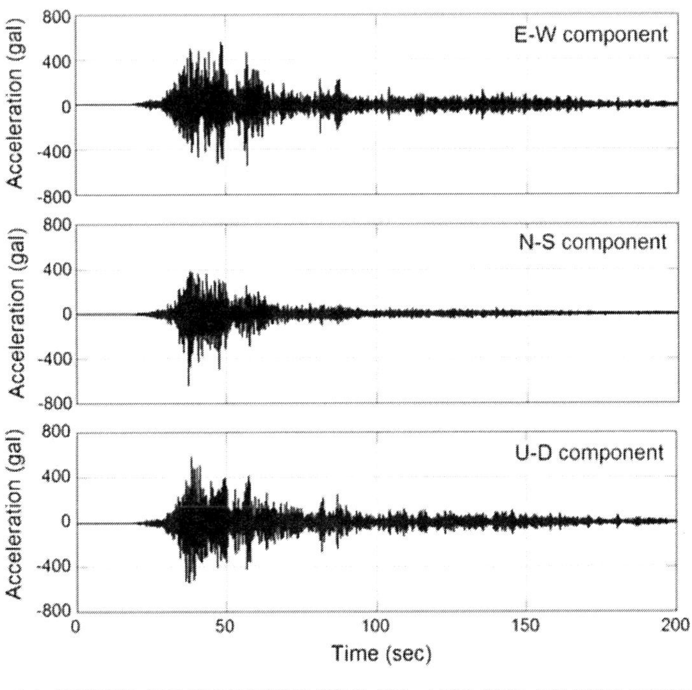


**Figure 11.** THREE COMPONENTS (EW, NS, AND UD) OF THE WENCHUAN EARTHQUAKE RECORDED AT SFB STRONG-MOTION STATION.

Generally, seismic intensity is attenuated by distance, and amplified by standing elevation and loose deposits. To apply the seismic record to Donghekou slide, modifications would have to be made. However, when considering the following factors, we assumed the seismic intensity at the Donghekou slide can be treated as the same as that recorded in SFB station. They are: 1) the SFB station is on the hanging wall of the secondary fault, and Beichuan fault is located near the source area of the Donghekou complex slide (Yuan et al. [2010]); and 2) the elevation difference between source area and toe of Donghekou slide is about 700 m. The source area of the Donghekou slide should have been affected by the seismic amplification. The SFB station is located in alluvium, so the recorded waves also should have been amplified. As both sites are located in the same seismic intensity zone of VIII, and the amplification by standing elevation in the Donghekou complex slide and the alluvium deposit at the SFB station can be treated as equivalent qualitatively. Although we know that the geological and topographic conditions and the distance to the source fault of the earthquake may cause amplification and attenuation of the seismic shaking in different site, the coefficient of the amplification and attenuation is very difficult to be determined. So for simplification, the seismic wave monitored at SFB station was used to form the seismic loading input signal for the simulation test, without any amplification and attenuation (Wang et al. [2000]).

Through directional transformation shown in Figure 12 with consideration of the initial slope condition of the Donghekou rockslide (slope angle and initial thickness), the earthquake induced incremental normal stress and the shear stress component acting on the sliding surface of the Donghekou rockslide could be calculated.

The seismic wave direction transformation procedure is presented below:

1. Determination of the horizontal seismic acceleration $a_{HR}$ (t) in the sliding direction of the landslide by means of the NS component $a_{NS}$ (t) and the EW component $a_{EW}$ (t) of the earthquake (Figure 12a), using Equation (1).

$$a_{HR}(t) = \sqrt{a_{NS}^2(t) + a_{EW}^2(t)\cos\delta} \tag{1}$$

2. Determination of the two components in parallel direction and normal direction to the sliding surface, by summing the horizontal component $a_{HR}$ (t) and the vertical component $a_{UD}$ (t) (Figure 12b). The component in the normal direction to the sliding surface $a_{NR}$ (t), and the component in the shear direction of the sliding surface, $a_{SH}$ (t), are obtained as shown in Equations (2) and (3). Where, $\theta$is the slope angle of the landslide.

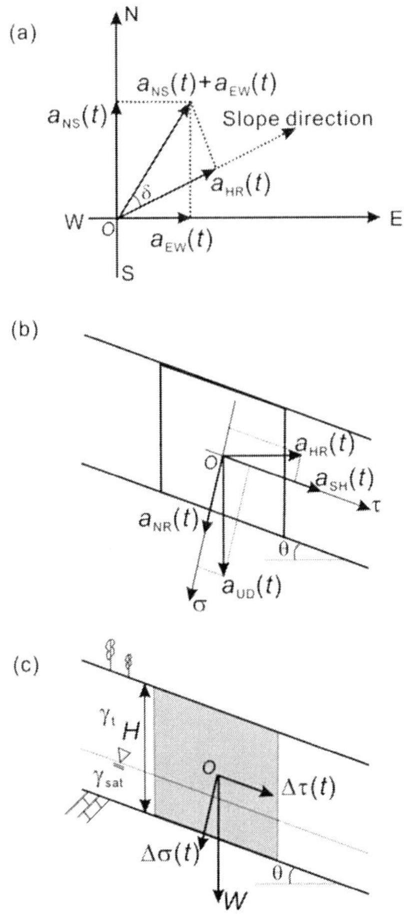


**Figure 12 . THE SYNTHESIZING PROCESS USING THE THREE COMPONENTS OF THE SEISMIC WAVE TO CALCULATE THE NORMAL STRESS INCREMENT AND SHEAR STRESS INCREMENT ACTING ON THE SLIDING SURFACE. (A)** Transfer of the horizontal (EW and NS) acceleration components ($a_{EW}$ (t) and $a_{NS}$ (t)) to slope direction in horizontal plane ($a_{HR}$ (t)); **(B)** Transfer of the horizontal acceleration component ($a_{HR}$ (t)) along the slope direction and the vertical acceleration component (UD) ($a_{UD}$ (t)) to normal stress component ($a_{NR}$ (t)) and shear stress component ($a_{SH}$ (t)) acting on the sliding surface; **(C)** Obtaining the initial normal stress and shear stress ($\sigma_0$, $\tau_0$), and the increments of normal stress and shear stress acting ($\Delta\sigma$(t), $\Delta\tau$(t)) on the sliding surface by multiplying acceleration and mass of the soil column. $\delta$ is the angle between slope direction and the horizontal shaking direction of the earthquake; $\theta$ is the slope angle; $\gamma_{sat}$ is unit weight of the saturated soil layer, and $\gamma_t$ is unit weight of the soil over the groundwater table; $H$ is the depth of sliding mass.

$$a_{NR}(t) = a_{UD}(t)\cos\theta - a_{HR}(t)\sin\theta \quad (2)$$

$$a_{SH}(t) = a_{UD}(t)\sin\theta + a_{HR}(t)\cos\theta \quad (3)$$

1. Calculation of the initial stress and the seismic stress increment on the sliding surface. A soil column with a unit width along the slope was considered as shown in Figure 12c. So, the weight of a unit column is given by Equations (4) and (5).

$$W = \gamma H\cos\theta \quad (4)$$

$$m = W/g \quad (5)$$

The initial stress condition can be determined using Equations (6) and (7).

$$\sigma_0 = W\cos\theta \quad (6)$$

$$\tau_0 = W\sin\theta \quad (7)$$

Moreover, the seismic stress increment can be determined from Equations (8) and (9).

$$\Delta\sigma(t) = ma_{NR}(t) \quad (8)$$

$$\Delta\tau(t) = ma_{SH}(t) \quad (9)$$

The initial normal stress is 295.7 kPa and the initial shear stress is 137.9 kPa at the source area of the Donghekou rockslide. These stress values were calculated based on a 25 degree average slope and a sliding mass thickness of 20 m with a unit weight of 18 kN/m$^3$. Figure 13 shows the synthesized input wave of normal stress and shear stress for the Donghekou slide. The initial stress condition of the initial slope before the earthquake, and the stress increment caused by the earthquake are also included. Because the earthquake occurred during a typical rainy season, the static pore-water pressure acting on the potential sliding surface was ignored, but the sliding surface itself was estimated to be in a fully saturated condition.

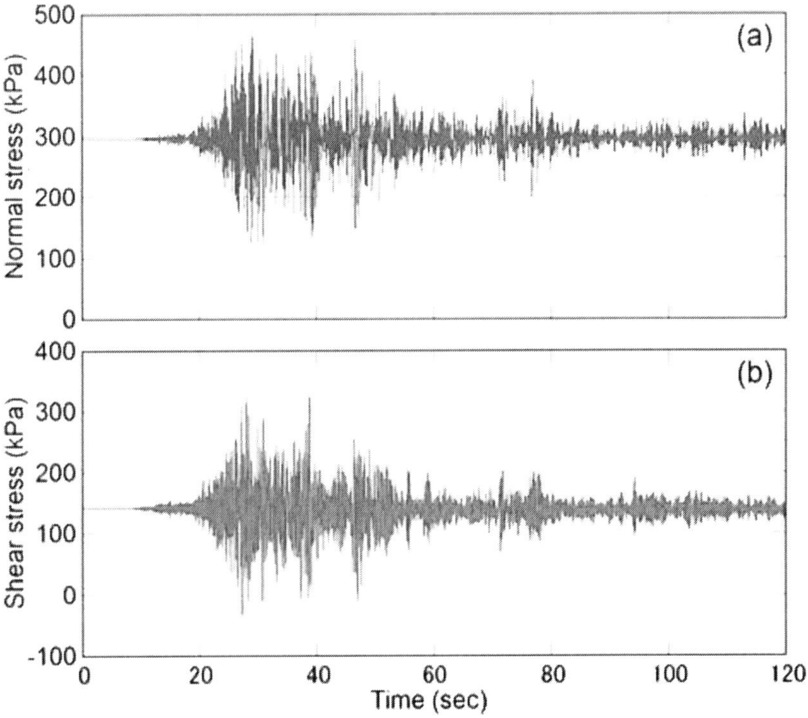


**Figure 13.** SYNTHESIZED INPUT SIGNAL FOR STRESSES DUE TO SEISMIC LOADING FOR THE SIMULATION TEST ON THE DONGHEKOU LANDSLIDE TO SHOW THE INITIATION MECHANISM OF THE LANDSLIDE TRIGGERED BY THE EARTHQUAKE. (A) normal stress and **(B)** shear stress.

The test was conducted with disturbed sample A from the source area. The sample was dried in an oven at 105°Cat first, and then separated into grains by gently applying a wooden hammer. By means of free fall deposition method (Ishihara [1993]), the dry sample grains were set in the shear box, and saturated by carbon dioxide gas and de-aired water. The saturation degree was confirmed by the $B_D$ value suggested by Sassa ([1988]). When the $B_D$ value is higher than 0.95, the sample is recognized as fully saturated. In this test, $B_D$ was 0.97, showing a fully saturated condition. After normal consolidation at 295.7 kPa, the initial shear stress 137.9 kPa was loaded gradually while keeping the sample in a drained condition to avoid excess pore-water pressure generation. At this moment, the void ratio of the soil sample reached 0.650. Then, the sample was changed to an undrained condition, and the input signals of seismic loading (Figure 13) were loaded

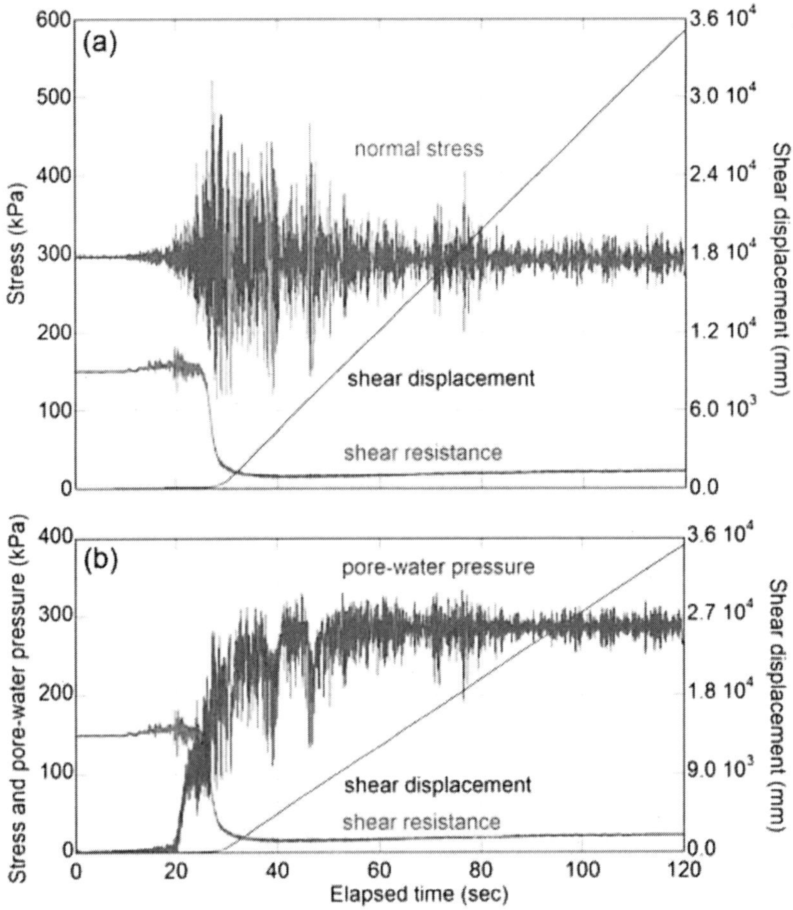


**Figure 14.** GEO-SIMULATION TEST RESULTS FOR THE DONGHEKOU ROCKSLIDE USING THE ACTUAL SEISMIC WAVE COMPONENTS AS THE TRIGGERING PROCESS. (A) normal stress, shear resistance and shear displacement **(B)** pore-water pressure, shear resistance and shear displacement.

to simulate the condition of the original slope when the earthquake motion was applied as the initiation factor.

Figure 14 shows the entire data series for the geo-simulation test on the Donghekou rockslide. A large decrease of the shear resistance and a rapid generation of the excess pore-water pressure with the application of the normal stress was observed. Figure 15 shows a close-up of the first interval of strong shaking, from 18 to 30 seconds. From the two figures, it can be seen that the sample failed at about 10 seconds from the beginning of the earthquake, and the shear displacement accelerated at about 25 seconds. At the same time (from 20 seconds), the pore-water pressure shows a rapid increase, and reached the same value as the normal stress at about 45 seconds, reaching the point of full liquefaction. Corresponding to the rapid increase of the pore-water pressure, the shear resistance decreased rapidly, and reached a very low value of about 15 kPa. At the end of the seismic loading, the shear displacement reached about 35 m with a high speed of 0.38 m/sec (the maximum speed of the apparatus at the selected gear).

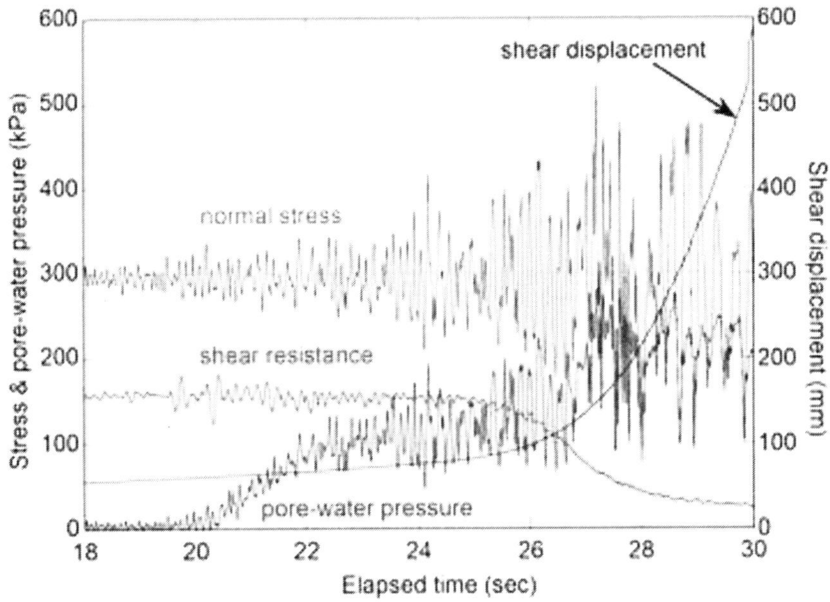


**Figure 15.** CLOSE-UP OF PART (18–30 SEC) OF THE GEO-SIMULATION TEST RESULTS FOR THE DONGHEKOU ROCKSLIDE.

In the field of liquefaction study, a loose structure of saturated sand layer is important for liquefaction when triggered by an earthquake. However, for a dense sandy soil layer, since the excess pore-water pressure related to the structure is limited or negative, a contribution from grain crushing is expected. Under undrained conditions, when a dense sandy soil layer is sheared, grain crushing will create a tendency for volume shrinkage, and result in an increase of excess pore-water pressure. Sassa et al. ([2004]) called this phenomenon "sliding surface liquefaction". Although it is recognized as a reasonable explanation for landslides with high mobility in dense sandy soil slopes, the evidence of grain crushing is limited in related case studies on rapid and long runout landslides. So, in this test, after the undrained shearing to 35 m was completed, the sample in the shear box was taken out, and the grain size distributions were analyzed (Figure 16). When compared with the original sample, it is evident that grain crushing occurred throughout the whole material of the samples. However, the material in the shear zone was finer than the upper and lower zones. Although migration of particles in the shear box can make the value slightly different from the reality, it is the small particles that were transported from the shear zone to the upper and lower sides because of high water pore-pressure at the shear zone (Sassa et al. [2004]). It is evident that grain crushing can occur under undrained conditions, and it may contribute to the "sliding surface liquefaction" of the soil during shearing.

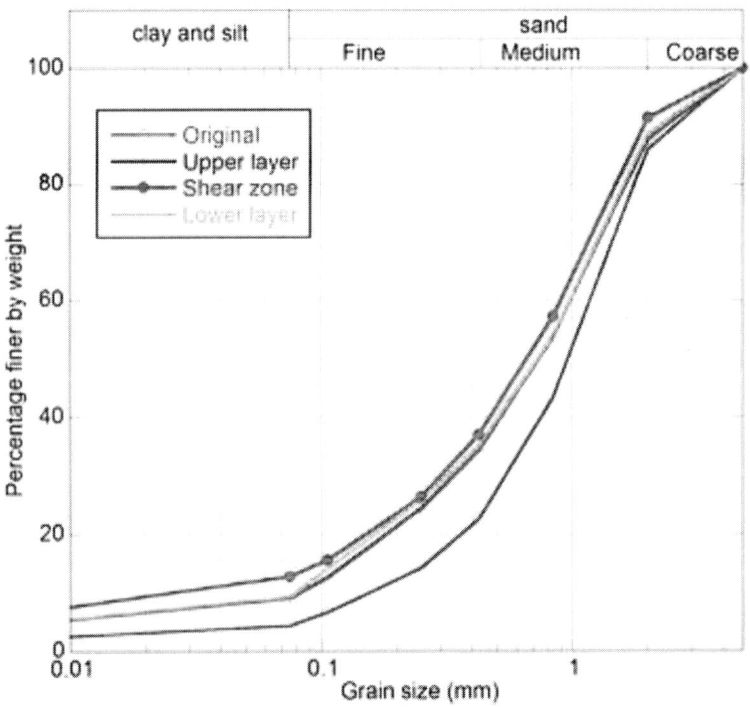


**Figure 16.** GRAIN SIZE DISTRIBUTION OF THE SOURCE AREA SAMPLE FROM THE DONGHEKOU COMPLEX SLIDE. The original sample, consists of soils from the upper layer of the shear box, the lower layer of the shear box, and the shear zone.

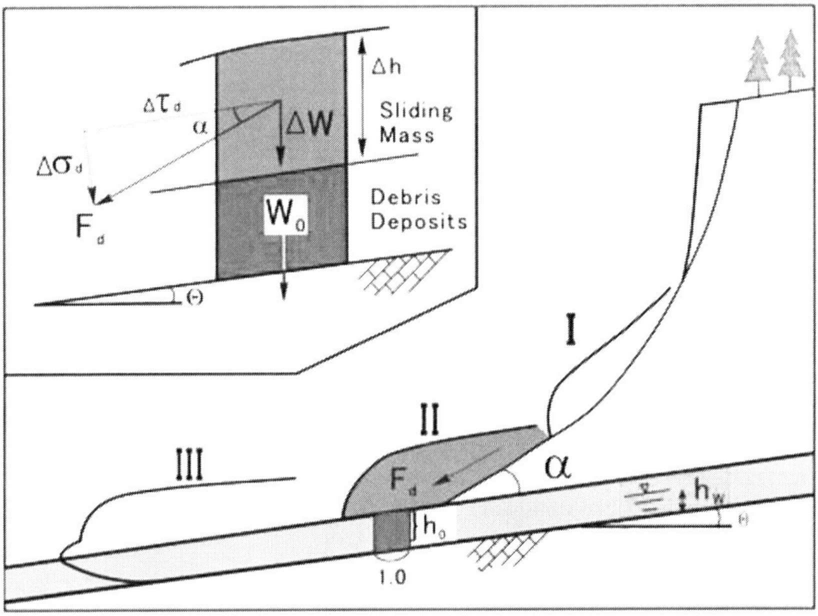


**Figure 17. MODEL OF THE LANDSLIDE TRIGGERED BY A DEBRIS FLOW (SASSA ET AL.**[1997]**).** The process can be divided into three stages. At stage I, the displaced sliding mass slid in the source area of the landslide. At stage II, the displaced sliding mass quickly moved onto a deposit layer (with a slope angle of $\theta$) in front of the landslide at an angle of α.$F_d$ is the impact force from the displaced sliding mass on the deposit layer.$h_0$ is the depth of the new sliding surface in the deposit layer,$h_W$ is the height of groundwater table over the new sliding surface. At stage III, the impact force disappeared, and the displaced sliding mass moved together with the deposit layer. The upper left corner is a close-up of a column in stage II.

### 2.2.2 Results of the simulation test showing the impact of failed slide mass loading into the valley deposit

To simulate the process of a debris flow triggered by a rockslide, the model proposed by Sassa et al. ([1997]) was applied (Figure 17). The slide mass moved down the slope (I), and applied load onto the valley deposits at the foot of the slope (II). When a surface-water stream or subsurface flow existed and some of the deposits were saturated, the valley deposit was sheared under undrained loading and transported downstream together with the sliding mass (III) (Sassa et al. [2004]). In this study, we change the saturated condition of valley deposits to three patterns - dry, partially saturated, and fully saturated - to observe the significance of the hydrogeological conditions along the travel path. A column of unit width, which is a part of the valley deposit, was considered. In the position (I) of the sliding mass, the weight of the column ($W_0$) was in effect. When the sliding mass moved onto the valley deposit (II) with a certain velocity, it provided dynamic loading on the column. Here, it is assumed that the applied stress on the valley deposits was the sum of the static stress, $W$ (load due to the weight of the sliding mass) and the dynamic (impact) stress, $F_d$, working in the direction of motion of the sliding mass.

The input parameters were determined by considering the actual slope condition at sampling point B and the capacity of the apparatus, as explained in the previous test. By using $\alpha = 17°$, $\theta = 6°$, $h_0 = 5$ m, $h_w = 0$, $\Delta h = 15$ m, $\gamma_t = 18$ kN/m$^3$, and $F_d = 1$, the impact stresses (normal stress and shear stress) for sample B were formed as shown in Figure 18a. The impact forces increased to peak in 5 seconds, and disappeared in 5 seconds. After the impaction, the valley deposits flowed down the gentle slope with the sliding mass from the source area of the rockslide.

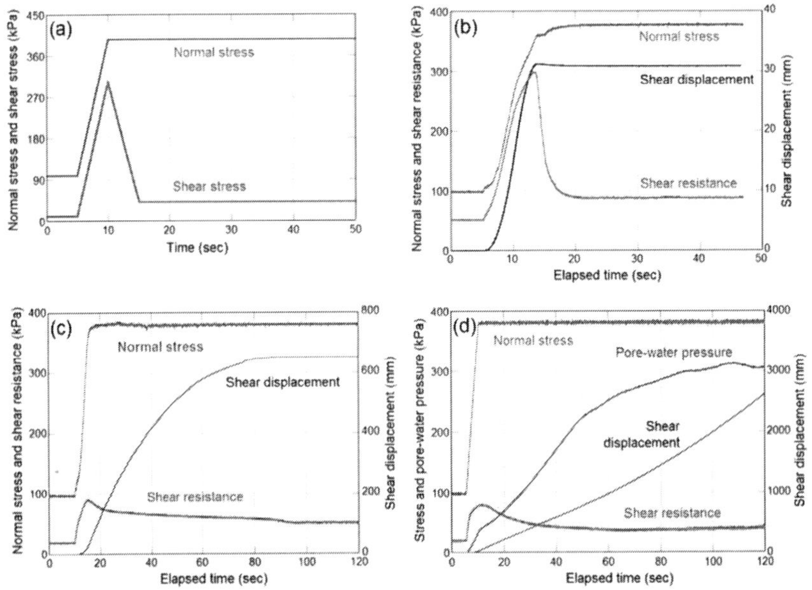


**Figure 18.** LOADED STRESS SIGNALS AND RESULTS OF THE DONGHEKOU COMPLEX SLIDE MOTION SIMULATION TEST USING SAMPLE B. Graph **(A)** shows loaded stress signals; Graph **(B)** shows results of dry condition; Graph **(C)** shows results of partially saturated condition; and graph **(D)** shows results of fully saturated condition.

Graphs b – d shown in Figure 18 illustrate the results for the three conditions (dry, partially saturated, fully saturated). When the failed sliding mass traveled down to the valley deposits, in a dry condition, only very limited shear displacement occurred, and when the valley deposits were fully saturated, a large travel distance together with high speed motion occurred. In the fully saturated condition, a high pore-water pressure was measured, and the shear resistance became small (about 45 kPa). This could be a possible reason for the long runout motion of more than 2 km of the Donghekou complex slide. Figure 19 shows the total stress path and effective stress path of the test with the fully saturated soil sample. Started from the same point, the total stress path (TSP) moved to the right side and then dropped down to a lower level of shear resistance, which means an increase in normal stress at first and a decrease in shear resistance, while the effective stress path (ESP) moved to the right at once, then returned to the left. This path shows the process of excess pore pressure generation. The apparent friction angle (the inclination of the line connecting the end of TSP and the original point) is only 6.0 degrees. It means that a high mobility of the sliding mass when it loaded on the saturated valley deposits can be expected. The small apparent friction angle will result in high velocity and long travel distance for the sliding mass.

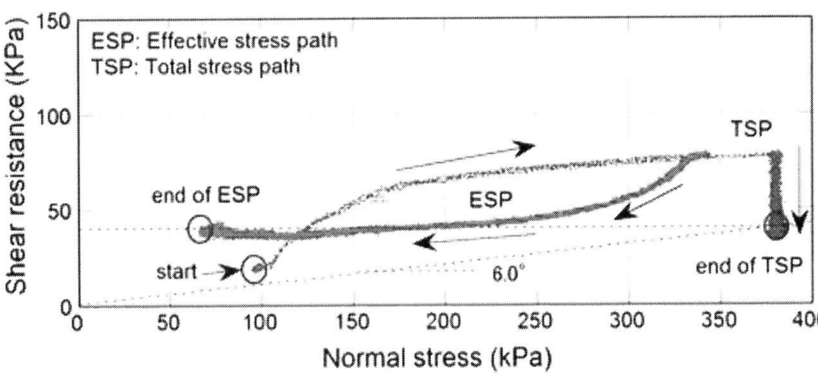


**Figure 19.** TOTAL STRESS PATH (TSP) AND EFFECTIVE STRESS PATH (ESP) OF THE SIMULATION TEST FOR FULLY SATURATED SAMPLE B.

As mentioned previously, the angle from the toe of the Donghekou complex slide and the source area is about 11 degrees. Caution is required to compare the actual apparent friction value with that obtained in ring shear tests. The most important difference is that the total normal stress in ring shear test is constant, while in the actual case, the normal stress generally decreases to lower level because the sliding mass always becomes thinner and thinner. Other reasons causing this difference may include (1) energy dissipation caused by the impact of the collision of the debris among the sliding mass, (2) some parts may not be in a fully saturated condition mobilizing a higher shear resistance. On the other hand, the test in laboratory was conducted in a simple and ideal situation with a fully saturated sample in undrained conditions. These factors may cause the difference between the actual apparent friction angle and test value. Moreover, it is clear that the test shows that the existence of the water is a very important factor for the rapid movement and long travel distance.

# 3 CONCLUSIONS

Through field investigation of the long runout landslides triggered by the Wenchuan earthquake and experimental study of the Donghekou complex landslide, in particular, it was found that the valley water and groundwater played key roles in the long runout and rapid landslide motion during the great earthquake. In reality, there is only limited information on the actual rainfall, valley water, and ground water conditions during the earthquake. It is suggested that in the source area, the landslide gained velocity when the "sliding surface liquefaction" phenomenon occurred as a result of undrained shearing in a saturated sliding zone. In turn, the displaced sliding mass can exert an impact on the valley deposit. When the valley deposit is also saturated, liquefaction can occur causing the valley deposit to work as a sliding zone with low shear resistance, transporting the sliding mass from the source area for a long distance. There may be other mechanisms for rapid and long runout landslides triggered by the earthquake, however, the hydrogeological condition in the source area and traveling path is certainly a key factor in controlling the shear resistance, and should be considered seriously in risk assessments for landslide disaster reduction. Impacts and losses depended upon whether the landslide moved but was abruptly stopped, and whether the landslides moved for a long distance with high speed. It would be useful if more earthquake-induced landslide case studies using methods outlined in this paper were to be analyzed. We could then be closer to concluding whether there are the same and/or additional variables that affect rate of movement, extent of movement and their relationships to the rates of death and damage.

# ACKNOWLEDGEMENTS

The field investigation and sampling was partially supported by Chinese State Key Fundamental Research Program Project (2008 CB425802, representative: Peng Cui). Dr. Laizhen Pei of Institute of Mountain Hazards and Environment, Chinese Academy of Sciences guided the field investigation. Dr. Hongshuai Liu of Institute of Engineering Mechanics, China Earthquake Administration supplied the seismic record of the Wenchuan earthquake. The authors deeply appreciated the discussions with Prof. Xiyong Wu of Southwest Jiaotong University, China, Prof. Masahiro Chigira of Kyoto University, Prof. Kazuo Konagai of University of Tokyo, and Prof. Satoshi Tsuchiya of Shizuoka University, Japan in the field investigation. The field investigation and ring shear tests were conducted when the first author worked in Disaster Prevention Research Institute of Kyoto University, Japan. The authors acknowledge Dr. Janusz Wasowski of CNR-IRPI, Italy, Dr. Julian Bommer of Imperial College London, and Beena Ajmera of California State University, Fullerton, USA for their helpful and constructive comments to improve the manuscript.

# COMPETING INTERESTS

The authors declare that they have no competing interests.

# AUTHORS' CONTRIBUTIONS

All authors participated the field investigations; FW & QC conducted the sampling; FW & PS conducted the ring-shear tests; FW & LH drafted the manuscript; all authors read and approved the final manuscript.

# REFERENCES

1. Bird JF, Bommer JJ: **EARTHQUAKE LOSSES DUE TO GROUND FAILURE.** Eng Geol 2004,**75**(2):147–179.
2. Burchfiel BC, Royden LH, Hilst RD, Hager BH, Chen Z, King RW, Li C, Lü J, Yao H, Kirby E: **A GEOLOGICAL AND GEOPHYSICAL CONTEXT FOR THE WENCHUAN EARTHQUAKE OF 12 MAY 2008, SICHUAN, PEOPLE'S REPUBLIC OF CHINA.** GSA Today 2008,**18**(7):4–11.

3. Chigira M, Wu XY, Inokuchi T, Wang G: **LANDSLIDES INDUCED BY THE 2008 WENCHUAN EARTHQUAKE, SICHUAN, CHINA.** Geophys J Roy Astron Soc 2010,**118**(3–4):225–238.
4. China Earthquake Administration (2008). (last access: 8 July 2014), [http://www.cea.gov.cn/manage/html/8a8587881632fa5c0116674a018300cf/sc/] China Earthquake Administration (2008). (last access: 8 July 2014)
5. Cui P, Zhu YY, Han YS, Chen XQ, Zhuang JQ: **THE 12 MAY WENCHUAN EARTHQUAKE-INDUCED LANDSLIDE LAKES: DISTRIBUTION AND PRELIMINARY RISK EVALUATION.** Landslides 2009,**6**(3):209–223.
6. Huang RQ, Li WL: **DEVELOPMENT AND DISTRIBUTION OF GEOHAZARDS TRIGGERED BY THE 5.12 WENCHUAN EARTHQUAKE IN CHINA.** Sci China Ser E-Tech Sci 2009,**52**(4):810–819.
7. Ishihara K: **LIQUEFACTION AND FLOW FAILURE DURING EARTHQUAKES.** Geologija 1993,**43**(3):351–451.
8. Kurahashi S, Irikura K: **CHARACTERIZED SOURCE MODEL FOR SIMULATING STRONG GROUND MOTIONS DURING THE 2008 WENCHUAN EARTHQUAKE.** Bull Seismol Soc Am 2010,**100**(5B):2450–2475.
9. Li XJ, Zhou ZH, Yu HY, Wen RZ, Lu DW, Huang M, Zhou YN, Cu JW: **STRONG MOTION OBSERVATIONS AND RECORDINGS FROM THE GREAT WENCHUAN EARTHQUAKE.** Earthquake Eng Eng Vibratio 2008, **7:** 235–246.
10. Lu M, Li XJ, An XW, Zhao JX: **A PRELIMINARY STUDY ON THE NEAR-SOURCE STRONG-MOTION CHARACTERISTICS OF THE GREAT 2008 WENCHUAN EARTHQUAKE IN CHINA.** Bull Seismol Soc Am 2010,**100**(5B):2491–2507.
11. Okuda S: **FEATURES OF DEBRIS DEPOSITS OF LARGE SLOPE FAILURES INVESTIGATED FROM HISTORICAL RECORDS.** Disaster Prev Res Inst Annuals B 1984,**27**(B-1):353–368.
12. Ouimet WB: **LANDSLIDES ASSOCIATED WITH THE MAY 12, 2008 WENCHUAN EARTHQUAKE: IMPLICATIONS FOR THE EROSION AND TECTONIC EVOLUTION OF THE LONGMEN SHAN.** Tectonophysics 2010, **491:** 244–252.
13. Qi SW, Xu Q, Zhang B, Zhou YD, Lan HX, Li LH: **SOURCE CHARACTERISTICS OF LONG RUNOUT ROCK AVALANCHES**

**TRIGGERED BY THE 2008 WENCHUAN EARTHQUAKE, CHINA.** J Asian Earth Sci 2011,**40**(4):896–906.

14. Sassa K: **GEOTECHNICAL MODEL FOR THE MOTION OF LANDSLIDES. SPECIAL LECTURE OF 5TH INTERNATIONAL SYMPOSIUM ON LANDSLIDES.** Landslides 1988, **1:** 37–55.

15. Sassa K: **A NEW INTELLIGENT-TYPE DYNAMIC-LOADING RING-SHEAR APPARATUS.** Landslide News 1997, **10:** 33.

16. Sassa K, Fukuoka H, Wang FW (1997) Mechanism and risk assessment of landslide- triggered-debris flows: lesson from the 1996.12.6 Otari debris flow disaster, Nagano, Japan. In: Cruden DM, Fell R (eds) Landslide Risk Assessment, Proceedings of the international workshop on landslide risk assessment. , Honolulu, pp 347–356, 19–21 February

17. Sassa K, Fukuoka H, Wang G, Ishikawa N: **UNDRAINED DYNAMIC-LOADING RING-SHEAR APPARATUS AND ITS APPLICATION TO LANDSLIDE DYNAMICS.** Landslides 2004,**1**(1):7–19.

18. Sato PH, Harp EL: **INTERPRETATION OF EARTHQUAKE-INDUCED LANDSLIDES TRIGGERED BY THE 12 MAY 2008, M7.9 WENCHUAN EARTHQUAKE IN THE BEICHUAN AREA, SICHUAN PROVINCE, CHINA USING SATELLITE IMAGERY AND GOOGLE EARTH.** Landslides 2009, **6:** 153–159.

19. Tang C, Zhu J, Liang J: **EMERGENCY ASSESSMENT OF SEISMIC LANDSLIDE SUSCEPTIBILITY: A CASE STUDY IN THE 2008 WENCHUAN EARTHQUAKE AREA.** Earthquake Eng Eng Vibration 2009, **8:** 207–217.

20. U.S. Geological Survey (2008). (last access: 8 July 2014), [http://earthquake.usgs.gov/earthquakes/eqinthenews/2008/us2008ryan/] U.S. Geological Survey (2008). (last access: 8 July 2014)

21. Wang FW, Sassa K, Fukuoka H: **GEOTECHNICAL SIMULATION TEST FOR THE NIKAWA LANDSLIDE INDUCED BY 1995.1.17 HYOGOKEN-NAMBU EARTHQUAKE.** Soils Found 2000,**40**(1):35–46.

22. Wang FW, Cheng QG, Highland L, Miyajima M, Wang HB, Yan CG: **PRELIMINARY INVESTIGATIONS OF SOME LARGE- SCALE LANDSLIDES TRIGGERED BY THE 2008 SICHUAN EARTHQUAKE.** Landslides 2009,**6**(1):47–54.

23. Wang D, Xie LL, Abrahamson NA, Li SY: **COMPARISON OF STRONG GROUND MOTION FROM THE WENCHUAN, EARTHQUAKE OF 12 MAY 2008 WITH THE NEXT GENERATION ATTENUATION (NGA) GROUND-MOTION MODELS.** Bull Seismol Soc Am 2010,**100**(5B):2381–2395.

24. Xu X, Wen X, Yu G, Chen G, Klinger Y, Hubbard J, Shaw J: **COSEISMIC REVERSE- AND OBLIQUE-SLIP SURFACE FAULTING GENERATED BY THE 2008MW7.9 WENCHUAN EARTHQUAKE, CHINA.** Geology 2009, **37:** 515–518.

25. Yi GX, Wen XZ, Wang SW: **STUDY ON FAULT SLIDING BEHAVIORS AND STRONG-EARTHQUAKE RISK OF THE LONGMENSHAN–MINSHAN FAULT ZONES FROM CURRENT SEISMICITY PARAMETERS.** Earthquake Res China 2006,**22**(2):117–125. (in Chinese) (in Chinese)

26. Yin YP, Wang FW, Sun P: **LANDSLIDE HAZARDS TRIGGERED BY THE 2008 WENCHUAN EARTHQUAKE, SICHUAN, CHINA.** Landslides 2009,**6**(2):139–152.

27. Yuan RM, Xu XW, Chen GH, Tan XB, Klinger Y, Xing HL: **EJECTION LANDSLIDE AT NORTHERN TERMINUS OF BEICHUAN RUPTURE TRIGGERED BY THE 2008 MW 7.9 WENCHUAN EARTHQUAKE.** Bull Seismol Soc Am 2010,**100**(5B):2689–2699.

# 8

# CHAPTER

# A REVIEW OF SEISMIC ISOLATION FOR BUILDINGS: HISTORICAL DEVELOPMENT AND RESEARCH NEEDS

**Gordon P. Warn** [1,*] **and Keri L. Ryan** [2]

[1] *Department of Civil and Environmental Engineering, Pennsylvania State University, 226B Sackett Building, University Park, PA 16802, USA*

[2] *Department of Civil and Environmental Engineering, University of Nevada, Reno/MS 0258, Reno, NV 89557, USA*

## ABSTRACT

Seismic isolation is a technique that has been used around the world to protect building structures, nonstructural components and content from the damaging effects of earthquake ground shaking. This paper summarizes current practices, describes widely used seismic isolation hardware, chronicles the history and development of modern seismic isolation through shake table testing of isolated buildings, and reviews past efforts to achieve three-dimensional seismic isolation. The review of current practices and past research are synthesized with recent developments from full-scale shake table testing to highlight areas where research is needed to achieve full seismic damage protection of buildings. The emphasis of this paper is on the application of passive seismic isolation for buildings primarily as practiced in the United States, though systems used in other countries will be discussed.

**Keywords:** isolation; elastomeric; sliding; seismic protection

# 1. INTRODUCTION

Seismic isolation is a technique to shift the fundamental natural period of a structure to the long period range, e.g., two to four seconds, by placing horizontally flexible isolation devices at the base of the structure to physically decouple it from the ground. For earthquake excitation this period shift translates into reduced floor acceleration and inter-story drift demands on the superstructure (structure above the isolation system) by comparison to the equivalent non-isolated structure. The reduced demands allow the superstructure to remain elastic, or nearly elastic, following a design level event. Furthermore, the reduced demands minimize the likelihood of damage to displacement sensitive and acceleration sensitive equipment, nonstructural components, and content. The period shift, however, does result in increased displacement demands concentrated at the isolation interface that must be accommodated by the seismic isolation devices. The simultaneous reduction in acceleration and drift demands achieved with seismic isolation makes it one of the most effective strategies to achieve "Operational" or "Fully Operational" performance following a large and infrequent earthquake event.

Though the concept of seismic isolation dates back more than one hundred years, it has only been practiced in the United States for the last three decades. The first record of seismic isolation is an 1870 U.S. Patent filed in San Francisco for a double concave rolling ball bearing, described as an "Earthquake-proof building" [1,2]. The 1870 double concave rolling ball bearing is remarkably similar to modern double concave Friction Pendulum™ bearings [2,3]. In 1985, one hundred and fifteen years later, the Foothill Communities Law and Justice Center in Rancho Cucamonga, California was the first seismically isolated building constructed in the United States [4]. The Law and Justice Center is isolated on 98 high damping natural rubber bearings. A thorough historical perspective and chronology of seismic isolation can be found in Naiem and Kelly [4]. Construction of seismically isolated buildings has increased at an almost exponential rate since the 1980s in Japan and China, while construction in the United States has remained modest [5] irrespective of the demonstrated effectiveness of this technology for protecting structures, nonstructural components, and content from horizontal earthquake ground shaking through more than thirty years of research. In addition to buildings, seismic isolation has been used for the protection of critical, non-building, structures such as bridges, liquefied natural gas (LNG) tanks, and offshore platforms [6]. However these applications are not discussed further as the scope of this paper is on seismic isolation of buildings.

A number of sources, both recent and dated, have provided comprehensive reviews of various aspects of the development, theory, and application of seismic isolation technology. Early reviews were excellent and thorough. For instance, Kelly [7] provided a historical perspective dating back to the rudimentary beginnings of seismic isolation technology, followed by a complete chronology of research and development efforts. Buckle and Mayes [8] also included a thoughtful historical discussion as well as a comprehensive list of the early applications that paved the way for acceptance and wider adoption. Taylor et al. [9] presented a review of the use of elastomers in seismic isolation bearings, with emphasis on their long-term behavior. A mid-1990s report provided information on several subtopics including theory, experiments, and application of sliding bearings, hybrid testing, and development and practice in several countries [10].

As the volume of information on seismic isolation has grown exponentially over the past 10–15 years, the attempts at a comprehensive review have diminished. Yet, several focused reviews have emerged. Kunde and Jangid [11] prepared a comprehensive review of research and application of seismic isolation to bridges, including analytical, experimental and parametric studies. Symans et al. [12] reviewed the development and application of seismic isolation and damping systems for wood frame structures, which are uniquely challenging to isolate due to the inherent flexibility of the framing system and relatively light mass. A recent primer, developed by the American Society of Civil Engineers (ASCE) [13], discusses the theory, hardware, analysis, design, and testing requirements specific to the United States. Finally, a collaboration of international experts acting as part of the International Council for Research and Innovation in Building and Construction (CIB) Task Group 44, prepared a report comparing devices, design codes, and current state of seismic isolation practice among countries that have been forward in adoption of seismic isolation technology [14].

Complete protection of a building system poses some unique challenges, which include: (1) protection of the nonstructural components and content, which are more sensitive to vertical excitation than the structural system; (2) mitigation of local uplift or tension demands in the isolation system that may be generated by overturning forces in moderate to slender structures; and (3) shifting of displacement demands from the isolation system to the structure in extreme events that can lead to superstructure yielding. While the vast majority of shake table tests have been conducted to demonstrate the effectiveness of seismic isolation to provide a linear elastic response of the structure under horizontal ground motion, only a few experimental studies have focused on the advanced aspects mentioned above.

The current volume of literature precludes a comprehensive review in a single paper. Instead, the aim of this paper is to summarize the current practice, describe widely used seismic isolation hardware, chronicle the history and development of modern seismic isolation through shake table testing of isolated buildings, review past efforts to achieve three-dimensional isolation, and discuss recently identified research needs. This article focuses on the application of passive seismic isolation for buildings primarily as is practiced in the United States, though systems used in other countries will be discussed. The paper is organized into the following major sections: seismic isolation hardware, experimental demonstration, past efforts to achieve 3D seismic isolation, and research needs.

# 2. SEISMIC ISOLATION HARDWARE

A variety of seismic isolation bearings have been developed and implemented around the world for the seismic protection of structures. In this section, a brief description of the basic construction, mechanical behavior and analytical modeling is presented for each type of bearing. In general, bearings are classified into two categories: (1) elastomeric and (2) sliding.

## 2.1. Elastomeric Bearings

Elastomeric bearings are composed of alternating layers of natural, or synthetic, rubber bonded to intermediate steel shim plates as shown in Figure 1. The bearings are constructed by placing un-vulcanized rubber sheets and steel shims in a mold, then subjecting the mold to elevated temperature and pressure to simultaneously vulcanize and bond the rubber. A rubber cover is provided to protect the internal rubber layers and steel plates from environmental degradation due to ozone attack [15] and corrosion, respectively. Elastomeric bearings can be categorized as: (1) low-damping natural or synthetic rubber; and (2) high-damping rubber. Low-damping natural rubber material exhibits nearly linear shear stress-strain behavior up to, approximately, 150% shear strain, after which the material stiffens. Natural rubber with type A durometer hardness of 50 is typically used for seismic applications having a shear modulus (G) that ranges from 0.65 MPa to 0.9 MPa. The equivalent damping ratio ξ, for low-damping natural rubber ranges between 2% and 3% at 100% shear strain. To control, or limit, displacements across the isolation interface, external supplemental damping devices such as yielding steel bars, plates, or viscous fluid dampers are typically used in parallel with low-damping natural rubber bearings. A higher level of damping is achieved through the addition of carbon black and other fillers to the raw rubber during the mixing process to produce high-damping rubber bearings [4]. The equivalent damping ratio of high-damping rubber bearings can range from 10% to 20% at 100% shear strain. Though the range of shear modulus for high-damping rubber is similar to that for low-damping rubber, fillers increase the hardness and thus the shear modulus of the rubber so that it can be difficult to achieve low shear modulus and high levels of damping.

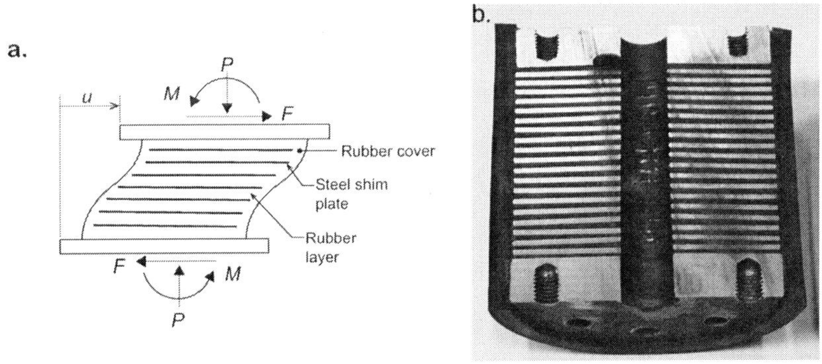


**Figure 1.** (**a**) Illustration of elastomeric bearing in the horizontally deformed configuration;(**b**) Photograph of elastomeric bearing cross-section.

The total thickness of rubber ($T_r$) provides the low horizontal stiffness needed to lengthen the fundamental natural period of the system, whereas the close spacing of the intermediate steel shim plates provides a large vertical stiffness and critical load capacity for a given G and bonded rubber area ($A_b$). However, the steel shim plates have no effect on the horizontal stiffness of the bearing, calculated as:

$$K_h = \frac{GA_b}{T_r} \tag{1}$$

The steel shims restrain the rubber at the bond interface and the spacing of the shims (or individual rubber layer thickness) controls the bulging around the perimeter and thus the compression modulus of the elastomeric layer [16,17]. For example, the compression modulus for an individual, solid, circular rubber layer, assuming the rubber is incompressible, is:

$$E_c = 6GS^2 \tag{2}$$

where S is the shape factor, a dimensionless geometric parameter, defined for a single rubber layer as:

$$S = \frac{\text{Loaded Area}}{\text{Area free to bulge}} \tag{3}$$

Although bulk compressibility was not considered in Equation 2 for simplicity, Chalhoub and Kelly [16] demonstrated that the assumption of incompressibility overestimates the compression stiffness of an elastomeric bearing, where the degree of over-estimation increases with increasing shape factor. Based on their work, Chalhoub and Kelly [16] concluded the effect of bulk compressibility should be considered in the analysis of seismic isolation. The close spacing of steel shim plates, i.e., thin rubber layers, produces a high shape factor that in turn results in a large vertical stiffness:

$$K_v = \frac{E_c A_b}{T_r} = \frac{6GS^2 A_b}{T_r} \tag{4}$$

Inspection of Equations 1,4 reveal the horizontal and vertical stiffness of the elastomeric bearing are related. For example, for circular bearings:

$$\frac{K_v}{K_h} = 6S^2 \tag{5}$$

Traditionally, elastomeric seismic isolation bearings have been designed with shape factors from 15 to 25 [18,19] and as high as 30 [20]. Such high shape factors have been used for the historically perceived benefits of mitigating rocking motion and enhancing stability [21,22], among others. However, according to Equation 5, this range of shape factors results in vertical stiffness that ranges from one to several thousand times greater than the horizontal stiffness. As a result, isolation systems composed of bearings detailed with high shape factors (15–30) typically have vertical isolation periods ranging from 0.03 to 0.15 second. Consequently, the vertical isolation frequencies can align with the dominant frequencies of the vertical spectrum, leading to significant amplification of the vertical acceleration. Therefore, isolation systems composed of elastomeric bearings detailed with high shape factors, e.g., 15–30, provide only horizontal isolation.

Figure 2 presents the shear force-horizontal deformation and vertical force-vertical deformation response of a low-damping natural rubber bearing [23]. While damping in natural rubber bearings is neither hysteretic nor viscous the shear force-horizontal deformation relationship is often modeled using either: (1) a linear viscous representation; or (2) a bilinear hysteretic representation. In Figure 2b, positive vertical force corresponds to tension. The vertical force-deformation behavior is highly nonlinear in tension due to softening or loss of stiffness resulting from multi-chain damage, damage of the micro-structure, and micro-void formation in cross-link polymers subjected to tensile strains [24,25]. Uplift, or tension, in elastomeric bearings is considered undesirable and efforts are made in the design process to avoid uplift on the isolation system. Therefore a simple linear force-deformation relationship is typically used to model the vertical force-deformation behavior where the stiffness is specified as the compression stiffness calculated using Equation 4, or similar, for different bearing plan geometries. If uplift cannot be avoided the vertical force-deformation relationship plotted in Figure 2b can be approximated with user-defined, multi-spring, models or, for example, a self-centering material model available in OpenSees [26], an open source earthquake simulation software.

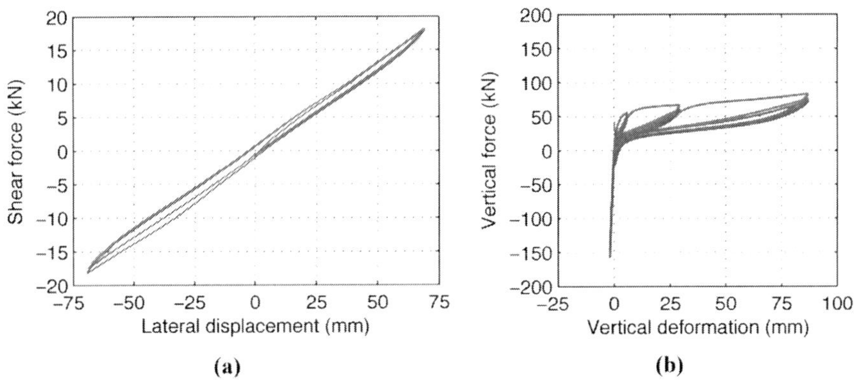


**Figure 2.** (**a**) Cyclic shear force-horizontal deformation response of low-damping natural rubber bearing; (**b**) Cyclic vertical force-deformation response of low-damping natural rubber bearing.

## 2.2. Lead-Rubber Bearings

Lead-rubber bearings were first introduced and used in New Zealand in the late 1970s [27,28]. Since then, lead-rubber bearings have been widely used for seismic isolation around the world including the United States and Japan [29]. From a construction perspective, lead-rubber bearings differ from low-damping natural rubber bearings only by the addition of a lead-plug that is press-fit into a central hole in the bearing. The lead-plug deforms plastically under shear deformation, enhancing the energy dissipation capabilities compared to the low-damping natural rubber bearing.

The horizontal force-deformation relationship of a lead-rubber bearing is characterized using bilinear behavior as shown in Figure 3. The zero-displacement force-intercept, $Q_d$, for a lead-rubber bearing is controlled by the yield strength of the lead in shear, $\sigma_L$, and the cross-sectional area of the lead-plug, $A_L$, or:

$$Q_d = \sigma_L A_L \tag{6}$$

The second-slope stiffness, $K_d$, is the stiffness of the elastomeric component of the bearing, determined using Equation 1. At a given horizontal displacement, d, the effective, or secant stiffness, of the lead-rubber bearing is:

$$K_{eff} = \frac{Q_d}{d} + K_d \tag{7}$$

The vertical stiffness of lead-rubber bearings is calculated from Equation 4 using the effective shear modulus, $G_{eff}$, calculated from the effective stiffness, $K_{eff}$, to approximately account for the increase in vertical stiffness from the lead-plug. Similar to elastomeric bearings, the vertical stiffness of a lead-rubber bearing is typically thousands of times larger than the horizontal stiffness so that isolation systems composed of lead-rubber bearings provide isolation only from the horizontal components of ground shaking.

The energy dissipation mechanism is primary hysteretic due to plastic deformation of the lead core. A Bouc-Wen or rate independent plasticity model [30] is typically used for analytically modeling the shear force-horizontal deformation response of lead-rubber bearings. Under bi-directional loading, the bearing model is coupled in the two orthogonal horizontal directions through a circular yield surface. The Bouc-Wen based hysteretic models, however, do not account for the effect of heating in the lead-core with repeated cycling that leads to degradation in the characteristic strength [2,31,32,33]. Theoretical models to account for the effects of heating in lead-rubber bearings have been developed and experimentally verified [31,32]. However, thermo-mechanical models [33] that account for heating and strength degradation in lead-rubber bearings have not been widely implemented for the analysis and design of lead-rubber isolation systems. Kalpakidis and Constantinou [34] developed a theory of scaling based on similitude and provide recommendations for testing reduced scaled bearings to properly account for the effects of heating.

The vertical force-deformation behavior is typically assumed to be linear with stiffness equal to the compressive stiffness of the bearing (Equation 4), though multi-linear models are possible as previously discussed. A model that includes the influence of vertical load on the effective horizontal stiffness and lead-core yield strength was developed by Ryan et al. [35] and has been implemented in OpenSees [26]. Models that account for the second-order effects due to vertical load at large horizontal displacement have been developed [36,37,38] and are capable of exhibiting zero or negative tangential horizontal stiffness as has been experimentally demonstrated [39]. However, these models have not been widely adopted due to the extensive experimental data required to calibrate the model parameters.

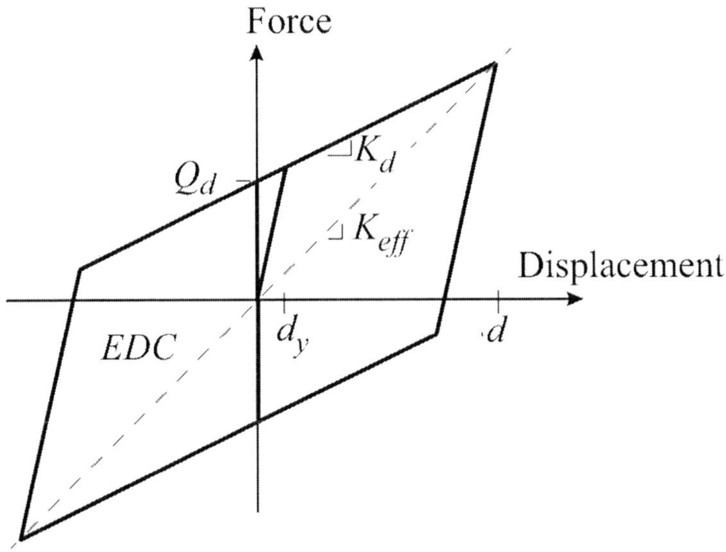


**Figure 3.** Bilinear horizontal force-displacement characterization of a seismic isolation bearing.

## 2.3. Sliding Bearings

Sliding bearings support the weight of the structure on a bearing that rests on a sliding interface. The sliding interface is designed with a low coefficient of friction, which limits the resistance to horizontal forces. Most sliding bearings use polytetrafluorethylene (PTFE) type material and stainless steel for the bearing material at the sliding interface. Restoring force is provided either by added springs or through geometry as with the Friction Pendulum™ (FP) bearing [40]. The FP bearing or derivatives such as the multi-spherical Friction Pendulum™ bearings [3,41,42,43] are among the most widely used seismic isolation bearings in the United States. The single FP bearing consists of a base-plate (ductile iron), an articulated slider (ductile iron with bonded PTFE type bearing material) and a spherical concave dish (cast steel with stainless steel overlay) as illustrated in Figure 4. As shown in Figure 4, under horizontal motion the spherical concave dish displaces horizontally relative to the articulated slider and base-plate. Friction between the PTFE type material and stainless steel surface provides frictional resistance and energy dissipation, whereas the radius of curvature of the spherical concave dish provides a restoring force. The FP bearing can be installed upside down from the configuration shown in Figure 4. However the upside down configuration results in the P-Δ moment being distributed to the structural element below the FP isolator rather than the element above the FP isolator as would occur with the configuration shown in Figure 4.

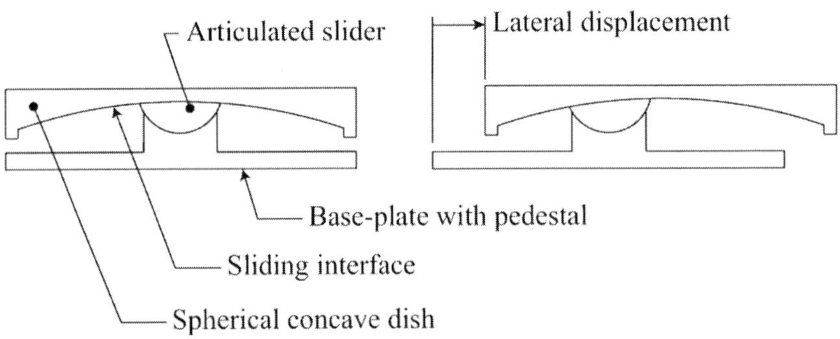


**Figure 4.** Illustration of Friction Pendulum™ bearing.

The shear force-horizontal deformation behavior of FP bearings is characterized using the bilinear relationship shown in Figure 3. The horizontal strength, or zero-displacement force-intercept, $Q_d$, is controlled by the bearing material and the weight W carried by the isolators, according to:

$$Q_d = \mu W \tag{8}$$

Where μ is the sliding coefficient of friction of the bearing interface. The sliding coefficient of friction for un-lubricated Teflon type material mated to stainless steel typically ranges between 0.07–0.18 depending on bearing pressure, peak velocity and material [44]. However, a bearing manufacturer reports that sliding coefficients of friction from 0.03 to 0.2 are possible [45]. The second-slope stiffness, $K_d$, of the FP bearing is controlled by weight acting on the isolator and the radius of curvature, R, of the spherical concave dish according to:.

$$K_d = \frac{W}{R} \tag{9}$$

The effective stiffness of an FP bearing can be determined by substituting Equations 8,9 into Equation 7. The FP bearing is unique in that the period based on the second-slope stiffness, $K_d$, is controlled only by the radius of the concave dish:

$$T_d = 2\pi\sqrt{\frac{R}{g}} \tag{10}$$

The significance of the weight independent property is that FP bearings can be effective for isolating light-weight structures. Furthermore, mass irregularities are naturally balanced by corresponding spatial variation in the restoring force, such that torsional response is minimal.

Similar to lead-rubber bearings, a bi-directionally coupled, Bouc-Wen or rate independent plasticity model is commonly used to analytically describe the force-deformation behavior of FP bearing undergoing horizontal planer motion [30]. The influence of bearing pressure and instantaneous velocity on the coefficient of sliding friction are considered using a model proposed by Constantinou et al. [46]. Similar to lead-rubber bearings, the characteristic strength of friction pendulum bearings is affected by heating [2]. The coefficient of friction is affected by the ambient, or bulk, temperature of the isolators as well as heating that develops at the isolation interface during high velocity sliding [2]. Heating due to sliding is directly related to the sliding velocity and must be carefully considered when designing reduced scale tests or extrapolating the results of characterization tests performed on reduced scale bearings to those of the prototype [2].

The vertical stiffness of FP bearings is several times larger than high shape factor elastomeric bearings due to the ductile iron, stainless steel and thin layer of PTFE used to construct the bearing. Typically the vertical stiffness of FP bearings is estimated using a simple axial stiffness equation where the modulus is that of ductile cast iron, the area is that of the articulated slider and the length is that of the combined height of the inner pedestal and articulated slider. The resulting vertical isolation period is around 0.03 seconds [45]. Therefore, isolation systems composed of FP bearings with vertical isolation periods around 0.03 seconds provide only horizontal isolation.

More recently, multi-spherical derivatives of the FP bearing have been developed, characterized [3,41,42,43] and implemented, e.g., [47]. The benefit of the multi-spherical sliding bearings over single FP bearings is that they allow an adaptive force-deformation behavior, whereby the stiffness and damping properties of the bearing can change at predetermined displacement amplitudes [41]. Variations of multi-spherical sliding bearings include the Double Pendulum™ (DP) and Triple Pendulum™ (TP), among others. Details of the DP bearing can be found in Fenz and Constantinou [3]. The TP bearing, consisting of four spherical sliding surfaces and three independent pendulum mechanisms, offers the most adaptable behavior of the multi-spherical sliding bearings [41,42,43]. A photograph of a TP bearing used in the NEES TIPS/E-Defense tests [48] is shown in Figure 5a with a cross-section view in Figure 5b. The internal pendulum mechanism with two concave plates and a rigid slider, denoted collectively as the "articulated slider" (Figure 5b), determines the response during low intensity shaking. The outer stainless steel concave surfaces, when designed with different curvatures and friction coefficients, provide two independent pendulum mechanisms that determine the response during medium to high intensity shaking. Details pertaining to the construction and force-deformation behavior of the TP bearing can be found in Fenz and Constantinou [41].

The force-deformation behavior of the TP bearing can be modeled analytically as three single FP elements in series [49]. Each of the individual series elements is modeled using the Bouc-Wen type plasticity model developed for the single FP bearing [30]. The vertical force-deformation behavior of the Triple FP bearings is assumed to be linear [49], and calculated according to the dimensions of the inner-most rigid slider.

(a)

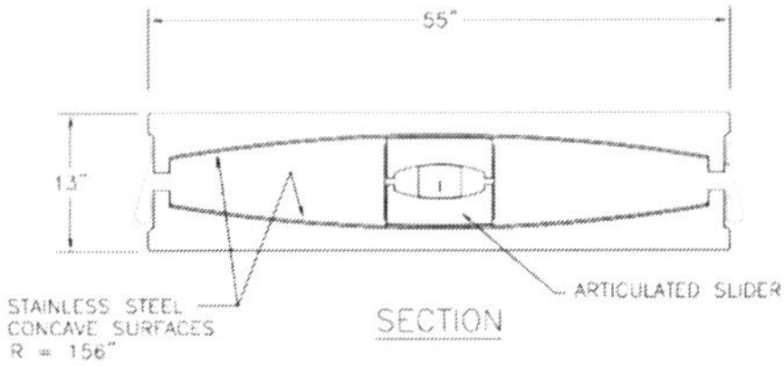


(b)

**Figure 5.** (**a**) photograph of Triple Pendulum™ bearing; (**b**) view of cross-section.

Both the single FP and the TP bearings provide no resistance to tensile forces and thus are free to uplift. In certain situations uplift in the bearings could occur, e.g., bearings on the perimeter of slender structures or those located under braced frames. For these situations resistance to uplift is considered desirable and a tension-capable sliding bearing, based on the principles of the single FP bearing, was developed [50,51]. The tension-capable device, referred to as the "XY-FP" isolator, consists of two orthogonal concave sliding rails connected through a sliding device that couples the rails in the vertical direction, allowing for the development of tensile forces in the bearing. The XY-FP isolator uniquely provides uplift resistance and an uncoupled bi-directional response [52] due to the separate orthogonal rails.

# 3. EXPERIMENTAL DEMONSTRATION

Extensive shake table testing of base-isolated buildings has been conducted over the past 30 plus years, which has paralleled the development of suitable isolation devices for large scale structures and the evolution of base isolation practice in the United States and other countries. The earliest tests focused on verification of different isolation devices and served mostly as proof-of-concept tests, without rigorous standards for evaluation of the building response. Elastomeric bearings matured more quickly than friction-based sliding systems. Several of the earliest systems were evaluated by shake table testing at the University of California's Earthquake Engineering Research Center (EERC). These systems included lightly damped elastomeric bearings, elastomeric bearings with steel dampers [53], and elastomeric bearings with enhanced damping through a friction fail-safe device [54,55]. The lead-rubber bearing was developed in New

Zealand [28], and evaluated also by shake table testing at EERC [56]. All systems were tested on a 5-story frame structure, allowing the development of higher mode response, which enabled the effectiveness of various implementation approaches for seismic isolation to be assessed. These research reports frequently observed that isolation systems with higher levels of damping, especially nonlinear damping, were effective at controlling isolator displacements, but increased floor accelerations and high frequency response.

While the concept of a friction-based system was very simple and attractive, the lack of a suitable restoring force delayed the development of sliding systems. Kelly and Chalhoub [57] tested an isolation system utilizing a combination of elastomeric bearings and flat sliders. The now mature FP system was first tested on a 2-story frame with the isolators located at the top of the first story columns; variations in structural height and aspect ratio were considered [40]. Around this time, several concepts for incorporating restoring force into friction base isolation systems were proposed [58], and an era of shake table testing of friction-based systems was launched at the State University of New York at Buffalo's National Center for Earthquake Engineering Research. A 6-story frame structure was used in evaluation tests of FP isolators [59,60,61] and teflon-disc bearings with helical steel springs [62,63]. Also, a hybrid elastomeric-friction device called resilient friction base-isolator (R-FBI) was tested at EERC [64,65]. All these systems were found to provide effective horizontal isolation.

Alongside the research and development in the United States, a major interest in base isolation developed in Japan, with research carried out mainly in the private sector by the most reputable construction companies [66]. Since Japan is an earthquake-prone region whose construction industry is dominated by a few companies, those companies have and continue to invest in the development of new technologies, including base isolation. Although some private companies and/or government agencies utilized shake table testing (e.g., Kajima, Mitsubishi, Nuclear Power Engineering Test Center), an alternative form of development adopted in Japan was the concept of a demonstration building. Kajima, Oiles, Shimizu, Obayashi, Takenaka, Taisei and Okumura all constructed base-isolation demonstration buildings, many of which were used as laboratory facilities on corporate research campuses. These demonstration buildings were continually monitored by forced vibration, free vibration, and multiple instances of seismic shaking every year, albeit mostly small intensity, as described and referenced in Kelly [66]. Most companies focused on natural rubber bearings with steel dampers, lead-rubber bearings, and eventually the application of rubber compounds with high damping fillers that could provide sufficient damping in a single device to eliminate the need for supplemental damping. Bridgestone emerged as one of the successful developers of the high damping rubber bearing, and also popularized the bolted connection detail, which was shown to improve stability and eventually replaced the doweled connection [20]. Taisei Corporation developed the friction based Taisei Shake Suppression System (TASS) that used TFE elastomeric bearings (bearings with a sliding interface) and Neoprene horizontal springs for a restoring force [67].

As the years have progressed, researchers have continued to invent new concepts for base isolation, many of them taken to the point of successful validation with shake table testing. While an exhaustive treatment of this subject is nearly impossible, a few examples are: (1) suspended pendulum isolator (SPI) [68]; (2) rolling ball isolation [69]; (3) almost lifted structure concept (ALSC) [70]; (4) stable unbonded fiber reinforced elastomeric isolator (SU-FREI) [71]; and (5) core suspended isolation system (CSI) [72]. Few of these systems have advanced to the point of widespread practical implementation. However, the new variations of the original FP isolator described earlier, including the tension capable bearing and the TP bearing, have achieved market acceptance and international implementation. The tension capable bearing was tested and validated by Roussis and Constantinou [50], and two comprehensive programs have evaluated the performance of multi-stage FP devices [73,74]. Some specifics of these test programs are mentioned below.

A few experimental studies focused specifically on evaluation of secondary system response [75,76,77,78]. The studies by Kelly [75] and Juhn et al. [77] designed oscillators (vertical cantilevers with mass) mounted on various floors of the frame to simulate the response of light equipment. Kelly [76] found that supplementary damping reduced the benefit of isolation on the secondary system response (floor acceleration and especially floor spectra), while Juhn et al. [77] found that the isolation system was very effective in reducing the response of an oscillator tuned to the natural frequency of the fixed-base structure, but could slightly increase the response of the high frequency detuned oscillator. Wolff and Constantinou [78] focused on secondary system response by evaluating floor accelerations and floor spectra for 8 different isolation systems, including combinations of low damping elastomeric bearings, lead-rubber bearings, friction pendulum bearings, and flat sliders with and without viscous dampers. The authors concluded that highly nonlinear systems incorporating friction pendulum bearings or flat sliders were not as detrimental to content response as suggested by earlier research, e.g., [29]. However, meaningful evaluation of the effect of nonlinearity is difficult when the isolation systems considered have significantly different isolation periods, thus providing different levels of force reduction, which was the case in these experiments. While useful, these studies lack a connection to the physical performance of realistic nonstructural components and content that appears in building structures.

Eventually, shake table experiments extended beyond basic validation to examine performance of the overall isolated building under more challenging loading conditions. One consideration was the possibility of uplift in slender structures under large horizontal loading. The effect of column uplift was examined in a 1/5 length scale 7-story R/C isolated structure [79] and a 1/4 length scale 9-story braced-steel frame isolated structure [80]. Different types of elastomeric bearings (natural rubber or lead-rubber) were designed for each test, and doweled connections were used to allow shear transfer while providing no restraint against uplift. The studies found that an uplift occurrence caused a vertical pulse to propagate through the structure, and likewise increased the higher frequency peaks in the horizontal floor spectra. The uplift occurrence sometimes produced a temporary instability in one or more bearing hysteresis loops, but this was a localized effect that was balanced by the other bearings and did not lead to global instability. In one case, the shear key disengaged from the bearing since the uplift of 1.25 cm, which exceeded the height of the shear key, occurred at the same time as the peak bearing displacement [79]. In tests utilizing softer bearings, a combination of moderate uplift and bearing roll out failure was observed, which occurred at a horizontal displacement less than the predicted roll out displacement. The tests suggested that uplift should be anticipated and accounted for in superstructure response and bearing connection details.

To minimize the effects of building overturning, researchers have developed and tested various uplift restraint mechanisms or mechanisms that carry tension. A thorough review of uplift restraint mechanisms studied to date is provided by Roussis [52]. Occasionally, uplift restraint has been provided by a mechanism independent of the isolation device. For example, Sumitomo Construction developed a mechanism—untested but implemented—to effectively tie the building down through interconnected steel arms attached to the structure and the foundation [52]. More commonly, uplift restraint mechanisms have been incorporated into the isolation device. An uplift restraint device embedded in the central hole of an annular elastomeric bearing was proposed and tested [81,82]. The restraint consisted of high strength bolts contained within a sleeve, wherein the bolts were detailed to allow vertical slip. The device could engage in tension and/or during large horizontal displacements to act as a fail-safe displacement restraint. The combined horizontal-vertical restraint provided enhanced functionality to the isolator, but ultimately limited the horizontal displacements that could be accommodated. An uplift restraint for a flat slider or FP isolator was tested by Nagarajaiah et al. [83]. The restraint was provided by L-shaped arms that extended from and engaged the underside of the isolator on two sides. The approach was demonstrated for unidirectional excitation, but is difficult to extend to bidirectional excitation and thus has limited application. The use of pre-stressing tendons was evaluated by Kasalanati and Constantinou [84]. Another solution is the tension capable XY-FP bearing mentioned earlier [50,51] that has been implemented in several applications. An alternative practical solution is to pair elastomeric bearings with tension capable cross-linear bearings [85]. The cross-linear bearings, developed in Japan, are essentially low friction sliders that use a double rail mechanism for tensile resistance. The friction coefficient in the cross-linear bearing is minimized by use of linear guides with recirculating ball technology. Shake table verification of this technology was a component of recent tests performed at E-Defense [48], that are described below.

Another project used shake table testing to engage the ultimate capacity of the isolated structure [86]. A three-story R/C building was designed and built to replicate the performance of a base-isolated demonstration building in Sendai, Japan (length scale factor = 0.4). The high damping rubber isolation system stiffened at large displacements, allowing demands to shift to the superstructure to develop the typical ductile degradation modes expected under large earthquakes. The experiment showed that design strategies can be adopted to ensure that the isolation system is not the weak link. In these tests, the isolators withstood significant tension due to the structure overturning.

At least two studies performed shake table tests where the intensity of the excitation was increased until the bearings ruptured [87,88]. Sato et al. [87] tested three different types of isolation devices (natural rubber bearings with steel dampers, lead rubber bearings, and high damping rubber bearings) to failure. Several of the tests were stopped only after all 4 bearings had failed; however, in each case the first failure was reported as rupture of the rubber compound under a combination of tension and shear. The failure shear strains in dynamic tests were found to be greater than the failure strains under static, monotonic loading. Takaoka et al. [88] performed shake table tests of a slender 1/9 length scale structure with lead rubber bearings, varying the structure slenderness ratio and the bearing aspect (diameter to height) ratio to engage different bearing failure modes. Bearing rupture was reported under extreme cases of buckling, tension, and combinations of tension and shear. Buckling rupture, accompanied by a "sinking" deformation of the bearing, occurred after the horizontal deformation had exceeded the diameter of the bearings. Large tension occurrences were accompanied by a notable increase in horizontal acceleration at the roof level when the uplifted structure landed on the bearings.

Limit state response of single and multi-spherical FP isolators has been observed in several of the previous test programs, e.g., [40,73,74], although engaging the limit states has never been a main focus of these programs. A short duration uplift occurrence of the bearings on one side of the building was observed in Morgan and Mahin [74], during which the slider assembly remained stable and the bearing hysteresis was not affected other than a localized (single bearing) reduction in horizontal shear due to the coupling of shear and supported axial load through friction. In Fenz and Constantinou [73], an accidental aggressive contact of the slider with the displacement restrainer in a DP bearing appeared to cause the building to rock off its supports on one side. The impact increased the base shear of the system, but not to the point of negating the effectiveness of base isolation relative to a non-isolated building; and the authors hypothesized that the model might have collapsed if a displacement restraining ring had not been present. The authors also suggested that an uplift occurrence could be particularly serious if the separation distance exceeds the height of the restrainer ring, causing the top concave plate to rise over the edge of the intermediate slider. However, such behavior has never been observed, and uplift occurences have typically been short duration, limiting the consequences of bearing separation.

In all tests described above, evaluation of the influence of vertical excitation was not a primary focus. However, many of the tests included a vertical component of excitation, and in some cases the test protocol included direct comparisons of shaking with and without vertical excitation. Most test programs have concluded that vertical excitation has no influence or very minor influence on the horizontal response and the effectiveness of base isolation. In tests with elastomeric bearings, Griffith et al. [79] noted a very minor (5%–11%) increase in horizontal floor accelerations with the inclusion of vertical excitation, and suggested that a very light coupling may have occurred, but the data was inconclusive. Responses due to horizontal only and horizontal plus vertical excitation were found to be similar in tests of the R-FBI system [65]. For multi-stage FP bearings, the influence of vertical excitation on the floor response spectra appeared to be negligible in Morgan and Mahin [74], and the influence of vertical excitation was also concluded to be "very minor" in Fenz and Constantinou [73]. Some instances of coupling between the system horizontal response and vertical ground excitation have been observed. For instance, in tests of a 3-story building isolated with lead-rubber and natural rubber bearings, Hwang and Hsu [89] found that for a structure with an asymmetric configured isolation system, vertical ground acceleration significantly increased the horizontal floor accelerations in the superstructure. No attempt was made to explain why the floor accelerations were amplified in the asymmetric configuration. Furthermore, data presented in Fenz and Constantinou [73] showed that peak horizontal floor accelerations were amplified by a factor of 2 in some of the tests with the largest peak vertical ground acceleration, which is in contrast to their conclusions.

All of the aforementioned studies involved reduced scale structural models and reduced scale isolation bearings. The multi-story structural models utilized in previous test programs employed length scale factors ranging from about 0.1 to 0.4, with a maximum supported structural weight on the order of 500 kN. Recently, two independent test programs on full scale isolated buildings with weights of 5,000–10,000 kN have been conducted at National Institute for Earth Science and Disaster Prevention (NIED) E-Defense shaking table of Japan. The first of these test programs evaluated the response of a 4-story R/C hospital building isolated with 4 high damping rubber bearings and 4 natural rubber bearings paired with steel dampers [90]. The second of these test programs, a NEES/E-Defense collaboration, evaluated the response of a 5-story steel moment frame building isolated with (1) 9 triple pendulum bearings (TPBs); and (2) 4 lead-rubber bearings (LRBs) paired with 5 cross linear bearings (CLBs). The results of the NEES/E-Defense collaboration have been reported only preliminarily to date [48].

Several features offered by full scale testing have led to significant new observations from the E-Defense tests even considering the wealth of data on reduced-scale systems. First and foremost, each full scale building served as a testbed for evaluation of physical nonstructural components and content. Thus, the effectiveness of seismic isolation to protect a variety of building content was evaluated directly, rather than by speculation of the influence of recorded responses, such as floor accelerations. In fact, the primary objective of the hospital test program [90] was to examine the performance of hospital equipment, and evaluation of the seismic performance of interacting partition walls, suspended ceilings, and sprinkler piping was a key component of the NEES/E-Defense tests [91]. Second, full scale models are required to realistically reproduce the effects of vertical vibration of the floor system and thus assess the influence of vertical excitation on building content. In most prior reduced scale tests, supplementary mass was bolted directly to a bare frame specimen, causing significant distortion of the vertical vibration modes and frequencies.

Based on observations from the hospital test program, functionality of base-isolated hospitals utilizing elastomeric isolation systems can be achieved following a near-fault motion. However, the same cannot be said for a long duration, long period ground motion generated from a subduction earthquake, due to significant motion of furniture items and medical appliances supported by casters. In such motions, velocities measured in the floors of the isolated building were larger than those in the comparable fixed-base building. The tests have generated serious discussion in Japan about how to maintain the function of a base-isolated hospital after a large, long-duration, earthquake. The influence of vertical excitation in this test program has not yet been reported [90].

The significant influence of vertical excitation was a dominant outcome of the NEES/E-Defense test program [48]. Input table excitations with peak vertical accelerations greater than about 0.5g resulted in some damage to suspended ceilings, causing ceiling panels to fall, or in the most extreme cases localized failures of the ceiling grid support system. The performance of the suspended ceilings was similar in all configurations (isolated with TPB, isolated with LRB/CLB, or fixed-base), and the damage seemed to be a direct result of the vertical acceleration recorded in the slabs, which was also similar in all configurations. In the building isolated with TPB, an occurrence of near simultaneous uplift of all 9 TPBs (i.e., bouncing of the entire building on the shake table) induced by a large vertical pulse with PGA = 1.3g. While the input sent a high frequency shock wave up through the columns, it was not transmitted to the slabs, as the slab vibration was dominated by single frequency vibration at the slab natural frequency. This suggests that the vertical motion transmitted to the nonstructural components and content is relatively insensitive to the base conditions (complete fixity, isolators with tensile restraint, or isolators free to uplift). Amplification of horizontal floor accelerations was observed in all configurations when subjected to combined horizontal/vertical table acceleration, although the extent to which this contributed to ceiling damage and content disruption is unclear. The amplification of horizontal acceleration was caused by at least two sources of horizontal-vertical coupling: coupling in the mode shapes due to building irregularities, and coupling in the TPB isolators (horizontal shear is proportional to vertical force). Though investigations are ongoing, the observations overwhelmingly suggest that mitigation of the effect of vertical component of excitation (e.g., 3D isolation, engineered floor systems, improved anchorage of nonstructural components and equipment) may be required to maintain functionality in essential base-isolated buildings.

# 4. DEVELOPMENT AND TESTING OF 3-DIMENSIONAL ISOLATION SYSTEMS

Efforts to combine horizontal seismic isolation with vertical vibration isolation date back to the early development of seismic isolation systems, driven by the power and utility industries focused on protection of large equipment. Multiple researchers have independently developed 3-dimensional isolation systems by modifying the design parameters of laminated rubber bearings. As mentioned previously, the bearing shape factor, controlled by the thickness of the individual rubber layers, determines the vertical stiffness of the system [92]. Hence, a natural approach to provide vertical isolation is to decrease the bearing shape factor (see Equation 4) in order to increase the vertical fundamental natural period. The earliest isolation system developed by Kajima in Japan utilized steel laminated natural rubber bearings that provided a vertical frequency = 5 Hz, which is substantially more vertically flexible than a typical modern design. Oil dampers were included to reduce vertical and rocking motions, and a 2-story R/C acoustic laboratory demonstration building was constructed using this approach [66]. The use of laminated rubber bearings for 3-dimensional isolation was also explored for the United States nuclear industry by Aiken et al. [93]. The research progressed to the point of designing a 3D isolation system for a nuclear facility and performing characterization tests of individual 1/4 length scale bearings [94]. The target horizontal and vertical periods of the system were 2 seconds and 0.33 seconds, respectively. The investigation concluded that elastomeric isolators could possibly be designed to effectively provide isolation in the horizontal and vertical directions [93,94]. Application of 3D isolation to elastomeric devices is by nature limiting due to stability issues. As the thickness of the rubber layers increases, the critical load capacity of the bearing $P_{cr}$ decreases proportionally. Furthermore, $P_{cr}$ has been experimentally demonstrated to reduce with increasing horizontal displacement [22,95,96,97]. For a lightweight building, balancing the design objectives for displacement capacity and stability in the deformed configuration can be extremely difficult for elastomeric bearings even without consideration of vertical isolation. Thus, successful implementation of 3D isolation using elastomeric bearings to any building would likely constrain the horizontal isolation period to be somewhat smaller than typically used today.

The GERB system for 3-dimensional earthquake protection of structures was developed by a company that specializes in vibration isolation [98]. The system utilizes helical springs—with similar flexibility in all three directions—and viscous dampers, and was developed for seismic and vibration isolation of diesel and turbo generators. Shake table testing was performed on a 5-story building model in Yugoslavia [98], and the system was also tested by Mitsubishi Industries in Japan [66]. The system can be designed with a vertical fundamental frequency as low as 1.2 Hz, leading to a flexible rocking mode around 0.65 Hz. The authors acknowledged that horizontal floor accelerations increased with the 3D isolation system compared to horizontal only seismic isolation, but argued that the responses were acceptable. A residential building in California isolated with the GERB system was shaken strongly in the 1994 Northridge Earthquake [99]. The maximum recorded floor acceleration was about 0.63g, relative to input PGA < 0.5g; thus the horizontal isolation was less effective than a typical isolation system. Observed vertical ground acceleration was not significant (PGA $\approx$ 0.1 g).

Prior to the disaster at Fukushima in the 2011 Great East Japan Earthquake and Tsunami, Japan actively pursued 3D seismic isolation approaches for nuclear facilities. Several approaches have been developed that use air springs or pressurized air in the vertical direction [100]. The rolling seal type air spring is a steel/concrete cylinder lowered into an air cavity and attached with a rolling rubber seal, and is configured in series with a laminated rubber bearing for horizontal isolation [101,102]. The cable reinforced air spring consists of an inner cylinder attached to the base and an outer cylinder attached to the structure separated by an air cavity bounded by a flexible rubber sheet [103,104]. The distance between the cylinders allows the device to move both horizontally and vertically and thus accommodate horizontal and vertical motion. Also used in series with laminated rubber bearings, the hydraulic system consists of load carrying hydraulic cylinders filled with nitrogen gas, to which fluctuating pressure can be transmitted by the attached accumulator units [105]. As the flexibility of the isolation system in the vertical direction increases, the structure tends to develop a rocking mode with a substantial modal mass participation factor due to increased coupling in the horizontal and vertical directions. The proposed systems have vertical isolation periods on the order of 1–2 seconds, and generally utilize dampers (oil dampers or viscous wall dampers) and rocking suppression devices to control both vertical and rocking displacements.

A commercial solution for 3D isolation is available through Shimizu Corporation in Japan, and has been implemented in at least one 3-story apartment building [106,107]. Each device consists of a laminated rubber bearing on a steel frame that transmits the loads to three air springs and three shear force transmitting vertical sliders. An oil damper system—two oil dampers connected by cross-coupled pipes—provides both vertical and rocking suppression. The system is quite complicated and costly.

# 5. CONCLUSIONS

This review of shake table testing of base-isolated building has shown that a multitude of devices have been experimentally validated to perform as intended, which is to provide high attenuation of the input excitation to allow the structural system to remain elastic under large horizontal ground shaking. Of the many devices that have been tested, a handful of devices (elastomeric bearings, lead-rubber bearings, Friction Pendulum™, and Triple Pendulum™) claim the majority of the market share in the United States and around the world. Sufficient research has been conducted to suggest that base-isolated buildings can be detailed to survive earthquake events larger than anticipated in design, through activation of one or more ultimate limit state behaviors such as: unrestrained uplift, rupture of elastomeric bearings under tension and shear, buckling of elastomeric bearings, engagement of a displacement restraint or large displacement hardening. Many of these ultimate behaviors of the isolation system will redistribute demands to the superstructure, which should ultimately be detailed for ductile response.

The recent shake table tests of full-scale base isolated buildings at E-Defense have provided insight into the performance of the LRB-CLB and TP isolation systems subjected to tri-directional earthquake ground shaking and have underscored existing knowledge gaps regarding the performance of secondary systems in isolated buildings during representative earthquake ground shaking. Specifically, nonstructural components and content were shown to be vulnerable to damage or disruption under long duration motions generating large floor velocities and motions with large vertical components of excitation. Damage and disruption to these systems can adversely affect the post-event functionality of critical base-isolated facilities, and prevent the performance objectives from being achieved. While the tests at E-Defense provided some data on the demands imposed on these systems, the data is limited due to the large cost associated with performing these tests.

The results from testing at E-Defense have several implications. First, improved understanding of the response of nonstructural components and sensitive equipment is needed for analysis/design of base-isolated buildings. Seismic fragilities for nonstructural systems has generally associated damage with a single demand parameter, such as horizontal floor acceleration. However, the testing suggests that observed damage may be more complex, relating to multiple factors such as peak horizontal acceleration, peak vertical acceleration, and frequency or spectral content of the accelerations. The demands imposed on nonstructural components in base-isolated buildings have different proportions of horizontal to vertical response and different frequency content than conventional buildings, and seismic fragility data of nonstructural systems should be developed in such a way that the results are broadly applicable to all types of buildings.

Second, for buildings that aim to provide post-event functionality following a large earthquake, it is clear that vertical ground acceleration needs to be considered in design. Currently, vertical ground acceleration is rarely incorporated into the response history analysis methods that are often used for the design of critical facilities. To effectively consider vertical excitation, the engineering profession is in need of guidance on: (1) selection and scaling of vertical ground motions that are compatible with the horizontal motions; (2) analytical modeling techniques that can be employed in commercial software to accurately reflect local vertical modes and frequencies related to vibration of the floor system; and (3) accurate analytical models for vertical force-deformation behavior. Floor system modeling techniques should be verified with vibration data of full-scale slabs in realistic structural systems under seismic demands. In particular, for steel framing with concrete floor slabs, the influence of slab to girder connection configuration (fully-composite, partially composite, and non-composite) should be considered. With regard to modeling of the bearings, analytical models for the horizontal force-displacement behavior have outpaced the analytical models for the vertical force-displacement behavior. For example, a simple linear model is typically used to represent the vertical force-displacement behavior of elastomeric bearings while the true behavior is highly nonlinear (see Figure 2b). For FP class of bearings, techniques for evaluating compressive stiffness should be verified and techniques to account for impact upon uplift should be developed and experimentally verified.

Finally, closely associated with the need to evaluate the influence of vertical excitation in the design process is the desire to mitigate the influence of vertical excitation through design. Multiple approaches may be possible. For instance, vertical floor vibrations may be reduced in part through improved slab design. A fundamental understanding of the influence of slab vibration characteristics on the vertical vibration demands and nonstructural component response is needed. Furthermore, limited evidence—described in this review—has suggested that structural configuration irregularities (torsional and vertical) have contributed to horizontal-vertical coupling phenomena observed in some tests.

An alternative is to develop a packaged approach to base isolation that can provide both horizontal and vertical attenuation. To support the revival of the United States nuclear industry, base isolation is being investigated as a viable solution to satisfy Nuclear Regulatory Commission (NRC) guidelines

[108] for seismic design based on a 100,000 year return period [109], and a packaged 3-dimensional base-isolation solution is considered highly desirable [109]. With vertical stiffness that is several thousand times the horizontal stiffness, current widely used seismic isolation devices provide isolation only in the horizontal direction, and may actually align the system with dominant frequencies of the vertical ground shaking. Past efforts to achieve 3-dimensional isolation have not produced a viable or cost-effective system. Stability considerations preclude low shape factor bearings from being a viable solution as only modest vertical periods, e.g., 0.33 seconds [94], can be achieved. While the Japanese systems might be effective, they are complex, have not been independently verified and their high cost and proprietary nature reduce the likelihood of adoption outside of Japan.

In summary, more research is needed to: (1) develop data on the 3-dimensional response of buildings isolated on current isolation systems; (2) evaluate existing analytical and numerical models of conventional isolation hardware for predicting demands imposed on the primary structural system as well as secondary and nonstructural systems; and (3) develop improved data on the seismic fragility of secondary and nonstructural systems under a variety of demands. Further research could show that complete protection of secondary and nonstructural systems using conventional isolation approaches might not be possible. Recognizing that the integrity of these systems is paramount to the post-event functionality of critical facilities, alternative or new systems to achieve complete 3-dimensional protection might be necessary.

# REFERENCES


1. Touaillon, J. Improvement in Buildings. U.S. Patent 99,973, 15 February 1870.
2. Constantinou, M.C.; Whittaker, A. Performacne of Seismic Isolation Hardware under Service and Seismic Loading; Technical Report MCEER-07-0012; Multidisciplinary Center for Earthquake Engineering Research, State University of New York at Buffalo: Buffalo, NY, USA, 2008.
3. Fenz, D.; Constantinou, M.C. Behaviour of the double concave Friction Pendulum bearing. Earthq. Eng. Struct. Dyn. 2006, 35, 1403–1424.
4. Naeim, F.; Kelly, J.M. Design of Seismic Isolated Structures: From Theory to Practice, 1st ed; John Wiley and Sons: Hoboken, NJ, USA, 1999.

5. Taylor, A.; Aiken, I. What's Happened to Seismic Isolation of Buildings in the U.S.? Structure, 11 March 2011. Available online: http://www.structuremag.org/article.aspx?articleID=1404 (accessed on 30 July 2012).

6. Clarke, C.S.J.; Buchanan, R.; Efthymiou, M. Structural platform solution for seismic arctic environments-Sakhalin II offshore facilities. In Proceedings of the Offshore Technology Conference, Houston, TX, USA, 2-5 May 2005.

7. Kelly, J.M. Aseismic base isolation: Review and bibliography. Soil Dyn. Earthq. Eng. 1986, 5, 202–216.

8. Buckle, I.B; Mayes, R.M. Seismic isolation: History, application, and performance—A world view. Earthq. Spectra 1990, 6, 161–201.

9. Taylor, A.W.; Lin, A.N.; Martin, J.W. Performance of elastomers in isolation bearings: A literature review. Earthq. Spectra 1992, 8, 279–303.

10. Soong, T.T.; Constantinou, M.C. Passive and Active Structural Vibration Control in Civil Engineering; Springer-Verlag: New York, NY, USA, 1994.

11. Kunde, M.C.; Jangid, R.S. Seismic behavior of isolated bridges: A-state-of-the-art review. Electron. J. Struct. Eng. 2003, 3, 140–170.

12. Symans, M.D.; Cofer, W.F.; Fridley, K.J. Base isolation and supplemental damping systems for seismic protection of wood structures: Literature review. Earthq. Spectra 2003, 18, 549–572.

13. Taylor, A.W.; Igusa, T. Primer on Seismic Isolation; ASCE and Task Committee on Seismic Isolation: Reston, VA, USA, 2004.

14. Higashino, M.; Okamoto, S. Response Control and Seismic Isolation of Buildings; SPON Press: London, UK, 2006.

15. Lake, G.J.; Lindley, P.B. Ozone Attack and Fatigue life of Rubber; Maclaren and Sons LTD: London, UK, 1967; pp. 56–71.

16. Chalhoub, M.S.; Kelly, J.M. Effect of bulk compressibility on the stiffness of cylindrical base isolation bearings. Int. J. Solids Struct. 1990, 26, 743–760.

17. Constantinou, M.C.; Kartoum, A.; Kelly, J.M. Analysis of compression of hollow circular elastomeric bearings. Eng. Struct. 1992, 14, 103–111.

18. Highway Innovative Technology Evaluation Center, Evaluation Findings for Skellerup Base Isolation Elastomeric Bearings; Civil Engineering Research Foundation: Washington, DC, USA, 1998.

19. Highway Innovative Technology Evaluation Center, Evaluation Findings for Dynamic Isolation Systems, Inc. Elastomeric Bearings; Civil Engineering Research Foundation: Washington, DC, USA, 1998.

20. Kelly, J.M. Dynamic Failure Characteristics of Bridgestone Isolation Bearings; Report No. UCB/EERC–91/04; University of California: Berkeley, CA, USA, 1991.

21. Kelly, J.M. Earthquake Resistant Design with Rubber, 2nd ed; Springer Verlag: New York, NY, USA, 1997.

22. Nagarajaiah, S.; Ferrell, K. Stability of elastomeric isolation bearings. J. Struct. Eng. 1999, 125, 946–954.

23. Warn, G.P.; Whittaker, A.S. A Study of the Coupled Horizontal-Vertical Behavior of Elastomeric and Lead-rubber Seismic Isolation Bearings; Technical Report MCEER-06-0011; Multidisciplinary Center for Earthquake Engineering Research: Buffalo, NY, USA, 2006.

24. Gent, A.N. Fracture of glassy polymers. ASME Appl. Mech. Div. 1976, 19, 55–68.

25. Kramer, E.J. Microscopic and molecular fundamentals of crazing. Adv. Polym. Sci. 1983, 52-53, 1–56.

26. Open System for Earthquake Engineering Simulation Framework, version 2.1.0; University of California: Berkeley: CA, USA, 2009. Available online: http://opensees.berkeley.edu/ (accessed on 20 July 2012).

27. Robinson, W.H. A lead-rubber shear damper. Bull. N. Z. Natl. Soc. Earthq. Eng. 1977, 3, 93–101.

28. Robinson, W.H. Lead-rubber hysteretic bearings suitable for protecting structures during earthquakes. Earthq. Eng. Struct. Dyn. 1982, 10, 593–604.

29. Skinner, R.I.; Robinson, W.H.; McVerry, G.H. An Introduction to Seismic Isolation; John Wiley and Sons Inc.: New York, NY, USA, 1993.

30. Nagarajaiah, S.; Reinhorn, A.M.; Constantinou, M.C. Nonlinear dynamic analysis of 3-D-base-isolated structures. J. Struct. Eng. 1991, 117, 2035–2054.

31. Kalpakidis, I.V.; Constantinou, M.C. Effects of heating on the behavior of lead-rubber bearings. I: Theory. J. Struct. Eng. 2009, 135, 1440–1449.

32. Kalpakidis, I.V.; Constantinou, M.C. Effects of heating on the behavior of lead-rubber bearings. II: Verification of theory. J. Struct. Eng. 2009, 135, 1450–1461.

33. Kalpakidis, I.V.; Constantinou, M.C.; Whittaker, A.S. Modeling strength degradation in lead-rubber bearings under earthquake shaking. Earthq. Eng. Struct. Dyn. 2010, 39, 1533–1549.

34. Kalpakidis, I.V.; Constantinou, M.C. Principles of scaling and similarity for testing of lead-rubber bearings. Earthq. Eng. Struct. Dyn. 2010, 39, 1551–1568.

35. Ryan, K.L.; Kelly, J.M.; Chopra, A.K. Nonlinear model for lead-rubber bearings including axial-load effects. J. Eng. Mech. 2005, 131, 1270–1278.

36. Iizuka, M. A macroscopic model for predicting large-deformation behaviors of laminated rubber bearings. Eng. Struct. 2000, 22, 323–334.

37. Yamamoto, S.; Kikuchi, M.; Ueda, M.; Aiken, I.D. A mechanical model for elastomeric seismic isolation bearings include the influence of axial load. Earthq. Eng. Struct. Dyn. 2009, 38, 157–180.

38. Kikuchi, M.; Nakamura, T.; Aiken, I.D. Three-dimensional analysis for square seismic isolation bearings under large shear deformations and high axial loads. Earthq. Eng. Struct. Dyn. 2010, 39, 1513–1531.

39. Sanchez, J.; Masroor, A.; Mosqueda, G.; Ryan, K. Static and dynamic stability of elastomeric bearings for seismic protection of structures. J. Struct. Eng. 2012. accepted for publication.

40. Zayas, V.; Low, S.; Mahin, S. The FPS Earthquake Resisting System; Report No. UCB/EERC-87/01; University of California: Berkeley, CA, USA, 1987.

41. Fenz, D.; Constantinou, M.C. Spherical sliding isolation bearings with adaptive behavior: Theory. Earthq. Eng. Struct. Dyn. 2008, 37, 163–183.

42. Fenz, D.; Constantinou, M.C. Spherical sliding isolation bearings with adaptive behavior: Experimental verification. Earthq. Eng. Struct. Dyn. 2008, 37, 185–205.

43. Becker, T.C.; Mahin, S.A. Experimental and analytical study of the bi-directional behavior of the triple friction pendulum isolator. Earthq. Eng. Struct. Dyn. 2012, 41, 355–373.

44. Mokha, A.; Constantinou, M.; Reinhorn, A. Teflon bearings in base isolation I: Testing. J. Struct. Eng. 1990, 116, 438–454.

45. Earthquake Protection Systems. Available online: http://www.earthquakeprotection.com/TechnicalCharacteristicsofFPBearngs.pdf (accessed on 20 April 2012).

46. Constantinou, M.; Mokha, A.; Reinhorn, A. Teflon bearings in base isolation II: Modeling. J. Struct. Eng. 1990, 116, 455–474.

47. Sarkisian, M.; Lee, P.; Hu, L.; Doo, C.; Zayas, V.; Constantinou, M.; Bachman, R. Property verification of triple pendulumtm seismic isolation bearings. In Proceedings of the 20th Analysis and Computation Specialty Track (2012 ASCE Structures Congress), Chicago, IL, USA, 29-31 March 2012.

48. Ryan, K.L.; Dao, N.D.; Sato, E.; Sasaki, T.; Okazaki, T. Aspects of isolation device behavior observed from full-scale testing of an isolated building at E-defense. In Proceedings of the 20th Analysis and Computation Specialty Track (2012 ASCE Structures Congress), Chicago, IL, USA, 29-31 March 2012.

49. Fenz, D.; Constantinou, M.C. Modeling triple friction pendulum bearings for response-history analysis. Earthq. Spectra 2008, 24, 1011–1028.

50. Rousis, P.C.; Constantinou, M.C. Uplift-restraining friction pendulum seismic isolation system. Earthq. Eng. Struct. Dyn. 2006, 35, 577–593.

51. Roussis, P.C.; Constantinou, M.C. Experimental and analytical studies of structures seismically isolated with an uplift-restraining friction pendulum bearing system. Earthq. Eng. Struct. Dyn. 2006, 35, 595–611.

52. Roussis, P.C. Study on the effect of uplift-restraint on the seismic response of base-isolated structures. ASCE J. Struct. Eng. 2009, 135, 1462–1471.

53. Kelly, J.M.; Skinner, M.S.; Beucke, K.E. Experimental Testing of an Energy Absorbing Seismic Isolation System; Report No. UCB/EERC-80/35; University of California,Berkeley: CA, USA, 1980.

54. Kelly, J.M.; Beucke, K.E.; Skinner, M.S. Experimental Testing of a Friction Damped Aseismic Base Isolation System with Fail-safe Characteristics; Report No. UCB/EERC-80/18; University of California: Berkeley, USA, 1980.

55. Kelly, J.M.; Beucke, K.E. A friction damped base isolation system withfail-safe characteristics. Earthq. Eng. Struct. Dyn. 1983, 11, 33–56.

56. Kelly, J.M.; Hodder, S.B. Experimental Study of Lead and Elastomeric Dampers for Base Isolation Systems; Report No. UCB/EERC-81/16; University of California: Berkeley, CA, USA, 1981.

57. Kelly, J.M.; Chalhoub, M.S. Earthquake Simulator Testing of a Combined Sliding Bearing and Rubber Bearing Isolation System; Report No. UCB/EERC-87/04; University of California: Berkeley, CA, USA, 1990.

58. Mokha, A.; Constantinou, M.C.; Reinhorn, A.M. Teflon Bearings in Aseismic Base Isolation: Experimental Studies and Analytical Modeling; Technical Report NCEER-88-0038; State University of New York at Buffalo: Buffalo, NY, USA, 1988.

59. Mokha, A.; Constantinou, M.C.; Reinhorn, A.M. Experimental Study and Analytical Prediction of Earthquake Response of a Sliding Isolation System with Spherical Surface; Technical Report NCEER-90-0020; State University of New York at Buffalo: Buffalo, NY, USA, 1990.

60. Mokha, A.; Amin, N.; Constantinou, M.C.; Zayas, V. Experimental study of friction pendulum isolation system. ASCE J. Struct. Eng. 1991, 117, 1201–1217.

61. Al-Hussaini, T.M.; Constantinou, M.C.; Zayas, V.A. Seismic Isolation of Multi-Story Frame Structures Using Spherical Sliding Isolation System; Technical Report NCEER-94-0007; State University of New York at Buffalo: Buffalo, NY, USA, 1994.

62. Constantinou, M.C.; Mokha, A.S.; Reinhorn, A.M. Experimental and Analytical Study of a Combined Sliding Disc Bearing and Helical Steel Spring Isolation System; Technical Report NCEER-90-0019; State University of New York at Buffalo: Buffalo, NY, USA, 1990.

63. Constantinou, M.C.; Mokha, A.S.; Reinhorn, A.M. Study of sliding bearing and helical-steel-spring isolation system. ASCE J. Struct. Eng. 1991, 117, 1257–1275.

64. Mostaghel, N.; Khodaverdian, M. Dynamics of resilient-friction base isolator (R-FBI). Earthq. Eng. Struct. Dyn. 1987, 15, 379–390.

65. Clark, P.W.; Kelly, J.M. Experimental Testing of the Resilient-Friction Base Isolation System; Report No. UCB/EERC-90/10; University of California: Berkeley, CA, USA, 1990.

66. Kelly, J.M. Base Isolation in Japan; Report No. UCB/EERC-88/20; University of California: Berkeley, CA, USA, 1988.

67. Nagashima, I.; Kawamura, S.; Kitazawa, K.; Hisano, M. Study on a base isolation system. In Proceedings of the 3rd Conference on Soil Dynamics and Earthquake Engineering, Princeton University, Princeton, NJ, 22-24 June 1987.

68. Bakhshi, A.; Araki, H.; Shimazu, T. Evaluation of the performance of a suspension isolation system subjected to strong ground motion. Earthq. Eng. Struct. Dyn. 1998, 27, 29–47.

69. Guerreiro, L.; Azevedo, J.; Muhr, A. Seismic tests and numerical modeling of a rolling-ball isolation system. J. Earthq. Eng. 2007, 11, 49–66.

70. Tashkov, L.; Manova, K.; Krstevska, L.; Garevski, M. Evaluation of efficiency of ALSC floating-sliding base-isolation system based on shake table test and floor response spectra. Bull. Earthq. Eng. 2010, 8, 995–1018.

71. Toopchi-Nezhad, H.; Tait, M.J.; Drysdale, R.G. Shake table study on an ordinary low-rise building seismically isolated with SU-FREIs (stable unbounded-fiber reinforced elastomeric isolators). Earthq. Eng. Struct. Dyn. 2009, 38, 1335–1357.

72. Nakamura, Y.; Saruta, M.; Wada, A.; Takeuchi, T.; Hikone, S.; Takahashi, T. Development of the core-suspended isolation system. Earthq. Eng. Struct. Dyn. 2011, 40, 429–447.

73. Fenz, D.M.; Constantinou, M.C. Development, Implementation, and Verification of Dynamic Analysis Models for Multi-spherical Sliding Bearings; Technical Report MCEER-08-0018; Multidisciplinary Center for Earthquake Engineering Research, State University of New York at Buffalo: Buffalo, NY, USA, 2008.

74. Morgan, T.A.; Mahin, S.A. The Use of Base Isolation Systems to Achieve Complex Seismic Performance Objectives; PEER Report No. 2011/06; Pacific Earthquake Engineering Research Center, University of California: Berkeley, CA, USA, 2011.

75. Kelly, J.M. The Influence of Base Isolation Systems on the Seismic Response of Light Secondary Equipment; Report No. UCB/EERC-81/17; Earthquake Engineering Research Center, University of California: Berkeley, CA, USA, 1981.

76. Kelly, J.M.; Tsai, H.C. Seismic response of light internal equipment in base-isolated structures. Earthq. Eng. Struct. Dyn. 1985, 13, 711–732.

77. Juhn, G.; Manolis, G.; Constantinou, M.C.; Reinhorn, A.M. Experimental investigation of secondary systems in a base-isolated structure. ASCE J. Struct. Eng. 1992, 118, 2204–2221.

78. Wolff, E.D.; Constantinou, M.C. Wolff, E.D.; Constantinou, M.C. Experimental Study of Seismic Isolation Systems with Emphasis on Secondary System Response and Verification of Accuracy of Dynamic Response History Analysis Methods; Technical Report MCEER-04-0001; Multidisciplinary Center for Earthquake Engineering Research, State University of New York at Buffalo: Buffalo, NY, USA, 2004.

79. Griffith, M.C.; Kelly, J.M.; Coveney, V.A.; Koh, C.G. Experimental Evaluation of Seismic Isolation of Medium-Rise Structures Subjected to Uplift; Report No. UCB-EERC 88-02; University of California: Berkeley, CA, USA, 1988.

80. Griffith, M.C.; Aiken, I.D.; Kelly, J.M. Experimental Evaluation of Seismic Isolation of a 9-story Braced Steel Frame Subjected to Uplift; Report No. UCB-EERC 88-05; University of California: Berkeley, CA, USA, 1988.

81. Kelly, J.M.; Griffith, M.C.; Aiken, I.D. A Displacement Control and Uplift Restraint Device for Base-Isolated Structures; Report No. UCB-EERC 87-03; University of California: Berkeley, CA, USA, 1987.

82. Griffith, M.C.; Aiken, I.D.; Kelly, J.M. Displacement control and uplift restraint for base-isolated structures. ASCE J. Struct. Eng. 1990, 116, 1135–1148.

83. Nagarajaiah, S.; Reinhorn, A.M.; Constantinou, M.C. Experimental study of sliding isolated structures with uplift restraint. ASCE J. Struct. Eng. 1992, 118, 1666–1682.

84. Kasalanati, A.; Constantinou, M.C. Testing and modeling of prestressed isolators. ASCE J. Struct. Eng. 2005, 131, 857–866.

85. Toniolo, R. THK:CLB crossed Linear Bearing Seismic Isolators. In Proceedings of the 2008 Seismic Engineering Conference: Commemorating the 1908 Messina and Reggio Calabria Earthquake, Reggio Calabria, Italy, 8-11 July 2008.

86. Clark, P.W.; Aiken, I.D.; Kelly, J.M. Experimental Studies of the Ultimate Behavior of Seismically Isolated Structures; Report No. UCB/EERC-97/18; University of California: Berkeley, CA, USA, 1997.

87. Sato, N.; Kato, A.; Fukushima, Y.; Iizuka, M. Shaking table tests on failure characteristics of base isolation system for a DFBR plant. Nuclear Eng. Des. 2002, 212, 293–305.

88. Takaoka, E.; Takenaka, Y.; Nimura, A. Shaking table tests and analysis method on ultimate behavior of slender base-isolated structure supported by laminated rubber bearings. Earthq. Eng. Struct. Dyn. 2011, 40, 551–570.

89. Hwang, J.-S.; Hsu, T.-Y. Experimental study of isolated building under triaxial ground excitations. ASCE J. Struct. Eng. 2000, 126, 879–886.

90. Sato, E.; Furukawa, S.; Kakehi, A.; Nakashima, M. Full-scale shaking table test for examination of safety and functionality of base-isolated medical facilities. Earthq. Eng. Struct. Dyn. 2011, 40, 1435–1453.

91. Soroushian, S.; Ryan, K.L.; Maragakis, M.; Sato, E.; Sasaki, T.; Okazaki, T.; Tedesco, L.; Zaghi, A.E.; Mosqueda, G.; Alvarez, D. Seismic response of ceiling/sprinkler piping nonstructural systems in NEES TIPS/NEES nonstructural/NIED collaborative tests on a full scale 5-story building. In Proceedings of the 2012 ASCE Structures Congress, Chicago, IL, USA, 29-31 March 2012.

92. Warn, G.P.; Whittaker, A.S.; Constantinou, M.C. Vertical stiffness of elastomeric and lead-rubber seismic isolation bearings. ASCE J. Struct. Eng. 2007, 133, 1227–1236.

93. Aiken, I.D.; Kelly, J.M.; Tajirian, F.F. Mechanics of Low Shape Factor Elastomeric Seismic Isolation Bearings; Report No. UCB/EERC-89/13; University of California: Berkeley, CA, USA, 1989.

94. Tajirian, F.F.; Kelly, J.M.; Aiken, I.D.; Veljovich, W. Elastomeric bearings for three-dimensional seismic isolation. In Proceedings of

the 1990 ASME PVP Conference, Nashville, TN, USA, 17-21 June 1990.

95. Buckle, I.G.; Liu, H. Critical loads of elastomeric isolators at high shear strain. In Proceedings of the 3rd U.S.-Japan Workshop on Earthquake Protective Systems for Bridges, Berkeley, CA, USA, 24-25 January 1994.

96. Buckle, I.G.; Liu, H. Experimental determination of critical loads of elastomeric isolators at high shear strain. NCEER Bull. 1994, 8, 1–5.

97. Buckle, I.G.; Nagarajaiah, S.; Ferrell, K. Stability of elastomeric isolation bearings: Experimental study. ASCE J. Struct. Eng. 2002, 128, 3–11.

98. Huffman, G.K. Full base isolation for earthquake protection by helical springs and viscous dampers. Nuclear Eng. Des. 1985, 84, 331–338.

99. Makris, N.; Deoskar, H.S. Prediction of observed response of base-isolated structure. ASCE J. Struct. Eng. 1996, 122, 485–493.

100. Inoue, K.; Morishita, M.; Fujita, T. Development of three-dimensional seismic isolation technology for next generation nuclear power plant in Japan. In Proceedings of the ASME/JSME 2004 Pressure Vessels and Piping Conference (PVP2004), San Diego, CA, USA, 25-29 July 2004.

101. Suhara, J.; Tamura, T.; Ohta, K.; Okada, Y.; Moro, S. Research on 3-D base isolation system applied to new power reactor 3-d seismic isolation device with rolling seal type air spring: Part 1. In Proceedings of the 17th International Conference on Structural Mechanics in Reactor Technology (SMiRT 17), Prague, Czech Republic, 17-22 August 2003.

102. Suhara, J.; Matsumoto, R.; Oguri, S.; Okada, Y.; Inoue, K.; Takahashi, K. Research on 3-D base isolation system applied to new power reactor, 3-d seismic isolation device with rolling seal type air spring: Part 2. In Proceeding of 18th International Conference on Structural Mechanics in Reactor Technology, Beijing, China, 7-12 August 2005.

103. Kageyama, M.; Iba, T.; Umeki, K.; Somaki, T.; Moro, S. Development of three-dimensional base isolation system with cable reinforcing air spring. In Proceedings of the 17th International Conference on Structural Mechanics in Reactor Technology (SMiRT 17), Prague, Czech Republic, 17-22 August 2003.

104. Kageyama, M.; Iba, T.; Umeki, K.; Somaki, T.; Hino, Y.; Moro, S.; Ikutama, S. Study on three-dimensional seismic isolation system for next generation nuclear power plant: independent cable reinforced rolling-seal air spring. In Proceedings of the 13th World Conference on Earthquake Engineering, Vancouver, BC, Canada, 1-6 August 2004.

105. Kashiwazaki, A.; Shimada, T.; Fujikawa, T.; Moro, S. Study on 3-dimensional base isolation system applying to new type power plant reactor (hydraulic 3-dimensional base isolation system: No.1). In Proceedings of the 17th International Conference on Structural Mechanics in Reactor Technology (SMiRT 17), Prague, Czech Republic, 17-22 August 2003.

106. Takahashi, O.; Aida, H.; Suhara, J.; Matsumoto, R.; Tsuyuki, Y.; Fujita, T. Construction of civil building using three dimensional seismic isolation system: Part 1, design of building using three dimensional seismic isolation system. In Proceedings of the 14th World Conference on Earthquake Engineering, Beijing, China, 12-17 October 2008.

107. Suhara, J.; Matsumoto, R.; Torita, H.; Tsuyuki, Y.; Kamei, T.; Takahashi, O.; Kunimatsu, Y.; Aida, H.; Fujita, T. Construction of civil building using three dimensional seismic isolation system: Part 2, tests for three dimensional seismic isolation system. In Proceedings of the 14th World Conference on Earthquake Engineering, Beijing, China, 12-17 October 2008.

108. USA Nuclear Regulatory Commission, Identification and characterization of seismic sources and determination of safe shutdown earthquake ground motion. In USA Nuclear Regulatory Commission Regulatory Guide 1.165; USA Nuclear Regulatory Commission: Rockville, MD, USA, 1997.

109. Malushte, S.R.; Whittaker, A.S. Survery of past base isolation applications in nuclear power plants and challenges to industry/regulatory acceptance. In Proceedings of the 18th International Conference on Structural Mechanics in Reactor Technology (SMiRT 18), Beijing, China, 7-12 August 2005.

# 9
# CHAPTER

# STABILITY AND RUN-OUT ANALYSIS OF EARTHQUAKE-INDUCED LANDSLIDES

**Yingbin Zhang**[1]

*[1] Department of Geotechnical Engineering, School of Civil Engineering, Southwest Jiaotong University, Chengdu, China*

## 1. INTRODUCTION

A large number of landslides can be caused by a strong earthquake and they have been the source of significant damage and loss of people and property. Therefore, it is very important to predict the stability of slope and the movement behaviors of a potential landslide under an earthquake loading, i.e., stability and run-out analysis (Figure 1).

Earthquake-induced landslides have been the source of significant damage and loss of people and property. One of the most serious event is the 1970 Peru earthquake. This event caused a huge rock avalanche that killed almost 54,000 people and buried two cities [143]. Another example is, in the 1920 Haiyuan earthquake, a large number of landslides caused widespread damage to infrastructure and buildings and killed at least 100,000 people, almost half of the total earthquake deaths [82].

Therefore, it is very important to predict the earthquake-induced landslides and to take countermeasures for potential landslides.

Main topics of earthquake-induced landslides are the following:

1. Investigation of recent and historical earthquake-induced landslides and their impacts so as to produce inventories of historical earthquake-induced landslides
2. Prediction of potential earthquake-induced landslides, including (i) failure mechanism and stability analysis of seismic slopes, (ii) movement mechanism and behaviours of earthquake-induced landslides, and (iii) Instrumentation and monitoring technologies for potential earthquake-induced landslides or post-earthquake landslides.
3. Preventive countermeasures for earthquake-induced landslides, including (i) Stabilization and disaster mitigation of earthquake-related landslides, (ii) risk assessment and management of earthquake-related landslides, and (iii) hazard map and early warning system for earthquake-related landslides

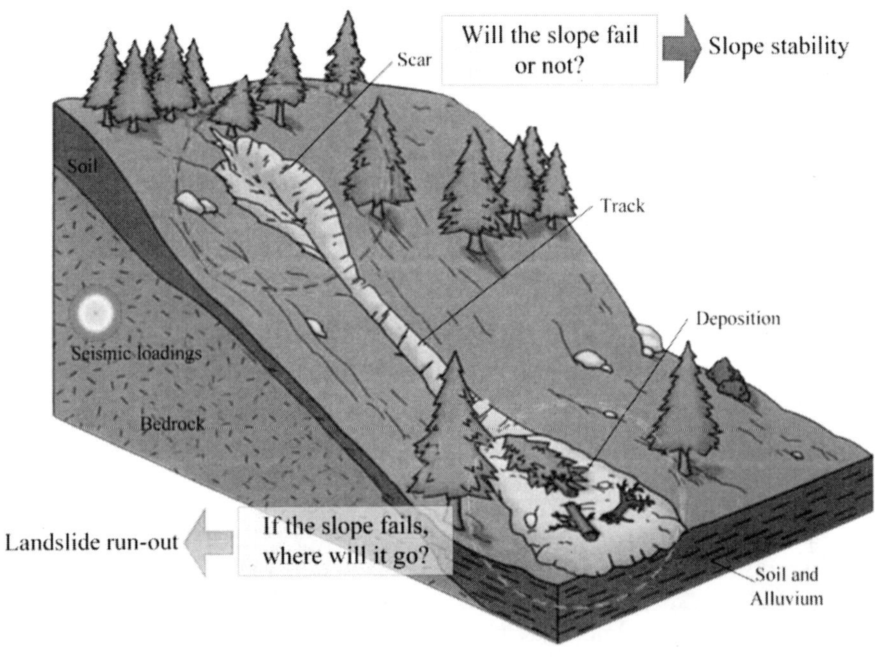


**Figure 1.** Stability and run-out analysis of earthquake-induced landslides

This chapter focuses on the prediction of potential earthquake-induced landslides. The prediction of potential landslide can be carried out using detailed geotechnical investigations and stability calculations. (i) Failure mechanism and stability analysis of seismic slopes, i.e. seismic slope stability analysis and (ii) movement mechanism and behaviours of earthquake-induced landslides, i.e. landside run-out analysis are outlined firstly, and then the merits and demerits of each method are clarified in this chapter.

# 2. SEISMIC SLOPE STABILITY ANALYSIS

So far, methods developed to analyze the stability of earthquake slopes can be divided into three types: (1) pseudo-static methods, (2) dynamic sliding block methods, and (3) stress-strain methods. These three types of methods can be applied in different cases due to each of them has merit and demerit [73].

## 2.1. Pseudo-static methods

[166] first presented the pseudo-static method, which is a simple method for evaluating of seismic stability of a slope. This type of method can be used to man-made or natural slopes based on either analytical method or numerical method. The earthquake force, acting on the an element or whole of the slope, is writed by a horizontal force and/or a vertical volum force equal to the gravitation force multiple a coefficient $k$, called the pseudo-static coefficient as shown in Figure 2 and Equation (1).

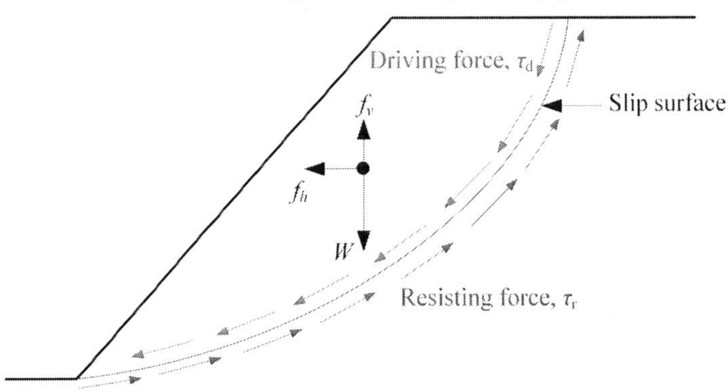


**Figure 2.** Forces acting on a slope in pseudo-static slope stability analysis

Thus, $k$ times the gravitational acceleration $g$, i.e. $a=kg$ forms the assumed seismic acceleration $a$. The assumed pseudo-static forces acting on a potential sliding mass of weight $W$ will be

$$f_h = \frac{a_h W}{g} = k_h W$$

$$f_v = \frac{a_v W}{g} = k_v W \qquad (1)$$

where $a_h$ and $a_v$ are horizontal and vertical pseudo-static accelerations, respectively, $k_h$ and $k_v$ are horizontal and vertical pseudo-static coefficients, respectively. The factor of safety(FOS) is represented as the ratio of the resisting force to the driving force, Equation (2).

$$\text{FOS}=\frac{\tau_r}{\tau_d} \qquad (2)$$

From Equation (1), the pseudo-static force is determined by the seismic coefficient. The key problem for the pseudo-static procedure is how to select an appropriate seismic coefficient under an acceptable FOS. There have been studies for determining the most appropriate pseudo-static coefficient by a matter of experience and judgment.

[166] classical paper made the original suggestion to use of $k_h$=0.1 for severe earthquakes, $k_h$=0.2 for violent and/or destructive earthquakes, and of $k_h$=0.5 for catastrophic earthquakes.

[103] presented a minimum pseudo-static FOS of 1.5 based on a slope material strength reduction factor (SRF) of 0.8 and the following acceleration values associated with two different earthquake magnitudes $M$. The same values of seismic coefficients for magnitude 6.5 and 8.25 earthquakes are recommended by [154], but with an acceptable FOS of 1.15.

$$a = 0.1g \text{ for } M = 6.5 \text{ implying } k = 0.1$$

$$a = 0.15g \text{ for } M = 8.25 \text{ implying } k = 0.15 \qquad (3)$$

[137] also presented the pseudo-static coefficient related to earthquake magnitude. In detail, for an 8.25, 7.5, 7.0 and 6.5 magnitude earthquakes, if the seismic coefficients equal to 1/2, 1/3, 1/4 and 1/5 of the PGA, respectively, the computed FOSs are larger than 1.0, the accumulated displacements of slope are likely to be acceptably small.

In the report published by the International Commission of Large Dams (ICOLD), [154] shows a list of the minimum FOS value and horizontal seismic coefficients for 14 large dams worldwide, in which the minimum FOSs range from 1.0 to 1.5 and the horizontal earthquake coefficients range from 0.1 to 0.15. The Corps of Engineers Manual recommended a earthquake coefficient of 0.1 or 0.15 for areas where major and great earthquake threats are estimated, respectively, and a FOS of no larger than 1.0 for all magnitude earthquakes.

Some references related the earthquake coefficient value to the peak ground acceleration (PGA) [10, 67, 108]. [108] related a pseudo-static coefficient of 1/3 to 1/2 of the PGA at the top of a double-side slope (a dam in the source reference), whereas [67] related a pseudo-static coefficient of 1/2 of the PGA of bedrock ($PGA_{rock}$) with a FOS of no larger than 1.0 and a SRF of 20%. And, [10] recommended the pseudo-static coefficient of 0.6 or 0.75 times of the PGA of bedrock (0.6 or $0.75PGA_{rock}$). It should be noted that the value given by [10] is conservative because the original study is designed for solid-waste landfills, where the allowable deformation are relatively small. [89] pointed that although engineering judgment is required for all cases, the criteria of [67] should be appropriate for most slopes.

[91] suggested one-half of PGA to use in an area of low seismicity (peak acceleration <0.15g) for the stability of earth embankments. This can be obtained from the peak horizontal motion (mean) from Modified Mercalli Intensity (MMI), magnitude-distance attenuation and the probability of a 50-year, 90% nonexceedance. However, in an area of moderate to strong seismicity ($0.15g \leq PGA \leq 0.40g$), PGA is obtained from the peak horizontal motion, from MMI, magnitude-distance attenuation and probability of 250-year, 90% nonexceedance.

[76] suggested a minimum FOS of 1.0, also based on a slope material SRF of 0.8 and the following values of pseudo-static coefficient: *a* equals to 0.17PGA or 0.5PGA for the dynamic response analysis is to be performed for the slope or earthquake structure or not.

[163] developed an expression for the earthquake coefficient in terms of characters of ground motion and magnitude of earthquake based on the data of [10].

It is almost common that only the horizontal acceleration is considered in evaluating the stability and deformation of a slope because the horizontal acceleration is the principal de-stabilizing force that acts on earth structures as well as the principal source of damage observed in earthquakes [4].

From Figure 2, the horizontal force clearly increases the driving force and decreases the FOS. The vertical pseudo-static force generally has less influence on the FOS than the horizontal pseudo-static force does because the vertical pseudo-static reduces both the driving and resisting forces. Hence, the effects of vertical seismic loading are frequently omitted in pseudo-static analysis [89].

Several investigators performed some analyses and have shown that the inclination of seismic loading have a significant influence on the seismic stability of slope by coupling the vertical and horizontal components of seismic force [20, 100].

In summary, pseudo-static method can be simply and directly used to identify the FOS and the critical seismic coefficient $k_c$. In addition, performance of slope is closely related to permanent displacement, but the results of pseudo-static method are difficult to interpret the performance of slope after a seismic event because this method provides no information about permanent displacement. Because the pseudo-static analysis method provides only a rough assessment of seismic slope, it should be only used for the preliminary procedures. More accurate methods can be used to the followed process [73, 163, 170].

## 2.2. Dynamic sliding block methods

Displacement-based dynamic slidng block method is another alternative approach to evaluate the seismic slope stability, as permanent displacement is a useful index of slope performance, especially for those man-made slopes constructed for special purposes such as dams, embankments et al. This method has been widely used in earthquake geotechnical engineering.

In 1965, [119] proposed the dynamic sliding block method for estmating the permanent displacement of embankment affected by a seismic loading. In this method, sliding would be induced once the seismic loading exceed the critical seismic force of a potential failure surface as shown in Figure 3. The sliding would be accumulated until the end of seismic loading. We can evaluate the accumulated permanent displacement to assess the seismic stability of a slope.

Newmark's method showes that the yield acceleration of a potential block is a function of the FOS and slope angle, as:

$$a_c = (\mathrm{FOS} - 1) g \sin\alpha \tag{4}$$

where $a_c$ is in terms of the gravity acceleration $g$; FOS is the static factor of safety; and $\alpha$ is the slope angle.

Since then, the method has been numerous extensions and applications. The section 2.2.1 and section 2.2.2 will give reviews for these two aspects, respectively. In addition, a regional scale application of the dynamic sliding block method is reviewed in section 2.2.3.

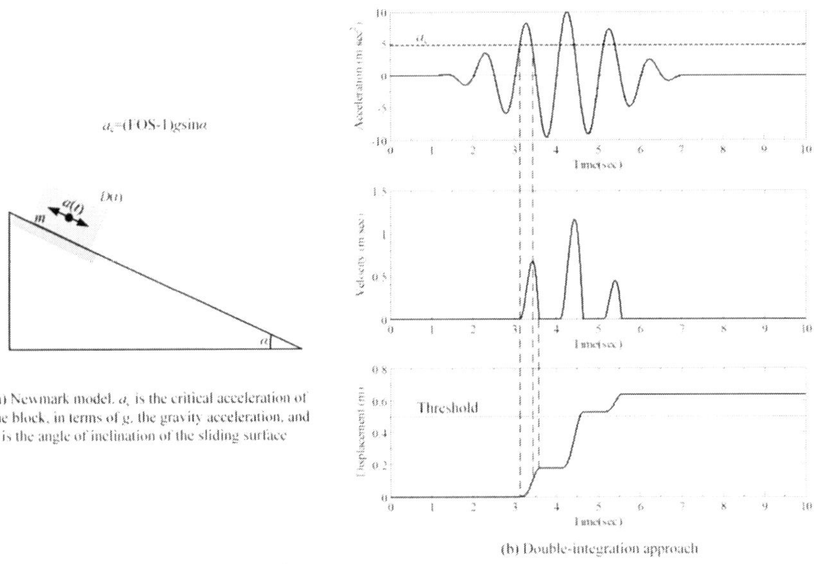


**Figure 3.** Illustration of the original Newmark's method

## 2.2.1. Extensions

More attention has been focused over the last decades on developing methods to more accurately analyze the seismic stability of a slope for dams, embankments or other important structures by modeling the dynamic response of a slope more rigorously.

After the first dynamic sliding rigid block method, [155 and 97] published more sophisticated methods to account for the un-rigid block. Similar studies also given by [103]. As the classification given by [73], methods for estimating the permanent displacement of a sliding system induced by earthquake loading can been grouped into: (1) rigid-block model [119], (2) decoupled model [10, 104], and (3) coupled model [11, 97, 139].

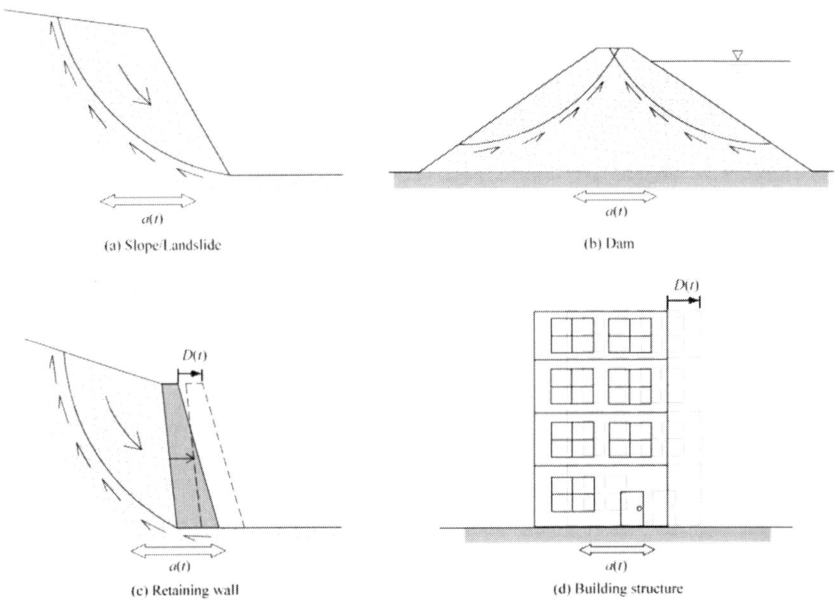


**Figure 4.** Applications of dynamic sliding block method in geotechnical engineering

### 2.2.2. Applications

Since the rigid-block method was published in 1965 by Newmark, it has seen numerous applications, four of which are shown in Figure 4. The applications in recent years include (1) the seismic deformation analysis of earth dams and embankments [1, 22, 48, 49, 89, 90, 97, 103, 138, 144, 145, 150, 155, 179, 180]; (2) the displacements associated with landslides [34, 53, 70, 171]; (3) the seismic deformation of landfills with geosynthetic liners [10, 181]; (4) the seismic settlement of surface foundations [141]; and (5) the potential sliding of concrete gravity dams [32, 47, 95]. The extension of the analogue by [140] to gravity retaining walls has met worldwide acceptance, and has found its way into seismic codes of practice. Several other generalised applications have also appeared (e.g. [2, 3, 45, 99, 139, 162, 169]).

### 2.2.3. Regional scale analysis

Except a single slope analysis, where the landslides are likely to occur and what kind of seismic conditions will cause it failure are two important topics in seismic hazard assessment, i.e.regional scale analysis [59].

For a regional scale analysis, slope stability analysis menthds will be not suitable [143, 168]. With the development of Geographic Information Systems (GIS) tools in recent years, regional scale analyses by the dynamic sliding block method have been proposed, in which ground shaking characteristic parameters, geotechnical material and topographic data are considered (e.g. [34, 71, 75, 106, 114, 151, 155]).

The Newmark analysis (which combines slope stability calculations with seismic ground-motion records) is widely used to evaluate the potential for landslides that could be triggered by earthquake shaking [70, 71, 72, 74, 113].

### 2.3. Stress-strain methods

With the developments of the simulation approach and computer technology in recent years, the stress-strain method is becoming increasingly used in seismic slope stability analysis. These methods can be grouped into continuous methods, e.g. finite element method (FEM) [21], finite difference method (FDM) [116], boundary element method (BEM) [12], and discontinuous methods, e.g. rigid block spring method (RBSM) [77; 80], discontinuous deformation analysis (DDA) [159, 160] and discrete element method (DEM) [31].

#### 2.3.1. Continuous methods

[21] developed and named FEM of engineering analysis, in which the studied system is meshed into small many elements. This method can be applied to estimate the slope stability including dynamic stability analysis.

Some applications of the continuous methods have been proposed, e.g. [89, 94, 153] and [156]. Recently, nonlinear in-elastic soil models have been developed and implemented in two-dimension (2-D) and three-dimension (3-D) models (e.g., [42, 50, 135, 164]). In addition, [93] and [183] studied the seismic slope stability by using FDM.

#### 2.3.2. Discontinuous methods

For the analysis of a potential failure mass consisting of multiple blocks as shown in Figure 5, the discontinuous methods are more applicable [120]. Some applications of RBSM and DEM can be found in some literature (e.g. [8, 52, 77, 79, 80, 85121, 127128, 129, 130, 131, 136, 182]).

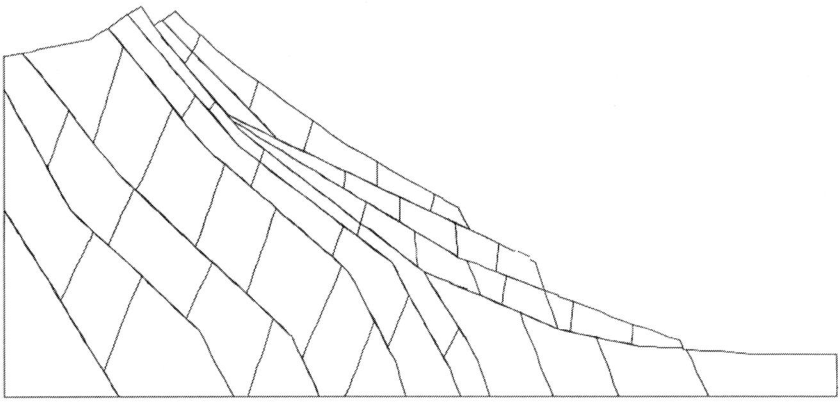

**Figure 5.** A jointed rock slope (modified from Bhasin and Kaynia, 2004)

DDA is also a discontinuous method developed for the modeling of the behaviors of multiple block systems. Since the novel formulation and the numerical code of DDA were presented, DDA draws more and more attention and many extensions and modifications to the original method have been proposed to overcome some limitations [19, 37, 38, 81, 87, 98] and make it more suitable, practical and efficient to seismic slope stability.

The DDA can be used both to static rock slope engineering (e.g. [17, 81, 102, 123, 176, 187]) and the seismic rock slope stability analysis [56, 57, 54].

In summary, stress-strain method represents a powerful alternative approach for seismic slope stability analysis which is accurate, versatile and requires fewer a priori assumptions, especially, regarding the shape of failure surface.

# 3. LANDSLIDE RUN-OUT ANALYSIS

It is important to estimate the movement behaviour of a potential landslide. For example, the movement distance is an important parameter in risk assessment and measure design. There are many run-out analysis methods, which can fall into four categories: (1) experimental methods, (2) empirical methods, (3) analytical methods, and (4) numerical simulation methods. The states of the art of these methods are reviewed in the following four subsections 3.1 - 3.4.

## 3.1. Experiment methods

Physical modelling typically involves using scale models to capture the motion of landslides. Physical experiments are usually preferred to models because models require more assumptions than direct measurements. But for landslides, direct experiment is difficult, dangerous, expensive, and of limited utility. Based on laboratory experiments and filed investigation data, there are many different available models developed for calculating run-out zones.

Some full-scale direct experiments with artificial landslides have been completed [118, 122124, 125, 126] and others). However, since landslides are frequently heterogeneous and single event cannot be repeated carefully through adjusting only one factor, direct experiment is difficult, dangerous, expensive and of limited utility. And observing conditions are complicated by the danger of being in close proximity to a landslide and the difficulty of measuring a material with properties that change when observed in-situ or when isolated for measurement. But laboratory experiments are still the first qualitative and quantitative observations on the obtained results became fundamental for a better understanding of movement runout behaviour.

## 3.2. Empirical methods

Several empirical methods for assessing landslide travel distance and velocity have been developed based on historical data and on the analysis of the relationship between parameters characterizing both the landslide, e.g. the volume of the landslide mass, and the path, e.g. local morphology, and the distance travelled by the failure mass [65]. Regression model-based methods and geomorphology-based methods are two kinds of common methods.

### 3.2.1. Regression model-based methods

The regression model-based methods are developed on an apparent inverse relationship between landslide volume and angle of reach (also called as fahrböschund by [58]). Several linear regression equations have been proposed [25, 96, 153]. Introduced by [58], the angle of reach is the inclination of the line connecting the crest of the source with the toe of the deposit, as measured along the approximate streamline of motion. The angle of reach is considered an index of the efficiency of energy dissipation, and so is inversely related to mobility. Similar correlations between volume and other simple mobility indices have been proposed [33, 60, 142]. Given estimated source location, volume and path direction, these methods provide estimates of the distal limit of motion [111].

Improved empirical model notable performing regressions on subsets with varying scopes were presented by [13, 25, 69] and others.

Regression model-based models play a valuable role in landslide run-out analysis due to the regression model-based methods are simple. But the regression model-based methods are difficult to apply in practice with a high degree of certainty. For example, the correlation coefficients for some of regression models are 0.7-0.8, while a value of larger than 0.95 generally indicates a strong correlation. And it is difficult in this method to take account of influences of the ground condition, the micro-topography, the degree of saturation of the landslide mass and et al. For this point, geomorphology-based method is another alternative approach to predict the run-out of landslide.

### 3.2.2. Geomorphology-based methods

Field work and photo interpretation are the main sources of the geomorphological analysis for determining the travel distance of landslides [65]. The outer margin of the landslide deposits give an appraisal of the maximum distances that landslides have been able to reach during the present landscape (Figure 6). Several authors have provided these studies (e.g. [23, 24, 26, 88]).

**Figure 6.** Boundary of the potential rockfall run-out area in Santa Coloma (Principality of Andorra), defined by the line that links the farthest fallen blocks observed in the field [24]. Arrows indicate historical rockfall paths and solid circles are large fallen boulders [65]

The geomorphological approach does not give any clue of the emplacement mechanism. Furthermore, the slope geometry and the circumstances responsible for past landslides might have changed. Therefore, results obtained in a given place cannot be easily exported to other localities.

In summary, empirical methods, both regression model-based methods and geomorphology-based methods, typically predict travel distances, while the deformation characteristics or the slide velocities of the landslide are not predicted. These models may be applied to establish initial hazard characteristics for preliminary run-out analysis, which may be later refined by other models.

## 3.3. Analytical methods

In contrast to empirical methods, analytical methods are based on mechanics and involve the solution of motion equations [111]. The simplest analytical model is the classical sliding block model as shown in Figure 7, which is based on work-energy theory [6, 9, 43, Müller-Bernet in 58, 63, 83, 84, 132, 147]. Internal deformation and it's associated energy dissipation are neglected and the landslides is treated as a lumped mass. At any position along the path, the sum of the energies including the potential energy, kinetic energy and net energy loss equals the initial potential energy. This energy balance can be visualized using the concept of energy grade lines, as shown in Figure 9. The concept of energy grade lines is useful for visualizing the energy balance. $v$ is the velocity of the block, $g$ is the vertical acceleration due to gravity and $v^2/2g$ is known as the velocity head, which is the kinetic energy of the block normalized by the product of its mass and $g$. The same normalization of net energy loss is known as head loss. Note that the positions of the energy lines are referenced to the centre of mass of the block and that the true energy line and mean energy line do not necessarily coincide. Given the initial position of the center of mass and a suitable relationship to approximate the energy losses, the position and velocity of the block can be determined at any given time.

Three-dimensional analysis for investigating runout of a slope were also proposed [36, 40, 51, 92 and 109]. These models require a high resolution Digital Elevation Model (DEM).

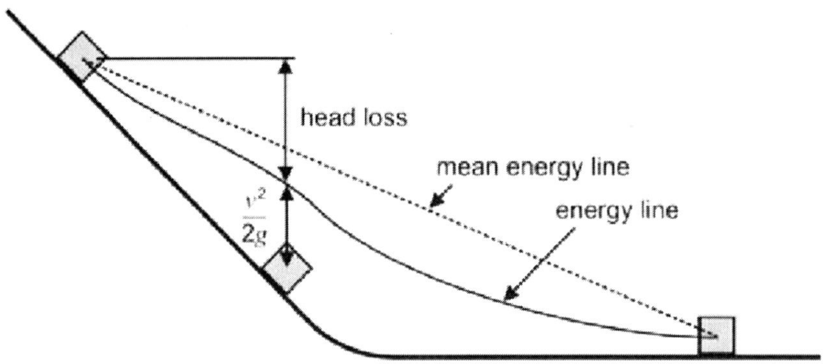


**Figure 7.** The classical sliding block model, based on work-energy theory [111]

Generally speaking, the use of analytical methods is somewhat motivated by the limitations of purely empirical methods, as the unique geometry and materials involved in each case can be accounted for explicitly and a statistically-significant database of previous events is not necessarily required. The simplicity of a lumped mass allows analytical solutions, fast and effectively [66]. However, because the landslide is reduced to a single point, lumped mass models cannot provide the exact maximum runout distance, but only the displacement concerning the centre of mass [44, 62].

## 3.4. Numerical simulation methods

The single-block model should be only applied to the motion of the center of mass of a rigid body, but more complex continuum deformable mass or multi-block system is often appeared in practice. Some numerical simulation methods have been developed to account explicitly for deformation during motion.

### 3.4.1. Continuous methods

When considering that the dimensions of a typical particle is much smaller than the depth and length of the debris, the debris mass is treated as continuum. According to depth averaged Saint Venant approach, the material is assumed to be incompressible and the mass and momentum equations are written in a depth-averaged form. Many numerical methods now exist to investigate the run-out process of landslide (e.g. [18, 27, 28, 29, 30, 35, 62, 111, 112, 133, 149, 161, 165]). These methods are usually based on continuum mechanics and assume that the avalanche thickness is very much smaller than its extent parallel to the bed, i.e. thin layer depth-averaged models. The primary differences are their representation of basal resistance force and the constitutive relations describing the mechanical behaviour of the considered material. These models can accurately take account of detailed topography effects, shown to be significant, with a reasonable computational time, making it possible to perform sensitivity studies of the parameters used in the model. They can provide effective properties that make it possible to roughly reproduce not only the deposit shape but also the dynamic as shown in [46) and 117] for examples. However, conventional continuum approaching models, which neglects the contact between rocks, makes it impossible to trace the position of individual rock during a landslide.

### 3.4.2. Discontinuous methods

When the landslide mass consists of large fragments and boulders, the run-out mass is modelled as an assembly of blocks moving down a surface. Some authors take circular shape models in their run-out analysis to evaluate maximum runout and final deposit position of past or potential events (e.g., [134]). Although polygonal shapes have the disadvantages due to the complexity of the contact patterns and penalty in computational time, methods using non-circular shapes will be required for more real-world problems. It is more appropriate when problems are limited in finite blocks. Discontinuous numerical simulation methods are powerful tools in simulation of failure and run-out process of rock avalanche controlled by weakness surface. DEM [31] and DDA [159, 157] are two of the most commonly used methods.

Both DEM and DDA employ the equations of dynamic motion which are solved at finite points in time, in a series of time steps, but there are some subtle but significant differences in their formulations of the solution schemes and contact mechanics. In the solution schemes, equations of motion in DDA are derived using the principle of minimization of the total potential energy of the system, while the equations of motion as implemented in DEM are derived directly from the force balance equations, which still resultant unbalanced force after a time step and damping is necessarily used to dissipate energy. In the contact mechanics, the DDA used a penalty method in which the contact is assumed to be rigid. No overlapping or interpenetration of the blocks is allowed as the same as real physical cases, whereas soft contact approach is used in DEM. The soft contact approach requires laboratory or field measured joint stiffness, which may be difficult to obtain in many cases. Many comparisons of basic models (sliding, colliding and rolling models) between the DEM and DDA were carried out and show that the results from DDA are more close to the analytical values than that from DEM [188]. Compared to DEM, DDA has a simpler and more straightforward physical meaning [172].

Applications of DEM can be found in some literatures, such as [85, 128, 129, 131, 136, 182].

DDA can be used for estimating the affected area of an earthquake-induced landslide.

[55] first validated the applicability of DDA for the dynamic behaviour of block sliding on an slope. Based on the same inputs model of seismic loadings, [7576886105, 158, 167, 178] studied the dynamic response or/and stability analysis of tunnel, slope, dam, foundation or ancient masonry structure by using DDA. Alternatively, the seismic loadings also can be applied to the base block [146, 147], which is different from the original DDA. Later, [173, 174, 175, 177] applied DDA to simulate the kinematic behavior of sliding rock blocks in the Tsaoling landslide and the Chiu-fen-erh-shan landslide induced by the 1999 Chi-Chi earthquake. Recently, [184] applied newest DDA program to simulate the largest landslides induced by the 2008 Wenchuan earthquake.

# 4. COMPARISONS OF VARIOUS METHODS

The studies in the field of the earthquake-induced landslides are generally reviewed. Two parts of contents, i) seismic stability analysis and ii) run-out analysis are reviewed and compared. Some conclusions can be drawn:

1. Three categories methods can be used to analyse the seismic stability of a slope. Each of these types of methods has strengths and weaknesses and each can be appropriately applied in different situations. In detail, pseudo-static methods can simply and directly determine the FOS and the critical coefficient $k_c$ of a slope, while the widely used Newmark's methods and its extensions can determine the co-seismic deformation of a slope. And the Newmark's methods can be used to estimate where earthquake-induced landslides are likely to occur and what kind of shaking conditions will trigger them based on the GIS technology. More sophisticated analysis for real dynamic process of a seismic slope should be carried out by stress-strain methods, including both continuous methods and discontinuous methods.

2. Four kinds of methods can be used to analyse the run-out of a landslide. In detail, experiment method can provide the qualitative and quantitative observations on the obtained results although this method is difficult, dangerous, expensive and of limited utility. Empirical method can be directly used for assessing landslide travel distance and velocity based on historical data and on the analysis of the relationship between parameters characterizing both landslide and the path. Analytical method can be more directly used without the need of statistically-significant database of previous events. Numerical simulation method can be used to provide more information for the landslide composed by the complex continuum deformable mass or multi-blocks.

# 5. A CASE STUDY: THE DAGUANGBAO LANDSLIDE [185, 186]

## 5.1. Background

The Daguangbao landslide is located in the hanging wall only 6.5 km away from the Yingxiu-Beichuan fault. It is a typical bedding landslide.

Figure 8 gives a pre- and post-earthquake 3-D topographies, from which cross-section of the Daguangbao landslide can be obtained (Figure 9). The extent of the damage caused by the Daguangbao landslide is reflected in the following statistics [61]:

1. The affected area covered 7.3~10 $km^2$;
2. The accumulation body width is 2.2 km;
3. Estimated volume of collapsed rock mass is 750~840 million $m^3$;
4. The failure zone is more than 1 km;
5. The failure mass moved about 4.5 km;
6. Formed a 600m high landslide dam.

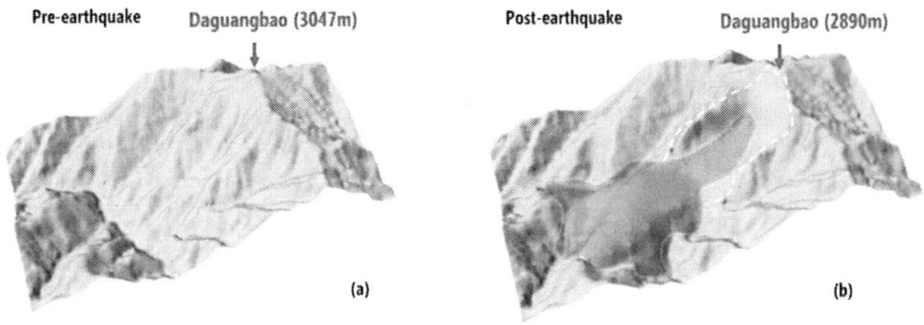


**Figure 8.** Pre- and post-earthquake 3-D topographies of the Daguangbao landslide

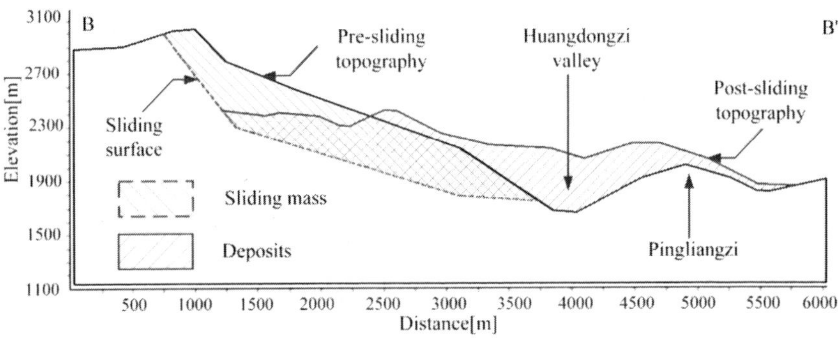


**Figure 9.** Cross-section of the Daguangbao landslide before and after the Wenchuan earthquake.

## 5.2. Material properties and ground motion

### 5.2.1. Material properties

The Daguangbao landslide is so huge that the size effect must be considered. To account for this discrepancy, experience equations based on Hoek-Brown failure criterion, which size effect can be considered, is used to back calculate the material strength. Table 1 lists the material properties of the Daguangbao landslide.

**Table 1.** Material properties of the Daguangbao landslide in FLAC$^{3D}$ and DDA

| | Material 1 | Material 2 | Materia 3 | |
|---|---|---|---|---|
| | | | FLAC$^{3D}$ | DDA |
| Density ($\rho$): g/cm$^3$ | 2.5 | 2.6 | 2.6 | 260000 |
| Unit weight of rock ($\gamma$) : kN/m$^3$ | 25 | 26 | 26 | 0 |
| Elastic modulus ($E$) : Gpa | 1.86 | 2.63 | 14.76 | |
| Poisson's ratio ($\nu$) | 0.2 | 0.2 | 0.1 | |
| Friction angle of discontinuities ($\varphi$) : º | 10.8 | 12.18 | 23.53 | |
| Cohesion of discontinuities ($c$) : Mpa | 1.276 | 1.576 | 4.052 | |
| Tensile strength of discontinuities ($\sigma_t$) : kPa | 12 | 32 | 556 | |

### 5.2.2. Ground motion

The horizontal earthquake wave is the projection combination in the main sliding direction (N60ºE) use the MZQP acceleration records in E-W and N-S directions as Equation (5). The inputted vertical earthquake wave is the MZQP acceleration records in U-D direction.

$$a_H = a_{E-W} \cdot \sin 60° + a_{N-S} \cdot \cos 60° \tag{5}$$

Figure 10 shows the input combined acceleration records. Velocity and displacement time histories can be obtained by first and second integration from acceleration record. The duration of earthquake wave is 60s.

## 5.3. Numerical simulations-Run-out analysis

Seismic DDA can successfully simulate the movement of earthquake induced landslide. Two main features determine the Daguangbao landslide is a unique case, one is near-fault location (≈6.5 km) and the other one is huge scale (≈800×10$^6$ m$^3$). The near-fault location determines the Daguangbao landslide must be shocked by the extreme ground motion from the strong Wenchuan earthquake. And the Daguangbao landslide located on the meizoseismal area where the vertical seismic component is very large. In addition, the landslide is so huge that the size effects must be considered. The friction coefficient measured in the laboratory is no longer suitable for stability and run-out analysis.

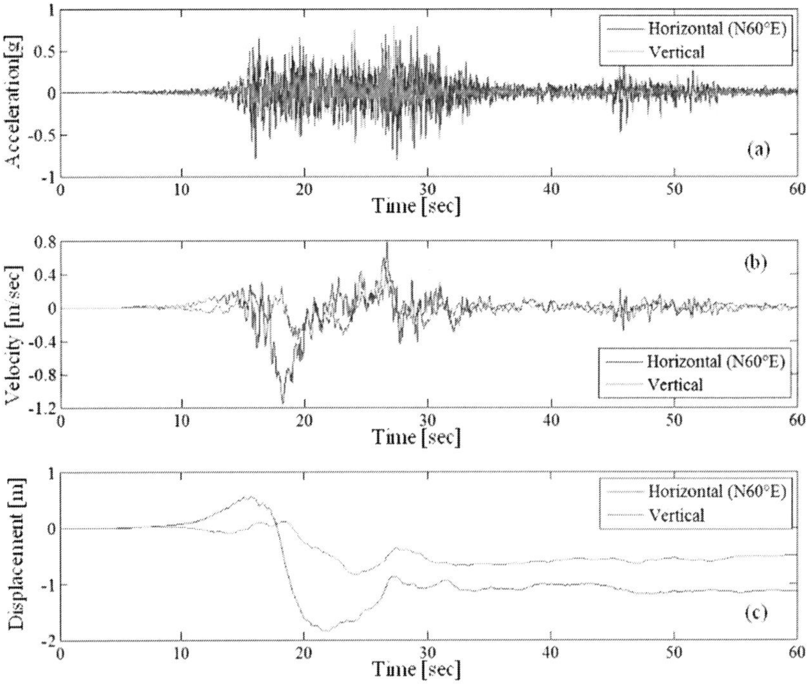


**Figure 10.** Input post-corrected horizontal and vertical ground records projected to N60ºE direction

To these two features, the Daguangbao landslide is simulated by the newest seismic DDA code in which multi-direction seismic forces can be applied in the base block directly, and experience equations based on Hoek-Brown failure criterion is applied to back-calculate the material strength by trying to consider the size effect.

**Table 2.** Control parameters for DDA

| Parameter | Value |
|---|---|
| Assumed maximum displacement ratio ($g_2$) | 0.001 |
| Total number of time steps | 20,000 |
| Time step ($g_1$) | 0.005s |
| Contact spring stiffness ($g_0$) | 5.0×10$^8$ kN/m |
| Factor of over-relaxation | 1.3 |

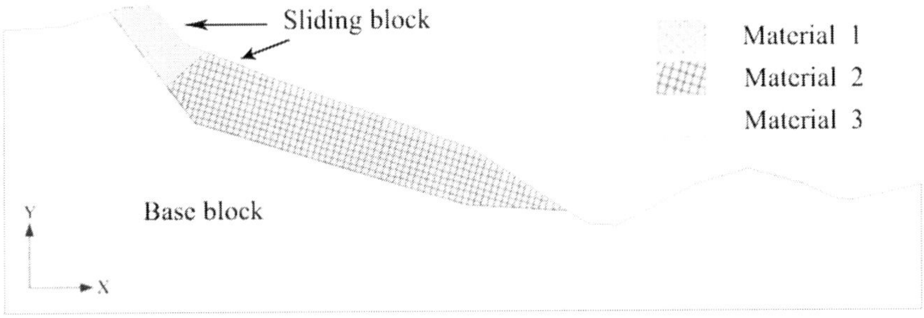


**Figure 11.** DDA model of the Daguangbao landslide

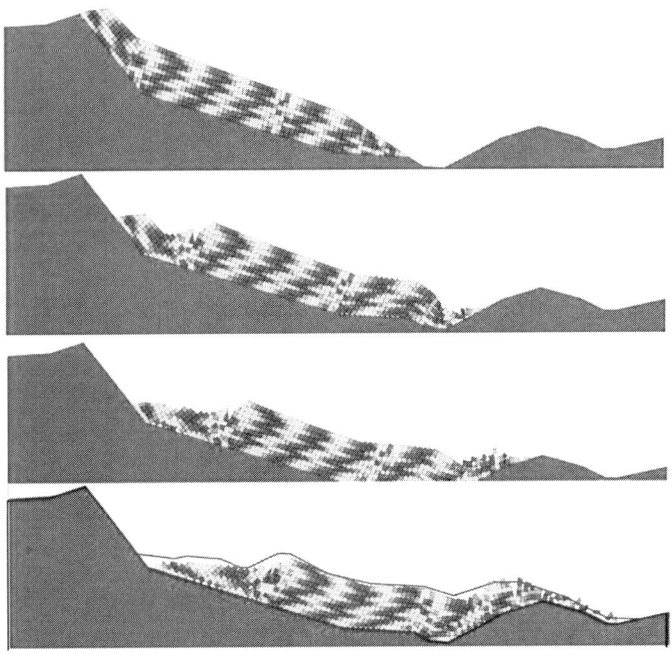

**Figure 12.** Simulation results of the Daguangbao landslide

### 5.3.1. Geometry of sliding blocks

The main sliding direction of the Daguangbao landslide, N60ºE, is selected as analysis profile. The DDA model is depicted in Figure 11. In this simulation, based on the shape of failure surface and the character of slope topography, the whole slope is divided into three parts: base block, upper sliding mass, and lower sliding mass. Then two sliding masses are divided into the smaller discrete deformable blocks based on pre-existing discontinuities.

### 5.3.2. Results

Figure 15 shows the post-failure behavior of the Daguangbao landslide simulated by the seismic DDA code. Simulated results show that the sliding blocks climb over the Pingliangzi. After overlapping the final step of DDA calculation with the topographic cross-section at the Daguangbao landslide, the deposit pattern of the simulated Daguangbao landslide under horizontal-and-vertical situation coincides well with local topography.

## 6. CONCLUSIONS

Five cases are performed using finite difference program $FLAC^{3D}$, under the real seismic waves near the study site. The results show that the seismic conditions cause a significant reduction in factor of safety than static situation. It also found that the vertical seismic has a significant influence on tension failure of block, although it has an insignificant influence on change of the factor of safety. Another important conclusion is the effect of vertical seismic force on relative displacement of potential sliding mass is significant. In addition, large area of tension failure caused by the combined seismic forces at back edge of the slope applies the evidence of effect of vertical seismic force on failure mechanism of slope stability.

A comparison of simulation results from three situations, static, only-horizontal and horizontal-and-vertical, is carried out. Seismic force has a significant influence on the arrival distance, and shape of post-failure. Arrival distance from horizontal-and-vertical situation is larger than that from only-horizontal situation. In addition, the deposit pattern of the simulated Daguangbao landslide under horizontal-and-vertical situation coincides well with local topography. The vertical seismic force should be considered for landslide assessment and management, especially in the situation that the studied site located on the meizoseismal area during the earthquake.

# 7. ACKNOWLEDGEMENTS

The authors gratefully acknowledge the financial support provided by the National Natural Science Foundation of China (No. 51408511), the Scientific Research Foundation for the Returned Overseas Chinese Scholars, State Education Ministry and the opening fund of State Key Laboratory of Geohazard Prevention and Geoenvironment Protection (Chengdu University of Technology) (No. SKLGP2014K015).

# REFERENCES

1. Ambraseys, N., and Sarma, S. 1967. The response of earth dams to strong earthquakes. Geotechnique, 17(3): 181-213.
2. Ambraseys, N., and Menu, J. 1988. Earthquake-induced ground displacements. Earthquake Engineering & Structural Dynamics, 16(7): 985-1006.
3. Ambraseys, N., and Srbulov, M. 1994. Attenuation of earthquake-induced ground displacements. Earthquake Engineering & Structural Dynamics, 23(5): 467-487.
4. Anderson, D., and Kavazanjian Jr, E. 1995. Performance of landfills under seismic loading. In Proc., 3rd Int. Conf. on Recent Advances in Geotechnical Earthquake Engineering and Soil Dynamics. Univ. of Missouri Rolla, MO, Vol.3, pp. 277-306.
5. Army, U.C.o.E. 1960. Stability of Earth and Rockfill Dams, EM 1110-2-1902.
6. Azzoni, A., La Barbera, G., and Zaninetti, A. 1995. Analysis and prediction of rockfalls using a mathematical model. In International Journal of Rock Mechanics and Mining Sciences & Geomechanics Abstracts. Elsevier, Vol.32, pp. 709-724.
7. Bakun-Mazor, D., Hatzor, Y., and Glaser, S. 2012. Dynamic sliding of tetrahedral wedge: The role of interface friction. International Journal for Numerical and Analytical Methods in Geomechanics, 36(3): 327-343.
8. Bhasin R, Kaynia AM. 2004. Static and dynamic simulation of a 700-m high rock slope in western Norway. Engineering Geology, 71(3-4): 213-226.

9. Bozzolo, D., and Pamini, R. 1986. Simulation of rock falls down a valley side. Acta Mechanica, 63(1-4): 113-130.

10. Bray, J.D., and Rathje, E.M. 1998. Earthquake-induced displacements of solid-waste landfills. Journal of Geotechnical and Geoenvironmental Engineering, 124(3): 242-253.

11. Bray, J.D., and Travasarou, T. 2007. Simplified procedure for estimating earthquake-induced deviatoric slope displacements. Journal of Geotechnical and Geoenvironmental Engineering, 133(4): 381-392.

12. Brebbia, C.A., and Wrobel, L. 1980. The boundary element method. Computer methods in fluids.(A 81-28303 11-34) London, Pentech Press, Ltd., 1980: 26-48.

13. Cannon, S.H. 1993. An empirical model for the volume-change behavior of debris flows. In Hydraulic Engineering '93, San Francisco.

14. CDCDMG:California Department of Conversation, Division of Mines and Geology, 1997. Guidelines for evaluating and mitigating seismic hazards in California. CDMG Special Publication.

15. Chang, K.-T., Lin, M.-L., Dong, J.-J., and Chien, C.-H. 2011. The Hungtsaiping landslides: from ancient to recent. Landslides, 9(2): 205-214.

16. Chen, L., and Zhao, W. 1979. Longling earthquake, 1976. Earthquake press, Beijing.

17. Chen, G., and Ohnishi, Y. 1999. Slope stability analysis using Discontinuous Deformation Analysis method. Rock Mecganics for Industry: 535-541.

18. Chen, H., and Lee, C. 2000. Numerical simulation of debris flows. Canadian Geotechnical Journal, 37(1): 146-160.

19. Cheng, Y.M. 1998. Advancements and improvement in discontinuous deformation analysis. Computers and Geotechnics, 22(2): 153-163.

20. Chopra, A.K. 1966. The importance of the vertical component of earthquake motions. Bulletin of the Seismological Society of America, 56(5): 1163-1175.

21. Clough, R.W. 1960. The finite element method in plane stress analysis. In 2nd Conference on Electronic Computation, Pittsburgh, PA.

22. Constantinou, M., and Gazetas, G. 1987. Probabilistic seismic sliding deformations of earth dams and slopes. In Probabilistic Mechanics and Structural Reliability (1984). ASCE, pp. 318-321.

23. Copons, R., and Vilaplana, J.M. 2008. Rockfall susceptibility zoning at a large scale: From geomorphological inventory to preliminary land use planning. Engineering Geology, 102(3): 142-151.

24. Copons, R., Vilaplana, J.M., Corominas, J., Altimir, J., and Amigó, J. 2004. Rockfall Risk Management in High-Density Urban Areas. The Andorran Experience. Landslide hazard and risk: 675-698.

25. Corominas, J. 1996. The angle of reach as a mobility index for small and large landslides. Canadian Geotechnical Journal, 33(2): 260-271.

26. Costa, J.E. 1984. Physical geomorphology of debris flows. In Developments and applications of geomorphology. Springer. pp. 268-317.

27. Crosta, G., and Agliardi, F. 2003a. A methodology for physically based rockfall hazard assessment. Natural Hazards and Earth System Science, 3(5): 407-422.

28. Crosta, G., and Frattini, P. 2003b. Distributed modelling of shallow landslides triggered by intense rainfall. Natural Hazards and Earth System Science, 3(1/2): 81-93.

29. Crosta, G.B., Frattini, P., and Fusi, N. 2007. Fragmentation in the Val Pola rock avalanche, Italian Alps. Journal of Geophysical Research: Earth Surface (2003–2012), 112(F1).

30. Crosta, G., Imposimato, S., Roddeman, D., Chiesa, S., and Moia, F. 2005. Small fast-moving flow-like landslides in volcanic deposits: the 2001 Las Colinas Landslide (El Salvador). Engineering Geology, 79(3): 185-214.

31. Cundall, P. 1971. A computer model for simulating progressive, large scale movements in blocky rock system. In In Symposium of International Society of Rock Mechanics, Nancy, France, pp. 11–18.

32. Danay, A., and Adeghe, L. 1993. Seismic-induced slip of concrete gravity dams. Journal of Structural Engineering, 119(1): 108-129.

33. Davies, T.R. 1982. Spreading of rock avalanche debris by mechanical fluidization. Rock Mechanics, 15(1): 9-24.

34. Del Gaudio, V., Pierri, P., and Wasowski, J. 2003. An approach to time-probabilistic evaluation of seismically induced landslide

hazard. Bulletin of the Seismological Society of America, 93(2): 557-569.

35. Denlinger, R.P., and Iverson, R.M. 2001. Flow of variably fluidized granular masses across three-dimensional terrain: 2. Numerical predictions and experimental tests. Journal of Geophysical Research: Solid Earth (1978–2012), 106(B1): 553-566.

36. Descoeudres, F., and Zimmermann, T. 1987. Three-dimensional dynamic calculation of rockfalls. In 6th ISRM Congress.

37. Doolin, D.M. 2005. Unified displacement boundary constraint formulation for discontinuous deformation analysis (DDA). International Journal for Numerical and Analytical Methods in Geomechanics, 29(12): 1199-1207.

38. Doolin, D.M., and Sitar, N. 2004. Time Integration in Discontinuous Deformation Analysis. Journal of Engineering Mechanics, 130(3): 249-258.

39. Dorren, L.K. 2003. A review of rockfall mechanics and modelling approaches. Progress in Physical Geography, 27(1): 69-87.

40. Dorren, L., and Heuvelink, G.B. 2004. Effect of support size on the accuracy of a distributed rockfall model. International Journal of Geographical Information Science, 18(6): 595-609.

41. Dreyfus, D., Rathje, E.M., and Jibson, R.W. 2013. The Influence of Different Simplified Sliding-Block Models and Input Parameters on Regional Predictions of Seismic Landslides Triggered by the Northridge Earthquake. Engineering Geology.

42. Elgamal, A.-W.M., Scott, R.F., Succarieh, M.F., and Yan, L. 1990. La Villita dam response during five earthquakes including permanent deformation. Journal of Geotechnical Engineering, 116(10): 1443-1462.

43. Evans, S., and Hungr, O. 1993. The assessment of rockfall hazard at the base of talus slopes. Canadian Geotechnical Journal, 30(4): 620-636.

44. Evans, S., Hungr, O., and Enegren, E. 1994. The Avalanche Lake rock avalanche, Mackenzie mountains, northwest territories, Canada: description, dating, and dynamics. Canadian Geotechnical Journal, 31(5): 749-768.

45. Fardis, M.N. 2009. Seismic design, assessment and retrofitting of concrete buildings: based on EN-Eurocode 8. Springer.

46. Favreau, P., Mangeney, A., Lucas, A., Crosta, G., and Bouchut, F. 2010. Numerical modeling of landquakes. Geophysical Research Letters, 37(15): L15305.

47. Fenves, G., and Chopra, A.K. 1986. Simplified analysis for earthquake resistant design of concrete gravity dams. Earthquake Engineering Research Center, University of California.

48. Franklin, A.G., and Chang, F.K. 1977. Permanent displacements of earth embankments by Newmark sliding block analysis.

49. Gazetas, G., and Uddin, N. 1994. Permanent deformation on preexisting sliding surfaces in dams. Journal of Geotechnical Engineering, 120(11): 2041-2061.

50. Griffiths, D., and Prevost, J.H. 1988. Two-and three-dimensional dynamic finite element analyses of the Long Valley Dam. Geotechnique, 38(3): 367-388.

51. Guzzetti, F., Malamud, B.D., Turcotte, D.L., and Reichenbach, P. 2002. Power-law correlations of landslide areas in central Italy. Earth and Planetary Science Letters, 195(3): 169-183.

52. Hamajima, R., Kawai, T., Yamashita, K., and Kusabuka, M. 1985. Numerical analysis of cracked and jointed rock mass. In the 5th International Conference on Numerical Methods in Geomechanics, Nagoya, Japan, pp. 207-214.

53. Harp, E.L., and Jibson, R.W. 1995. Inventory of landslides triggered by the 1994 Northridge, California earthquake. US Geological Survey.

54. Hatzor, Y.H. 2003. Fully Dynamic Stability Analysis of Jointed Rock Slopes. In Proceedings of the 10th ISRM Congress, pp. 503-514.

55. Hatzor, Y.H., and Feintuch, A. 2001. The validity of dynamic block displacement prediction using DDA. International Journal of Rock Mechanics & Mining Sciences.

56. Hatzor, Y.H., Arzi, A.A., and Tsesarsky, M. 2002. Realistic dynamic analysis of jointed rock slopes using DDA. In 5th Int. Conf. on Analysis of Discontinuous Deformation - Stability of rock structures, Abingdon, Balkema, Rotterdam, The Netherlands, pp. 47–56.

57. Hatzor, Y., Arzi, A.A., Zaslavsky, Y., and Shapira, A. 2004. Dynamic stability analysis of jointed rock slopes using the DDA method: King

Herod's Palace, Masada, Israel. International Journal of Rock Mechanics and Mining Sciences, 41(5): 813-832.

58. Heim, A. 1932. Bergsturz und Menschenleben, Zurich: Fretz and Wasmuth Verlag.

59. Hsieh, S.-Y., and Lee, C.-T. 2011. Empirical estimation of the Newmark displacement from the Arias intensity and critical acceleration. Engineering Geology, 122(1-2): 34-42.

60. Hsü, K.J. 1975. Catastrophic debris streams (sturzstroms) generated by rockfalls. Geological Society of America Bulletin, 86(1): 129-140.

61. Huang, R., Pei, X., Fan, X., Zhang, W., Li, S., and Li, B. 2012. The characteristics and failure mechanism of the largest landslide triggered by the Wenchuan earthquake, May 12, 2008, China. Landslides, 9(1): 131-142.

62. Hungr, O. 1995. A model for the runout analysis of rapid flow slides, debris flows, and avalanches. Canadian Geotechnical Journal, 32(4): 610-623.

63. Hungr, O., and Evans, S. 1988. Engineering evaluation of fragmental rockfall hazards. In Proceedings of the Fifth International Symposium on Landslides, Lausanne, AA Balkema, Rotterdam, Netherlands, pp. 685-690.

64. Hungr, O., and Evans, S. 2004. Entrainment of debris in rock avalanches: An analysis of a long run-out mechanism. Geological Society of America Bulletin, 116(9-10): 1240-1252.

65. Hungr, O., Corominas, J., and Eberhardt, E. 2004. Estimating landslide motion mechanism, travel distance and velocity.

66. Hürlimann, M., Rickenmann, D., Medina, V., and Bateman, A. 2008. Evaluation of approaches to calculate debris-flow parameters for hazard assessment. Engineering Geology, 102(3): 152-163.

67. Hynes-Griffin, M.E., and Franklin, A.G. 1984. Rationalizing the seismic coefficient method. Defense Technical Information Center.

68. Ishikawa, T., Sekine, E., and Ohnishi, Y. 2002. Shaking table tests of coarse granular materials with discontinuous analysis. Proc. of ICADD-5, BALKEMA: 181-187.

69. Jakob, M., and Hungr, O. 2005. Debris-flow hazards and related phenomena. Springer.

70. Jibson, R.W. 1993. Predicting earthquake-induced landslide displacements using Newmark's sliding block analysis. Transportation Research Record: 9-9.

71. Jibson, R. 2000. A method for producing digital probabilistic seismic landslide hazard maps. Engineering Geology.

72. Jibson, R.W. 2007. Regression models for estimating coseismic landslide displacement. Engineering Geology, 91(2-4): 209-218.

73. Jibson, R.W. 2011. Methods for assessing the stability of slopes during earthquakes—A retrospective. Engineering Geology, 122(1-2): 43-50.

74. Jibson, R.W., and Jibson, M.W. 2003. Java programs for using Newmark's method and simplified decoupled analysis to model slope performance during earthquakes. US Department of the Interior, US Geological Survey.

75. Jibson, R.W., Harp, E.L., and Michael, J.A. 1998. A Method for Producing Digital Probabilistic Seismic Landslide Hazard Maps: An Example from the Los Angeles, California, Area.

76. Kavazanjian, E., and Consultants, G. 1997. Design Guidance: Geotechnical Earthquake Engineering for Highways. Design Principles. Federal Highway Administration.

77. Kawai, T. 1977. A new discrete analysis of nonlinear solid mechanics problems involving stability, plasticity and crack. In the Symposium on Applications of Computer Methods in Engineering, Los Angeles, USA, pp. 1029-1038.

78. Kawai, T. 1978. New discrete models and their application to seismic response analysis of structures. Nuclear Engineering and Design, 48(1): 207-229.

79. Kawai, T., Takeuchi, N., and Kumeta, T. 1981. New discrete models and their application to rock mechanics. In ISRM International Symposium.

80. Kawai, T., Kawabata, Y., Kumagai, K., and Kondou, K. 1978. A new discrete model for analysis of solid mechanics problems. Numerical methods in fracture mechanics: 26-37.

81. Ke, T.C. 1996. The issues of rigid-body rotation in DDA. In First international forum on discontinuous deformation analysis (DDA) and simulations of discontinuous media, Berkeley, USA, pp. 318-325.

82. Keefer, D.K. 2000. Statistical analysis of an earthquake-induced landslide distribution—the 1989 Loma Prieta, California event. Engineering Geology, 58(3): 231-249.

83. Kirby, M.J., and Statham, I. 1975. Surface stone movement and screen formation. Journal of Geology, 83(3): 349-362.

84. Kobayashi, Y., Harp, E., and Kagawa, T. 1990. Simulation of rockfalls triggered by earthquakes. Rock Mechanics and Rock Engineering, 23(1): 1-20.

85. Komodromos, P., Papaloizou, L., and Polycarpou, P. 2008. Simulation of the response of ancient columns under harmonic and earthquake excitations. Engineering Structures, 30(8): 2154-2164.

86. Kong, X., and Liu, J. 2002. Dynamic failure numeric simulations of model concrete-faced rock-fill dam.. Soil Dynamics and Earthquake Engineering 22(9–12): 1131–1134.

87. Koo, C.Y., and Chern, J.C. 1998. Modification of the DDA method for rigid block problems. International Journal of Rock Mechanics & Mining Sciences, 35: 683-693.

88. Kostaschuk, R. 1987. Identification of debris flow hazard on alluvial fans in the Canadian Rocky Mountains. Debris flows/avalanches: process, recognition, and mitigation, 7: 115.

89. Kramer, S.L. 1996. Geotechnical earthquake engineering. Prentice-Hall Civil Engineering and Engineering Mechanics Series, Upper Saddle River, NJ: Prentice Hall,| c1996, 1.

90. Kramer, S.L., and Smith, M.W. 1997. Modified Newmark model for seismic displacements of compliant slopes. Journal of Geotechnical and Geoenvironmental Engineering, 123(7): 635-644.

91. Krinitzsky, E.L. 1993. Fundamentals of earthquake-resistant construction. Wiley. com.

92. Lan, H., Derek Martin, C., and Lim, C. 2007. RockFall analyst: A GIS extension for three-dimensional and spatially distributed rockfall hazard modeling. Computers & Geosciences, 33(2): 262-279.

93. Latha, G.M., and Garaga, A. 2010. Seismic Stability Analysis of a Himalayan Rock Slope. Rock Mechanics and Rock Engineering, 43(6): 831-843.

94. Lee, K.L. 1974. Seismic permanent deformation in earth dams, University of California, Los Angeles, CA.

95. Leger, P., and Katsouli, M. 1989. Seismic stability of concrete gravity dams. Earthquake Engineering & Structural Dynamics, 18(6): 889-902.

96. Li, T. 1983. A mathematical model for predicting the extent of a major rockfall. Zeitschrift Fur Geomorphologie, 24: 473-482.

97. Lin, J.S., and Whitman, R.V. 1983. Decoupling approximation to the evaluation of earthquake-induced plastic slip in earth dams. Earthquake Engineering & Structural Dynamics, 11(5): 667-678.

98. Lin, C.T., Amadei, B., Jung, J., and Dwyer, J. 1996. Extensions of discontinuous deformation analysis for jointed rock masses. international Journal of Rock Mechanics and Mining Sciences & Geomechanics Abstracts, 33(7): 671-694.

99. Ling, H.I. 2001. Recent applications of sliding block theory to geotechnical design. Soil Dynamics and Earthquake Engineering, 21: 189-197.

100. Ling, H., and Leshchinsky, D. 1998. Effects of vertical acceleration on seismic design of geosynthetic-reinforced soil structures. Geotechnique, 48(3): 347-373.

101. Lomnitz, C. 1970. Casualties and behavior of populations during earthquakes. Bulletin of the Seismological Society of America, 60(4): 1309-1313.

102. Luan, M., Li, Y., and Yang, Q. 2000. Discontinuous deformation computational mechanics model and its application in stability analysis of rock slope. Chinese Journal of Rock Mechanics and Engineering, 3: 006.

103. Makdisi, F.I., and Seed, H.B. 1977. Simplified procedure for estimating dam and embankment earthquake-induced deformations. In ASAE Publication No. 4-77. Proceedings of the National Symposium on Soil Erosion and Sediment by Water, Chicago, Illinois, December 12-13, 1977.

104. Makdisi, F.I., and Seed, H.B. 1978. Simplified procedure for estimating dam and embankment earthquake-induced failures. Journal of the Geotechnical Division, ASCE, 104: 849–861.

105. Makris, N., and Roussos, Y. 2000. Rocking response of rigid blocks under near-source ground motions. Geotechnique, 50(3): 243-262.

106. Mankelow, J.M., and MURPHY, W. 1998. Using GIS in the probabilistic assessment of earthquake triggered landslide hazards. Journal of Earthquake Engineering, 2(4): 593-623.

107. Marcuson, W. 1981. Moderator's report for session on Earth Dams and Stability of Slopes under Dynamic Loads. In Proceedings, International Conference on Recent Advances in Geotechnical Earthquake Engineering and Soil Dynamics, Vol.3, p. 1175.

108. Marcuson III, W.F., and Franklin, A.G. 1983. Seismic Design, Analysis, and Remedial Measures to Improve Stability of Existing Earth Dams, DTIC Document.

109. Masuya, H., Amanuma, K., Nishikawa, Y., and Tsuji, T. 2009. Basic rockfall simulation with consideration of vegetation and application to protection measure. Natural Hazards and Earth System Science, 9(6): 1835-1843.

110. McDougall, S. 2006. A new continuum dynamic model for the analysis of extremely rapid landslide motion across complex 3D terrain, University of British Columbia.

111. McDougall, D. 2006. The distributed criterion design. Journal of Behavioral Education, 15(4): 236-246.

112. McDougall, S., and Hungr, O. 2004. A model for the analysis of rapid landslide motion across three-dimensional terrain. Canadian Geotechnical Journal, 41(6): 1084-1097.

113. Meunier, P., Hovius, N., and Haines, A.J. 2007. Regional patterns of earthquake-triggered landslides and their relation to ground motion. Geophysical Research Letters, 34(20).

114. Miles, S.B., and Ho, C.L. 1999. Applications and issues of GIS as tool for civil engineering modeling. Journal of Computing in Civil Engineering, 13(3): 144-152.

115. Miles, S.B., and Keefer, D.K. 2000. Evaluation of seismic slope-performance models using a regional case study. Environmental & Engineering Geoscience, 6(1): 25-39.

116. Mitchell, A.R., and Griffiths, D.F. 1980. The finite difference method in partial differential equations(Book). Chichester, Sussex, England and New York, Wiley-Interscience, 1980. 281 p.

117. Moretti, L., Mangeney, A., Capdeville, Y., Stutzmann, E., Christian Huggel, C., Schneider, D., and Francois Bouchut, F. 2012. Numerical modeling of the Mount Steller landslide flow history and of the

generated long period seismic waves. Geophys. Res. Lett., 39(L16402).

118. Moriwaki, H., Yazaki, S., and Oyagi, N. 1985. A gigantic debris avalanche and its dynamics at Mount Ontake caused by the Nagano-ken-seibu earthquake, 1984. In Proc. 4th Int. Conf. Field Workshop on Landslides, pp. 359-364.

119. Newmark, N.M. 1965. Effects of earthquakes on dams and embankments. Géotechnique, 15: 139-159.

120. Ning, Y., and Zhao, Z. 2012. A detailed investigation of block dynamic sliding by the discontinuous deformation analysis. International Journal for Numerical and Analytical Methods in Geomechanics: 1-21.

121. Niwa, K., Kawai, T., Ikeda, M., and Takeda, T. 1984. Application of a new discrete method to fracture analysis of brittle materials. In the 3rd International Conference on Numerical Methods in Fracture Mechanics, Swansea, U.K., pp. 13-27.

122. Ochiai, H., Okada, Y., Furuya, G., Okura, Y., Matsui, T., Sammori, T., Terajima, T., and Sassa, K. 2004. A fluidized landslide on a natural slope by artificial rainfall. Landslides, 1(3): 211-219.

123. Ohnishi, Y., Chen, G., and Miki, S. 1995. Recent development of DDA in rock mechanics. Proc. ICADD, 1: 26-47.

124. Okura, Y., Kitahara, H., and Sammori, T. 2000. Fluidization in dry landslides. Engineering Geology, 56(3): 347-360.

125. Okura, Y., Kitahara, H., Sammori, T., and Kawanami, A. 2000. The effects of rockfall volume on runout distance. Engineering Geology, 58(2): 109-124.

126. Okura, Y., Kitahara, H., Ochiai, H., Sammori, T., and Kawanami, A. 2002. Landslide fluidization process by flume experiments. Engineering Geology, 66(1): 65-78.

127. Pal, S., Kaynia, A.M., Bhasin, R.K., and Paul, D.K. 2011. Earthquake Stability Analysis of Rock Slopes: a Case Study. Rock Mechanics and Rock Engineering.

128. Papaloizou, L., and Komodromos, P. 2009. Planar investigation of the seismic response of ancient columns and colonnades with epistyles using a custom-made software. Soil Dynamics and Earthquake Engineering, 29(11-12): 1437-1454.

129. Papantonopoulos, C., Psycharis, I.N., Papastamatiou, D.Y., Lemos, J.V., and Mouzakis, H.P. 2002. Numerical prediction of the earthquake response of classical columns using the distinct element method. Earthquake Engineering & Structural Dynamics, 31(9): 1699-1717.

130. Pekau, O.A, and Yuzhu, C. 2004. Failure analysis of fractured dams during earthquakes by DEM. Engineering Structures, 26(10): 1483-1502.

131. Pekau, O.A, and Yuzhu, C. 2004. Seismic collapse behaviour of Damaged dams. In 13 WCEE: 13 th World Conference on Earthquake Engineering Conference Proceedings.

132. Pfeiffer, T.J., and Bowen, T. 1989. Computer simulation of rockfalls. Bulletin of the Association of Engineering Geologists, 26(1): 135-146.

133. Pirulli, M. 2005. Numerical modelling of landslide runout. A continuum mechanics approach, Politecnico di Torino.

134. Poisel, R., Preh, A., and Hungr, O. 2008. Run Out of Landslides–Continuum Mechanics versus Discontinuum Mechanics Models. Geomechanics and Tunnelling, 1(5): 358-366.

135. Prevost, J.H. 1981. DYNA-FLOW: a nonlinear transient finite element analysis program. Princeton University, Department of Civil Engineering, School of Engineering and Applied Science.

136. Psycharis, I., Lemos, J., Papastamatiou, D., Zambas, C., and Papantonopoulos, C. 2003. Numerical study of the seismic behaviour of a part of the Parthenon Pronaos. Earthquake Engineering & Structural Dynamics, 32(13): 2063-2084.

137. Pyke, R. 1991. Selection of Seismic Coefficients for Use in Pseudo-Static Slope Stability Analyses. http://www.tagasoft.com/Discussion/article2_html.

138. Rathje, E.M., and Bray, J.D. 1999. An examination of simplified earthquake-induced displacement procedures for earth structures. Canadian Geotechnical Journal, 36(1): 72-87.

139. Rathje, E.M., and Bray, J.D. 2000. Nonlinear coupled seismic sliding analysis of earth structures. Journal of Geotechnical and Geoenvironmental Engineering, 126(11): 1002-1014.

140. Richards, R., and Elms, D.G. 1979. Seismic behavior of gravity retaining walls. Journal of the Geotechnical Engineering Division, 105(4): 449-464.

141. Richards, J., Elms, D., and Budhu, M. 1993. Seismic bearing capacity and settlements of foundations. Journal of Geotechnical Engineering, 119(4): 662-674.

142. Rickenmann, D. 1999. Empirical relationships for debris flows. Natural Hazards, 19(1): 47-77.

143. Rodriguez, C.E., Bommer, J., and Chandler, R.J. 1999. Earthquake-induced landslides 1980-1997. soil Dynamics and Earthquake Engineering, 18: 325-346.

144. Sarma, S.K. 1975. Seismic stability of earth dams and embankments. Geotechnique, 25(4): 743-761.

145. Sarma, S.K. 1981. Seismic displacement analysis of earth dams. Journal of the Geotechnical Engineering Division, 107(12): 1735-1739.

146. Sasaki, T., Hagiwara, I., Sasaki, K., Ohnishi, Y., and Ito, H. 2007. Fundamental studies for dynamic response of simple block structures by DDA. In In Proceedings of the eighth international conference on analysis of discontinuous deformation: fundamentals and applications to mining & civil engineering, Beijing, China, pp. 141–146.

147. Sasaki, T., Hagiwara, I., Sasaki, K., Yoshinaka, R., Ohnishi, Y., and Nishiyama, S. 2004. Earthquake response analysis of rock-fall models by discontinuous deformation analysis. In In Proceedings of the ISRM international symposium 3rd ARMS, Kyoto, Japan, pp. 1267–1272.

148. Sassa, K. 1988. Motion of Landslides and Debris Flows: Prediction of Hazard Area: Report for Grant-in-aid for Scientific Research by Japanese Ministry on Education, Science and Culture (project No. 61480062). Disaster Prevention Research Institute.

149. Savage, S., and Hutter, K. 1989. The motion of a finite mass of granular material down a rough incline. Journal of Fluid Mechanics, 199(1): 177-215.

150. Sawada, T., Chen, W.F., and Nomachi, S.G. 1993. Assessment of seismic displacements of slopes. Soil Dynamics and Earthquake Engineering, 12: 357-362.

151. Saygili, G., and Rathje, E.M. 2009. Probabilistically based seismic landslide hazard maps: An application in Southern California. Engineering Geology, 109(3): 183-194.

152. Scheidegger, A.E. 1973. On the prediction of the reach and velocity of catastrophic landslides. Rock Mechanics, 5(4): 231-236.

153. Seed, H.B. 1973. Analysis of the Slides in the San Fernando Dams During the Earthquake of Feb. 9, 1971: Report to State of California Department of Water Resources, Los Angeles Department of Water and Power, National Science Foundation. College of Engineering, University of California.

154. Seed, H.B. 1979. Considerations in the earthquake-resistant design of earth and rockfill dams. Geotechnique, 29(3): 13-41.

155. Seed, H.B., and Martin, G.R. 1966. The seismic coefficient in earth dam design. Journal of Soil Mechanics & Foundations Div, 92(Proc. Paper 4824).

156. Serff, N. 1976. Earthquake induced deformations of earth dams. College of Engineering, University of California.

157. Shi, G.-H. 1988. Discontinuous Deformation Analysis A New Numerical Model for the Statics and Dynamics of Block Systems, University of California, Berkeley.

158. Shi, G. 2002. Single and multiple block limit equilibrium of key block method and discontinuous deformation analysis. In Proceedings of the 5th International Conference on Analysis of Discontinuous Deformation. Rotterdam: AA Balkema, pp. 3-43.

159. Shi, G.-H., and Goodman, R.E. 1985. Two dimensional discontinuous deformation analysis. International Journal for Numerical and Analytical Methods in Geomechanics, 9: 541-556.

160. Shi, G.-H., and Goodman, R.E. 1989. Generalization of two-dimensional discontinuous deformation analysis for forward modelling. International Journal for Numerical and Analyrucal Methods in Geomechanics, 13: 359-380.

161. Sousa, J., and Voight, B. 1991. Continuum simulation of flow failures. Geotechnique, 41(4): 515-538.

162. Stamatopoulos, C. 1996. Sliding system predicting large permanent co-seismic movements of slopes. Earthquake Engineering & Structural Dynamics, 25(10): 1075-1093.

163. Stewart, J.P., Blake, T.F., and Hollingsworth, R.A. 2003. A Screen Analysis Procedure for Seismic Slope Stability. Earthquake Spectra, 19(3): 697.

164. Taiebat, M., Kaynia, A.M., and Dafalias, Y.F. 2011. Application of an Anisotropic Constitutive Model for Structured Clay to Seismic Slope Stability. Journal of Geotechnical and Geoenvironmental Engineering, 137(5): 492.

165. Takahashi, T., Momiyama, A., Hirai, K., Hishinuma, F., and Akagi, H. 1992. Functional correlation of fetal and adult forms of glycine receptors with developmental changes in inhibitory synaptic receptor channels. Neuron, 9(6): 1155-1161.

166. Terzaghi, K. 1950. Theoretical Soil Mechanics.

167. Tsesarsky, M., Hatzor, Y., and Sitar, N. 2005. Dynamic displacement of a block on an inclined plane: analytical, experimental and DDA results. Rock Mechanics and Rock Engineering, 38(2): 153-167.

168. Varnes, D.J., Landslides, t.I.A.E.G.C.o., and Slopes, O.M.M.o. 1984. Landslide hazard zonation: a review of principles and practice.

169. Wartman, J., Asce, M., Bray, J.D., and Seed, R.B. 2003. Inclined Plane Studies of the Newmark Sliding Block Procedure. Journal of Geotechnical and Geoenvironmental Engineering, 129(8): 673-684.

170. Wasowski, J., Keefer, D.K., and Lee, C.-T. 2011. Toward the next generation of research on earthquake-induced landslides: Current issues and future challenges. Engineering Geology, 122(1-2): 1-8.

171. Wilson, R.C., and Keefer, D.K. 1983. Dynamic analysis of a slope failure from the 6 August 1979 Coyote Lake, California, earthquake. Bulletin of the Seismological Society of America, 73(3): 863-877.

172. Wu, J.-H. 2003. Numerical analysis of discontinuous rock masses using discontinuous deformation analysis, Kyoto University, Kyoto, Japan.

173. Wu, J.-H. 2010. Seismic landslide simulations in discontinuous deformation analysis. Computers and Geotechnics, 37(5): 594-601.

174. Wu, J.-H., and Chen, C.-H. 2011a. Application of DDA to simulate characteristics of the Tsaoling landslide. Computers and Geotechnics, 38(5): 741-750.

175. Wu, J.-H., and Tsai, P.-H. 2011b. New dynamic procedure for back-calculating the shear strength parameters of large landslides. Engineering Geology.

176. Wu, A., Ren, F., and Dong, X. 1997. A study on the numerical model of DDA and its preliminary application to rock engineering. Chinese Journal of Rock Mechanics and Engineering, 16(5): 411-417.

177. Wu, J., Lin, J., and Chen, C. 2009. Dynamic discrete analysis of an earthquake-induced large-scale landslide. International Journal of Rock Mechanics and Mining Sciences, 46(2): 397-407.

178. Yagoda-Biran, G., and Hatzor, Y.H. 2010. Constraining paleo PGA values by numerical analysis of overturned columns. Earthquake Engineering & Structural Dynamics, 39(4): 463-472.

179. Yegian, M.K. 1991a. Seismic risk analysis for earth dams. ASCE.

180. Yegian, M.K., Marciano, E.A., and Ghahraman, V.G. 1991b. Earthquake-induced permanent deformations: probabilistic approach. Journal of Geotechnical Engineering, 117(1): 35-50.

181. Yegian, M., Harb, J., and Kadakal, U. 1998. Dynamic response analysis procedure for landfills with geosynthetic liners. Journal of Geotechnical and Geoenvironmental Engineering, 124(10): 1027-1033.

182. Zhang, C., Pekau, O.A., Jin, F., and Wang, G. 1997. Application of distinct element method in dynamic analysis of high rock slopes and blocky structures. soil Dynamics and Earthquake Engineering, 16: 385-394.

183. Zhang, Y., Chen, G., Zheng, L., Wu, J., and Zhuang, X. 2012a. Effects of vertical seismic force on the initiation of the Daguangbao landslide induced by the Wenchuan earthquake. In The 8th Annual Conference of International Institute for Infrastructure, Renewal and Reconstruction, Kumamoto, Japan, pp. 530-539.

184. Zhang, Y., Chen, G., Zheng, L., and Li, Y. 2012b. Numerical analysis of the largest landslide induced by the Wenchuan earthquake, May 12, 2008 using DDA. In International Symposium on Earthquake-induced Landslides, Kiryu, Japan.

185. Zhang, Y., G. Chen, L. Zheng, Y. Li and J. Wu. 2013. Effects of near-Fault Seismic Loadings on Run-out of Large-Scale Landslide: A Case Study. Engineering Geology 166, 216-236.

186. Zhang, Y., G. Chen, L. Zheng, Y. Li, X. Zhuang. 2014. Effects of vertical seismic force on initiation of the Daguangbao landslide induced by the 2008 Wenchuan earthquake, Soil dynamics and earthquake engineering (In press).

187. Zhao, S.L., Salami, M.R., and Rahman, M.S. 1997. Discontinuous Deformation Analysis Simulation of Rock Slope Failure. In 9th International Conference on Computer Methods and Advances in Geomechanics, Wuhan, China

188. Zheng, L. 2010. Development of new models for landslide simulation based on discontinuous deformation analysis, Kyushu University, Fukuoka, Japan.

# INDEX

## N



## P



## R



## S



## T



## W